Mass Spectrometry

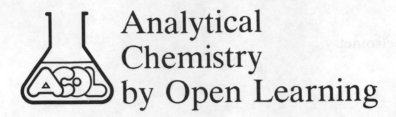

Analytical Chemistry by Open Learning

Titles in Series:

Mass Spectrometry

Analytical Chemistry by Open Learning

Authors:
REG DAVIS
Kingston Polytechnic, UK

MARTIN FREARSON
Hatfield Polytechnic, UK

Editor:
F. ELIZABETH PRICHARD

on behalf of ACOL

Published on behalf of ACOL, Thames Polytechnic, London
by
JOHN WILEY & SONS
Chichester · New York · Brisbane · Toronto · Singapore

Published by permission of the Controller of
Her Majesty's Stationery Office

Library of Congress Cataloging in Publication Data:

Davis, Reg.
  Mass spectrometry.

  (Analytical chemistry by open learning)
  Bibliography: p.
  1. Mass spectrometry—Programmed instruction.
2. Chemistry, Analytic—Programmed instruction.
I. Frearson, Martin.   II. Prichard, F. Elizabeth.
III. Title.   IV. Series.
QD96.M3D38   1987   543'.0873   87-2029
ISBN 0 471 91388 X
ISBN 0 471 91389 8 (pbk.)

British Library Cataloguing in Publication Data:

Davis, Reg
  Mass spectrometry.—(Analytical
  chemistry).
  1. Mass spectrometry
  I. Title   II. Frearson, Martin
  III. Prichard, F. Elizabeth   IV. ACOL
  V. Series
  543'.0873   QD96.M3

  ISBN 0 471 91388 X
  ISBN 0 471 91389 8 Pbk

Printed and bound in Great Britain

Analytical Chemistry

This series of texts is a result of an initiative by the Committee of Heads of Polytechnic Chemistry Departments in the United Kingdom. A project team based at Thames Polytechnic using funds available from the Manpower Services Commission 'Open Tech' Project has organised and managed the development of the material suitable for use by 'Distance Learners'. The contents of the various units have been identified, planned and written almost exclusively by groups of polytechnic staff, who are both expert in the subject area and are currently teaching in analytical chemistry.

The texts are for those interested in the basics of analytical chemistry and instrumental techniques who wish to study in a more flexible way than traditional institute attendance or to augment such attendance. A series of these units may be used by those undertaking courses leading to BTEC (levels IV and V), Royal Society of Chemistry (Certificates of Applied Chemistry) or other qualifications. The level is thus that of Senior Technician.

It is emphasised however that whilst the theoretical aspects of analytical chemistry can be studied in this way there is no substitute for the laboratory to learn the associated practical skills. In the U.K. there are nominated Polytechnics, Colleges and other Institutions who offer tutorial and practical support to achieve the practical objectives identified within each text. It is expected that many institutions worldwide will also provide such support.

The project will continue at Thames Polytechnic to support these 'Open Learning Texts', to continually refresh and update the material and to extend its coverage.

Further information about nominated support centres, the material or open learning techniques may be obtained from the project office at Thames Polytechnic, ACOL, Wellington St., Woolwich, London, SE18 6PF.

How to Use an
Open Learning Text

Open learning texts are designed as a convenient and flexible way of studying for people who, for a variety of reasons cannot use conventional education courses. You will learn from this text the principles of one subject in Analytical Chemistry, but only by putting this knowledge into practice, under professional supervision, will you gain a full understanding of the analytical techniques described.

To achieve the full benefit from an open learning text you need to plan your place and time of study.

- Find the most suitable place to study where you can work without disturbance.

- If you have a tutor supervising your study discuss with him, or her, the date by which you should have completed this text.

- Some people study perfectly well in irregular bursts, however most students find that setting aside a certain number of hours each day is the most satisfactory method. It is for you to decide which pattern of study suits you best.

- If you decide to study for several hours at once, take short breaks of five or ten minutes every half hour or so. You will find that this method maintains a higher overall level of concentration.

Before you begin a detailed reading of the text, familiarise yourself with the general layout of the material. Have a look at the course contents list at the front of the book and flip through the pages to get a general impression of the way the subject is dealt with. You will find that there is space on the pages to make comments alongside the text as you study—your own notes for highlighting points that you feel are particularly important. Indicate in the margin the points you would like to discuss further with a tutor or fellow student. When you come to revise, these personal study notes will be very useful.

∏   When you find a paragraph in the text marked with a symbol such as is shown here, this is where you get involved. At this point you are directed to do things: draw graphs, answer questions, perform calculations, etc. Do make an attempt at these activities. If necessary cover the succeeding response with a piece of paper until you are ready to read on. This is an opportunity for you to learn by participating in the subject and although the text continues by discussing your response, there is no better way to learn than by working things out for yourself.

We have introduced self assessment questions (SAQ) at appropriate places in the text. These SAQs provide for you a way of finding out if you understand what you have just been studying. There is space on the page for your answer and for any comments you want to add after reading the author's response. You will find the author's response to each SAQ at the end of the text. Compare what you have written with the response provided and read the discussion and advice.

At intervals in the text you will find a Summary and List of Objectives. The Summary will emphasise the important points covered by the material you have just read and the Objectives will give you a checklist of tasks you should then be able to achieve.

You can revise the Unit, perhaps for a formal examination, by re-reading the Summary and the Objectives, and by working through some of the SAQs. This should quickly alert you to areas of the text that need further study.

At the end of the book you will find for reference lists of commonly used scientific symbols and values, units of measurement and also a periodic table.

Contents

Study Guide

This unit is intended to provide you with a good understanding of mass spectrometry. It covers the basic theory of ion formation and behaviour, the instrumentation and the interpretation of the spectra of organic molecules. There are also sections covering combined gas chromatography – mass spectrometry and liquid chromatography – mass spectrometry, which are two of the most powerful techniques available to the analytical chemist.

Mass spectrometry has undergone a great deal of development as an empirical subject. That is, many useful approaches to the analysis and identification of organic molecules have been developed without a detailed understanding of theory of ion behaviour. This is also the way in which the subject is usually taught and it is the approach adopted here. That does not mean to say that an understanding of ion behaviour is not important – it is, but that understanding is still being developed and we wish to use the technique now.

Understanding the fragmentations of ions formed from organic molecules is greatly aided by an understanding of the basic concepts of organic chemistry which you will have met while studying that subject. Section 9 depends on this knowledge, although some of the most important concepts are highlighted in the text. It is assumed, throughout, that the reader has an understanding of chemistry equivalent to that of a student holding an HNC in Chemistry and a knowledge of physics and mathematics to at least GCE (O-level).

The course is constructed to be used at two levels – by those studying for the LRSC qualification or its equivalent and by those who wish to obtain a very detailed knowledge of mass spectrometry, much beyond that required for LRSC. Those using the text as preparation for their LRSC examinations are not expected to study the whole unit. They are advised to omit the following sections:

3; 4.3, 4.4; 6.2; 7.4; 8.2; 9.3, 9.4, 9.5.2 to 9.5.7 inclusive, 9.8, 9.9, 9.10; 10.2.

You may find that the coverage given in text-books clarifies some aspects of this text. For this reason a bibliography is provided. We hope you will look at at least one of these books sometime during studying this unit. This will provide balance in the approach to the subject.

Bibliography

1. General Analytical and Spectroscopy books. Most analytical chemistry text books have a chapter on mass spectrometry. These include:

(*a*) F. W. Fifield and D. Kealey, *Principles and Practice of Analytical Chemistry*, International Textbook Co Ltd, 2nd Edn 1983.

(*b*) H. H. Willard, L. L. Merritt, J. A. Dean and F. A. Settle, *Instrumental Methods of Analysis*, Van Nostrand, 6th Edn, 1981.

(*c*) R. L. Pecsok, L. D. Shields, T. Cairns and I. G. McWilliam, *Modern Methods of Chemical Analysis*, J. Wiley and Sons, 2nd Edn 1976.

Spectroscopy books have in some cases just a brief description of mass spectrometry but all have examples for analysis and show how mass spectrometry is used in conjunction with other spectroscopic techniques.

(*d*) D. L. Pavia, G. M. Lampman and G. S. Kriz, *Introduction to Spectroscopy*, Saunders, 1979.

(*e*) R. Davis and C. H. J. Wells, *Spectral Problems in Organic Chemistry*, International Textbook Co Ltd, 1984.

2. Books specialising in Mass Spectrometry.

(*a*) J. H. Beynon and A. G. Brenton, *Introduction to Mass Spectrometry*, Univ Wales Press, 1982.

(*b*) I. Howe, D. H. Williams and R. D. Bowen, *Mass Spectrometry, Principles and Applications*, 2nd Ed, McGraw Hill, 1981.

(*c*) M. E. Rose and R. A. W. Johnstone, *Mass Spectrometry for Chemists and Biochemists*, Cambridge U.P., 1982.

(*d*) J. R. Majer, *The Mass Spectrometer*, Wykeham, 1977.

(*e*) B. S. Middleditch, *Practical Mass Spectrometry*, Plenum, 1979.

(*f*) G. M. Message, *Practical Aspects of Gas Chromatography – Mass Spectrometry*, Wiley–Interscience, 1984.

(*g*) B. J. Millard, *Quantitative Mass Spectrometry*, Heyden, 1978.

(*h*) J. R. Chapman, *Computers in Mass Spectrometry*, Academic Press, 1978.

(*i*) J. P. Payne, J. A. Bushman and D. W. Hill, *The Medical and Biological Application of Mass Spectrometry*, Academic Press, 1979.

(*j*) H. E. Duckworth, R. C. Barber and V. S. Venkatasubramanian, *Mass Spectroscopy*, 2nd Ed, Cambridge University Press, 1986.

Acknowledgements

Figures 3.1c, 3.1g, 3.7a, 5.1a and 6.2c are redrawn from Rose and Johnstone, *Mass Spectrometry for Chemists and Biochemists*, Cambridge University Press, 1982 with permission.

Figures 3.2a, 3.3b and 4.2g are redrawn from D. H. Williams and I. Howe, *Mass Spectrometry, Principles and Applications*, McGraw Hill, 1972.

Figure 3.2b is redrawn from S. B. Martin, *Biomed. Mass Spectrom.*, **1**, 320, 1974 with permission from John Wiley & Sons Ltd.

Figure 3.4a is redrawn from N. J. Haskins et al, *Biomed. Mass Spectrom.*, **1**, 423, 1974.

Figures 3.5b, c and d are redrawn from *Fast Atom Bombardment Spectra*, published by Kratos Analytical, by permission of the publishers.

Figure 4.3a is redrawn from Millard, *Quantitative Mass Spectrometry*, Heyden, 1979 with permission of John Wiley & Sons Ltd.

Figure 4.3b is redrawn from J. R. Chapman, *Computers in Mass Spectrometry*, Academic Press, 1978 with permission.

Figure 7.3b, 7.3c, 9.6e, 9.6g and 9.6h are redrawn from R. Davies and C. H. J. Wells *Spectral Problems in Organic Chemistry*, International Textbook Company, Blackie Group, 1984. Permission has been requested.

Figures 9.6b, 9.10c and 9.10d are redrawn from H. Budzikiewicz, C. Djerassi and D. H. Williams, _Mass Spectrometry of Organic Compounds_, Holden-Day Inc, 1967. Permission has been requested.

Figures 9.6f and 9.7d are mass spectra taken from D. L. Pavia, G. M. Lampman and G. C. Kris Jr, _Introduction to Spectroscopy_, Saunders and Co, 1979 with permission from Holt, Rinehart and Winston, Inc.

Figures 10.1a, 10.2d and 10.2e are taken from instrument manuals published by V G Instruments Group Ltd and reproduced with permission.

Figures 10.2b and c are redrawn from _European Spectroscopy News_, 63, 22, 1985 with permission of John Wiley & Sons Ltd.

Figure 10.1h is a cartoon by Birch first published in _Chemistry in Britain_, September 1985 and reproduced with permission of the Royal Society of Chemistry.

1. Introduction to Mass Spectrometry

This part of the Unit will introduce you to some of the basic principles on which mass spectrometry is based. It will discuss the phenomena of ionisation and fragmentation of organic molecules. Some basic terminology will be defined and exemplified by examination of a mass spectrum.

1.1. ANALYSIS BY MASS SPECTROMETRY

Mass spectrometry is unique among the molecular spectrometric methods of analysis in terms of both the principle on which it is based and the instrumentation required to perform the experiments.

∏   Think about any other common technique used for molecular spectroscopic analysis with which you have some familiarity. Without going into too much detail, try to recall the principle on which it is based.

The common techniques used for analysing molecules, other than mass spectrometry, are infrared, ultraviolet/visible and nuclear magnetic resonance spectrometry. They are all based on the same basic principle. This is that a molecule exists in a ground state energy level and when irradiated with electromagnetic radiation of the correct frequency, if the electromagnetic radiation is absorbed the molecule is promoted to an excited state energy level. Absorption of infrared radiation is accompanied by vibrational changes within the molecule and absorption of ultraviolet or visible radiation is accompanied by electronic excitation. In nuclear magnetic resonance spectrometry, when molecules are placed in a magnetic field, and irradiated with radio waves absorption of energy is accompanied by changes in the nuclear spin energy states. Thus in all these and in other methods, electromagnetic radiation is absorbed in exciting the molecule. In each case, the molecule will eventually return to its ground state energy level and the process can be repeated. These forms of spectrometry are, therefore, non-destructive methods of analysis.

In the mass spectrometer, molecules are ionised and it is these ions that are subsequently examined in detail. The first thing to say is that once the ions have been formed, one cannot reverse the process and recover the sample. Mass spectrometry is thus a destructive means of analysis.

1.2.   IONISATION OF MOLECULES

Let us now begin our discussion of the ionisation of molecules in a mass spectrometer. Almost invariably, positive ions are formed by simply removing one electron from the original molecule. For the moment do not worry how this is done experimentally. For a molecule M the process can be represented by Eq. 1.2a.

$$M \rightarrow M^+ + e \tag{1.2a}$$

The ion M^+ is known as the *molecular ion*. Two of the most important properties of an ion are its charge (z) and its mass (m). The mass spectrometer measures the *mass to charge ratio* (m/z) for ions. Wherever ions are formed with a single positive charge, then z will be $+1$ and thus the mass to change ratio will be equivalent to the mass of the ion. The mass of the ion is, of course, related to the relative molecular mass of the molecule.

SAQ 1.2a

> Given that 1 atomic mass unit (amu) = 1.661 $\times 10^{-27}$ kg and that the mass of the electron = 9.110×10^{-31} kg, express the mass of the electron in atomic mass units and use these data to calculate the mass difference between: (i) CH_4 and CH_4^+, and (ii) $C_{20} H_{42}$ and $C_{20} H_{42}^+$, if each mass is required to the nearest whole number.

SAQ 1.2b

> Calculate the mass difference, to five places of decimal, between the following pairs:
>
> (i) CH_4 and CH_4^+;
>
> (ii) $C_{20} H_{42}$ and $C_{20} H_{42}^+$.

You will see from the response to SAQ 1.2a that removing one (or even two or three electrons) has no effect on a determination of the relative molecular mass of a molecule by mass spectrometry when measured to the nearest whole number. It only introduces an error of 0.00002 amu when measured to five places of decimals, see SAQ 1.2b.

This represents an error of $1.25 \times 10^{-4}\%$ for CH_4 and $7.08 \times 10^{-6}\%$ for $C_{20}H_{42}$. In practice, even the most accurate mass spectrometer cannot measure to this degree of precision and so masses of molecular ions determined by mass spectrometry can be directly related to relative molecular masses of the parent molecules.

Let us now return to Eq. 1.2a and take a slightly closer look at it. We have, in fact made a small but very significant omission in writing that equation. It is shown in its correct form in Eq. 1.2b.

$$M: \rightarrow M^{\ddot{+}} + e \qquad\qquad (1.2b)$$

The difference is that we have now given the proper designation of the electron configuration of the positive ion. Virtually all stable organic molecules have an even number of electrons. (Thus, we have represented the molecule as M:). These electrons occupy either bonding or non-bonding orbitals. Thus, if on ionisation one electron is removed, the resultant molecular ion will have an odd electron configuration – it will be left with one unpaired electron. It is thus designated as $M^{\ddot{+}}$. That is, it is a radical cation. We will return to this aspect of electronic configuration shortly, but at this stage it is worth making the point that *molecular ions always have an odd electron configuration.*

1.3. FRAGMENTATION

Let us now look briefly at the energetics of positive ion formation. In order to form a molecular ion from a neutral molecule, we must, of course, reach the first ionisation potential of the molecule. The value of this ionisation potential will be determined by the nature of the highest occupied molecular orbital, as it is from this orbital that an electron is usually removed.

∏ What types of molecular orbitals are commonly found in both saturated and unsaturated molecules?

In organic molecules electrons may occupy σ-bonding, (σ), σ-antibonding (σ^*), π-bonding (π), π-antibonding (π^*) and non-bonding (n) orbitals.

SAQ 1.3a

> On a qualitative energy level diagram show the order in which the above mentioned molecular orbitals appear. Having done this, use the diagram to decide the order of ionisation energies for alkanes, alkenes and compounds containing a carbonyl group.

Most organic molecules have first ionisation potentials in the range 8–15 eV (cf 5.2 eV for the first ionisation potential of sodium) (1 eV $= 9.649 \times 10^4$ J mol^{-1}).

In mass spectrometry, ionisation can be achieved in a number of ways most of which involve using an energy greater than that required just to achieve ion formation. Typically, using a method of ionisation called electron impact, in which a beam of electrons is directed at a vaporised sample, the ionising electrons may have energies in the range 20 to 75 eV and most often an energy of 70 eV is used. This is clearly far in excess of the energy required simply to ionise the molecule. In the ionisation process some of this extra energy is transferred to the ion as it is formed. The ion is thus formed with excess internal energy and this may be partitioned in a number of ways.

∏   List as many ways as you can in which the excess energy may be partitioned.

It may lead to the ion existing in an electronically excited state; having excess rotational and/or vibrational energy; or it may appear as translational energy.

The most important of these, as far as mass spectrometry is concerned, are the ions existing in an electronically excited state and/or having excess vibrational energy. Ions existing in excited electronic states are often unstable and prone to dissociation. Furthermore, if the excess vibrational energy is sufficiently large, it will lead to vibrations of amplitude greater than the elastic limits of at least some of the bonds, again leading to dissociation.

Thus ions will be prone to dissociation or as it is more commonly called, *fragmentation* and these fragmentation processes almost invariably yield a new ion and a neutral particle.

∏   Consider the molecular ion, M^+ fragmenting into a new ion and a neutral particle. For each molecular ion there are two different reaction pathways that can be followed. Taking into account the need to balance equations representing fragmentation in terms of both the positive charge and the odd electron, see if you can write equations representing these two processes.

The equations are:

$$M^{+\cdot} \rightarrow A^+ + N^\cdot \qquad (1.3a)$$

$$M^{+\cdot} \rightarrow B^{+\cdot} + N \qquad (1.3b)$$

Eq. 1.3a shows the odd electron molecular ion fragmenting into an even-electron fragment ion and an odd electron neutral. The neutral may be an atom, such as $H^\cdot$, or a radical, such as $CH_3^\cdot$.

Eq. 1.3b shows the molecular ion fragmenting into an odd electron fragment ion and an even electron neutral. In this case the neutral may be a stable molecule, such as C_2H_4 or CO, or an even electron species such as ketene, CH_2CO.

SAQ 1.3b   For the molecule, C_2H_6 represent three fragmentation processes for the molecular ion, each of which involves only C—H bond cleavage but which lead to the production of a neutral atom, a neutral radical and a neutral molecule respectively.

You will note that the answers given to SAQ 1.3b form two complementary pairs. The first pair is formation of either $C_2H_5^+$ or H^+ and the second pair is formation of either $C_2H_4^{+\cdot}$ or $H_2^{+\cdot}$. It is quite common in mass spectrometry to observe fragment ions which form such complementary pairs, but it is important to realise that each ion is formed by a separate fragmentation process.

∏   Examine the four fragmentation pathways of $C_2H_6^{+\cdot}$ shown in the response to SAQ 1.3b, and in each case, list how many bonds are broken and formed.

In both the pathways leading to the formation of $C_2H_5^+$ and H^+ only one bond is broken in each case and no new bonds are formed. In the pathway leading to formation of $H_2^{+\cdot}$ and C_2H_4, two bonds are broken and two bonds formed.

$$\begin{matrix} H_2C-H^{+\cdot} \\ | \\ H_2C-H \end{matrix} \quad \longrightarrow \quad \begin{matrix} H_2C \\ || \\ H_2C \end{matrix} \;+\; \begin{matrix} H^{+\cdot} \\ | \\ H \end{matrix}$$

Note one of the new bonds formed in this process converts the $C-C$ single bond to a $C=C$ double bond.

In the formation of $C_2H_4^{+\cdot}$ and H_2, the answer is not quite so clear cut, because $C_2H_4^{+\cdot}$ will not have sufficient electrons to form 4 $C-H$ single bonds and a $C=C$ double bond, but even so, it is usually represented this way:

$$\begin{matrix} H_2C-H^{+\cdot} \\ | \\ H_2C-H \end{matrix} \quad \longrightarrow \quad \begin{matrix} H_2C^{+\cdot} \\ || \\ H_2C \end{matrix} \;+\; \begin{matrix} H \\ | \\ H \end{matrix}$$

If the fragment ions formed in Eq. 1.3a and Eq. 1.3b, A^+ or $B^{+\cdot}$, have sufficient internal energy associated with them after their formation, they will, themselves, subsequently decompose with the formation of new fragment ions as shown below:

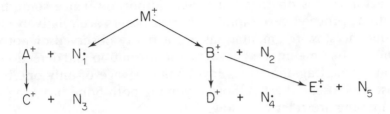

Again, these secondary fragmentation processes may involve loss of either even or odd electron neutrals and both molecular and fragment ions may decompose by competing, as well as unique pathways. Such fragmentation processes give rise to the range of ions which form the *mass spectrum* for any compound and the fragmentation pathways together form the *fragmentation pattern*, which is characteristic of that compound.

The complexity of the fragmentation pattern is usually related to the complexity of the original molecule, but this is not always the case. Some relatively simple molecules show a considerable number of competing and consecutive fragmentations and some very complex molecules show spectra dominated by only one process.

∏ The extent to which any fragmentation pathway is followed or dominates over others is determined by a number of factors. List as many of these as you can, but don't worry if you can't think of very many.

There are four factors which determine fragmentation pathways.

(*i*) The strengths of the bonds which are to be broken.

(*ii*) The stability of the products of fragmentation, both ions and neutrals.

(*iii*) The internal energy of the fragmenting ions.

(*iv*) The time interval between ion formation and ion detection.

I hope you will appreciate that points (*i*) (*ii*) and (*iii*) are important, even if you didn't manage to list them all. The last point, the

time factor, is less obvious but is also important, since some fragmentations proceed very rapidly whereas others are relatively slow.
Whether the slow fragmentations are actually observed or not will
depend on the time interval between ion formation and detection. In
general terms, fragmentations involving cleavage of only one bond
are fast processes, whereas those involving both bond breaking and
bond making are relatively slow.

∏   Think back to your previous courses on kinetics and mech-
       anism, and try to explain why the latter processes are slow.

Bond making and bond breaking in the same process requires the
ion to achieve the correct transition state for reaction. Only a proportion of the molecules will achieve the correct geometry and possess the correct energy for reaction. Thus, the number of reactions
per second will be relatively low compared with simple bond cleavages where molecules only have to achieve the correct energy.

From the above discussion on fragmentation you may now be wondering if molecular ions are ever detected in spectra obtained using
70 eV bombarding electrons, and if they are not detected, how mass
spectrometers measure relative molecular mass. In fact, for most
molecules, molecular ions are observed as well as a considerable
number of fragment ions. This arises because, in any one experiment, mass spectrometers usually produce molecular ions with a
spread of internal energies. Lets not worry about the experimental
reasons for this, but concentrate on the consequences of it. Each
molecular ion that is formed in a single experiment will fragment at
a rate partly determined by the internal energy gained at the instant
of ionisation. This is rather different from normal chemical processes and occurs because mass spectrometers operate under high
vacuum.

∏   Pause for a moment and see if you can think out the con-
       sequence of this difference between ions in a vacuum in a
       mass spectrometer and ions undergoing reactions in a solu-
       tion. Don't worry too much if you can't figure this one out.

In reactions in solution (a condensed phase), continual collisions
of reacting species lead to an equilibration of internal energies, so

they all have more or less the same energy for reaction. In the mass spectrometer, collisions between ions are rare because the disperse nature of the particles in the vacuum means they have a long mean free path. Thus there is no collisional equilibration of internal energies. Therefore whether an ion decomposes or not depends upon the initial internal energy gained at its formation. A hypothetical internal energy distribution is shown in Fig. 1.3a.

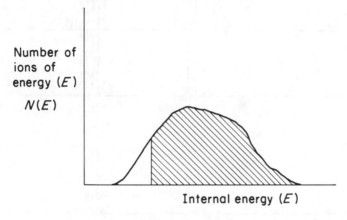

Fig. 1.3a. *Hypothetical internal energy distribution of an ion*

That proportion of ions in the unshaded area have insufficient internal energy to fragment within the time scale of the mass spectrometer and are thus detected as molecular ions. The proportions of ions in the shaded area have sufficient energy to fragment and, depending on the factors listed above, will form fragment ions.

1.4. INTERPRETATION OF MASS SPECTRA

Now, you may feel its been a long time coming, but lets examine our first mass spectrum. It is that of methanol and is shown in Fig. 1.4a as the photographic trace that is obtained directly from the recorders of many mass spectrometers. Three superimposed traces are shown, the middle one being obtained at three times the sensitivity of the bottom one, and the top one at ten times the sensitivity of the bottom one. This allows both very abundant ions and ions of very low abundance to be recorded on a single scan of the spectrometer.

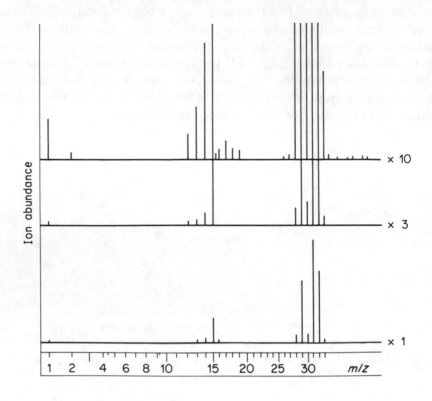

Fig. 1.4a. *Mass spectrum of CH_3OH*

Π   Examine the spectrum of methanol and list the m/z values
of the four most abundant ions. Attempt to assign formulae
to these ions.

The major ions are as follows:

m/z	32	$CH_3OH^{+\cdot}$
m/z	31	CH_2OH^+
m/z	29	CHO^+
m/z	15	CH_3^+

The ion, m/z 32 is of course the molecular ion. Apart from a very
small line at m/z of 33 it is the highest mass ion in the spectrum.
The other ions are fragment ions and arise as follows.

m/z 31 is formed by the loss of a hydrogen atom from the molecular ion. It is one of the C—H bonds which is broken in this process rather than the O—H bond. Hence its formulation as CH_2OH^+. Of course just looking at our mass spectrum of methanol does not allow us to distinguish between C—H and O—H bond cleavage, so don't be concerned if you wrote down CH_3O^+.

m/z 29 arises as a consequence of the CH_2OH^+ ion losing a molecule of hydrogen and can be formulated as CHO^+.

m/z 15 is the CH_3^+ ion formed by those molecular ions which have the correct energy for fragmentation of the C—O bond rather than C—H bond cleavage.

SAQ 1.4a

Draw up a fragmentation pattern for methanol based on the above assignments and underline the most intense ion in the spectrum.

The most intense ion in the methanol spectrum is at m/z 31, CH_2OH^+, and this is termed the *base peak* of the spectrum. Since in one common method of representing mass spectra in the form of line or bar diagrams, the most intense peak is assigned 100% relative abundance and this forms the base against which the relative abundance of other ions is measured.

SAQ 1.4b   By measuring the intensities of the four major peaks in Fig. 1.4a, setting the most intense to 100% relative abundance and normalising the others to it, construct a bar diagram for methanol.

1.4.1. Peaks of Low Relative Abundance

Let us now consider why there is a very small peak at m/z 33. This is an *isotope peak* and arises from the small proportion of methanol molecules in the sample analysed that contain either a ^{13}C atom (1.08% natural abundance) rather than a ^{12}C atom or a single ^2H atom (0.02% natural abundance) and three ^1H atoms rather than four ^1H atoms. Although such isotope peaks are often small, they can be important and the subject of isotopes in mass spectrometry is discussed in detail in Part 7. Lastly, if you again look carefully at the methanol spectrum in the region of m/z 15 on the photographic trace (Fig. 1.4a) you will see a small peak at m/z 15.5. This is not as unusual as it seems at first sight. It is in fact a *doubly charged ion* of mass 31 and therefore m/z 15.5. Firstly notice how weak it is compared to m/z 31. This is not perhaps surprising, since it must be more difficult to remove two electrons rather than one. Secondly, it is, of course, relatively easy to identify doubly charged ions of odd mass, since they have half integral m/z values. Identification of doubly charged ions of even mass is rather more difficult, since they will occur at the same position in the spectrum as singly charged ions of half the mass.

∏ The ion CH_3OH^{2+} has an m/z value of 16. What singly charged ions of this m/z value might you expect in the methanol spectrum?

The two possibilities are $CH_4^{+\cdot}$ formed by eliminating an oxygen atom from $CH_3OH^{+\cdot}$ and the O^+ ion.

1.4.2. Rules for Assignment of Fragments

In SAQ 1.4a you constructed your first fragmentation pattern. Although that one was relatively straightforward, deciding on whether to write a fragment ion with an odd electron ($^+_\cdot$) or even electron ($+$) configuration often causes difficulties to begin with. So helpful tips for doing this are.

(i) If the formula of the fragment ion corresponds with that of a stable neutral molecule or an isomer of such a molecule, the ion is always odd electron.

eg $C_2H_4^+$ CH_3COOH^+ $CH_2C\begin{smallmatrix} OH^+ \\ \\ OH \end{smallmatrix}$

(*ii*) If the formula of the ion is that of a stable neutral molecule minus an atom or radical, then the ion is even electron.

eg $C_6H_5^+$, $C_4H_9^+$, CH_2OH^+

(*iii*) If the formula of the ion is that of a stable neutral molecule minus a second stable neutral molecule, then the ion is odd electron.

eg $C_4H_4^+$ ($C_6H_6 - C_2H_2$)

SAQ 1.4c   Decide whether ions of the following formulae are odd or even electron.

(*i*) C_3H_8

(*ii*) CH_3CO

(*iii*) C_6H_4

(*iv*) $C_6H_5COOC_2H_5$

(*v*)   HCl

(*vi*) C_7H_{15}

(*vii*) C_7H_7

(*viii*) $C_2H_3OC_3H_7$

(*ix*) C_7H_{13}

(*x*) $C_6H_5NO_2$

SAQ 1.4c

SAQ 1.4d   Calculate m/z values (to the nearest whole number) for the following molecular ions:

    (*i*) $C_2H_6^{+\cdot}$

    (*ii*) $(CH_3)_2CO^{+\cdot}$

    (*iii*) $C_6H_6^{+\cdot}$

    (*iv*) $CH_3NH_2^{+\cdot}$

    (*v*) $CH_3COOH^{+\cdot}$

    (*vi*) $C_6H_5CH_2CN^{+\cdot}$

    (*vii*) $C_4H_9SH^{+\cdot}$

    (*viii*) $C_6H_4(NH_2)_2^{+\cdot}$

    (*ix*) $C_6H_3(NH_2)_3^{+\cdot}$

    (*x*) $(CH_3O)_3P^{+\cdot}$

    (*xi*) $C_6H_6^{2+}$

SAQ 1.4d

1.4.3.   The Nitrogen Rule

If you look at the answers to SAQ 1.4d again, you will perhaps notice something which is important in mass spectrometry. What you should have noticed is that all the singly charged molecular ions containing an odd number of nitrogen atoms have odd numbered masses and those with either an even number of nitrogen atoms, or none at all have even masses. This is, in fact, a universal rule, known as the *Nitrogen Rule* and can be extended from molecular ions to include fragment ions. In its general form the nitrogen rule can be stated as: all odd electron ions have even m/z values unless they contain an odd number of nitrogen atoms. (For the purposes of the rule, zero is an even number).

You may find this rule difficult to accept at first sight, but be assured it is universal, try as many examples as you wish. It arises because nitrogen is the only element commonly encountered in organic chemistry which has even relative atomic mass but an odd valency.

The following SAQs form a revision test and you should not proceed to the next part until you are sure you are able to answer all the questions. If you get any of the questions wrong, go back and revise that section.

SAQ 1.4e Calculate mass to charge ratios for singly charged ions of the following formulae:

(*i*) C_4H_{10};

(*ii*) C_6H_5;

(*iii*) CH_2COOH;

(*iv*) $C_6H_5NH_2$;

(*v*) Cl_2;

(*vi*) $(C_6H_5)_3P$;

(*vii*) $H_2NOCCONH_2$;

(*viii*) C_7H_7.

SAQ 1.4f	For each of the ions in SAQ 1.4e decide whether it should be classified as an even or an odd electron ion.

SAQ 1.4g	Examine the following fragmentation processes. Fill in the missing pluses ($+$) and dots (.) indicating ionic and radical character respectively. The ionic product is written first on the right hand side of each equation. (i) serves as an example.

(i) $\quad C_4H_{10}^{\stackrel{+}{\cdot}}$ $\qquad \rightarrow C_4H_8^{\stackrel{+}{\cdot}} + H_2$

(ii) $\quad C_5H_{11}CONH_2$ $\rightarrow C_5H_{11} + CONH_2$

(iii) $\quad C_7H_7$ $\qquad\qquad \rightarrow C_5H_5 + C_2H_2$

(iv) $\quad (CH_3)_2CHOH$ $\rightarrow CH_3CHOH + CH_3$

$\longrightarrow$

**SAQ 1.4g
(cont.)**

(v) C_2H_5 $\rightarrow$ $C_2H_3 + H_2$

(vi) C_2H_5SH $\rightarrow$ $C_2H_4 + H_2S$

(vii) $C_{10}H_{22}$ $\rightarrow$ $C_7H_{15} + C_3H_7$

$(viii)$ C_4H_9 $\rightarrow$ $C_3H_5 + CH_4$

(ix) $C_6H_5NO_2$ $\rightarrow$ $C_6H_5 + NO_2$

(x) $C_6H_5SO_2NH_2$ $\rightarrow$ $C_6H_5NH_2 + SO_2$

SAQ 1.4h

Given in Fig. 1.4b is the mass spectrum of ethanol, C_2H_5OH.

(i) Write down the m/z of the molecular ion.

(ii) Write down the m/z of the base peak.

(iii) The base peak is formed from the molecular ion in a single step. Write down an equation for this fragmentation process (nb refer back to the discussion of methanol for the formula of the base peak).

(iv) Suggest two fragmentation processes for production of ions of m/z 29 which may have different formulae (refer back to discussion of the methanol spectrum if you are stuck). $\longrightarrow$

SAQ 1.4h
(cont.)

(*v*)   Write down fragmentation processes for the stepwise decompositions,

m/z 46 → m/z 45 → m/z 43.

(*vi*) Assign a formula to m/z 18 and give a fragmentation process for its direct formation from the molecular ion.

(*vii*) Combine the fragmentation pathways discussed above to give the fragmentation pattern for ethanol.

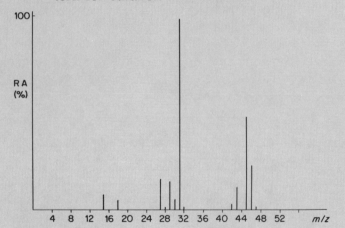

Fig. 1.4b.   *Mass spectrum of ethanol (C_2H_5OH)*

SAQ 1.4h

Summary

You have now been introduced to some of the basic ideas of mass spectrometry. These include the ionisation and fragmentation of organic molecules. You have also seen a mass spectrum presented as a photographic trace, a bar diagram and in the form of a fragmentation pattern. I hope you also appreciate the difference between odd-and even-electron ions and neutrals and understand the importance of the nitrogen rule.

Objectives

When you have completed this part of the unit you should be able to:

- explain the principle on which mass spectrometry is based and how ions are formed from molecules;

- explain the circumstances under which ions fragment;

- recognise the difference between, and the significance of, odd-and even-electron ions;

- explain what is meant by the terms fragmentation pathway and fragmentation pattern;

- define the terms molecular ion, fragment ion and base peak;

- represent mass spectra as bar diagrams;

- use the nitrogen rule.

2. The Mass Spectrometer

This part of the Unit introduces you to the instrumental side of mass spectrometry. It discusses the most commonly encountered type of mass spectrometer, explaining the function of the components and discussing the physical principles involved. It also explains the type of inlet systems used on all mass spectrometers.

2.1. THE MASS SPECTROMETER

Let's begin with an explanation of how one of the most common types of mass spectrometer works. But first a simple block diagram that introduces you to the principle components of this and most other mass spectrometers (Fig. 2.1a).

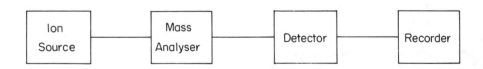

Fig. 2.1a. *Block diagram of mass spectrometer*

The role of the *ion source* is fairly obvious; it converts molecules into ions which usually carry a positive charge, although negative ions may be produced. The *mass analyser* isn't an absolutely accurate name, it should really be called the 'mass to charge ratio analyser'. As you might expect, this analyser distinguishes ions according to their mass to charge ratios. As most ions carry a single positive charge (ie $z = +1$) m/z is equivalent to m and thus it is legitimate to use the term *mass analyser*. When ion analysis has been achieved, the ions strike the *detector* and the information gained can be recorded in several ways.

The instrument we are going to examine is called a single focussing magnetic sector mass spectrometer with an electron impact ion source. Quite a mouthful, but it's well worth remembering, as it is probably the most commonly encountered instrument. Fig. 2.1b shows a diagram of this type of mass spectrometer.

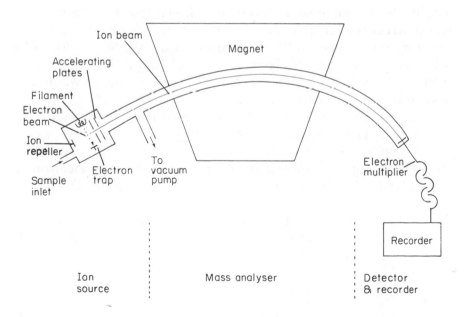

Fig 2.1b. *Single focussing mass spectrometer with an electron impact ion source*

I have annotated the diagram in terms of the segments we have previously discussed. Let's now deal with each in turn, but at the outset, note that the whole instrument is under vacuum, the pressure within the instrument is about 10^{-6} torr, (1 torr $= 133.325$ Nm^{-2}) – we'll see why as our journey through the mass spectrometer progresses.

Firstly the *ion source* – or rather not quite first – because we must have a means of getting a sample into the instrument and making sure the sample is in the vapour phase, but we'll return to that later. So assuming we have sample vapour in the ion source, ionisation is brought about by bombarding the sample molecules with high energy electrons. The electrons are emitted from a heated metal filament (usually made from tungsten or rhenium wire). On the opposite side of the ion source is a plate known as the *electron trap*. The electron trap has an electrical connection to the filament and is held at a positive potential with respect to the filament.

∏   What effect will this positive potential have on the electrons?

The negatively charged electrons are drawn towards the trap and travel across the ion chamber. Not shown on our diagram are some small magnets placed in this region of the source which are called the *collimating magnets*. They have the effect of drawing the electrons into a well-defined beam. Now for the crux of the whole process of ionisation. This is carried out by *electron impact*, but the term somewhat misleading as the electrons do not 'hit' the molecule in the classical sense. Rather, they either pass through or very close to the sample molecules. When these bombarding electrons have sufficient energy, the field they generate interacts with electrons of the molecule, leading to ionisation by removal of an electron.

∏   The process can be written in terms of an equation where the bombarding electron (e_b) reacts with the molecule (M). Complete the equation

$$e_b + M \rightarrow$$

The equation is

$$e_b + M \rightarrow M^{+\cdot} + e_m^- + e_b \qquad (2.1a)$$

where e_m represents an electron removed from the molecule. Thus, the molecular ion is formed by this process. The potential between the filament and the trap can be varied, but usually it is held at 70 volts producing 70 eV electrons.

SAQ 2.1a Can you recall from Part 1 why such a high value is used when most molecules are ionised at energies of 8–15 eV?

Note that, in Fig. 2.1b at the back of the ion source is another plate, which is labelled the *ion repeller* and on the opposite side of the source are yet another pair of plates, each with a hole in the middle, called the *accelerating plates*. These plates are also linked by electrical connections and the ion repeller is held at a positive potential with respect to the first accelerating plate. Because ions are positively charged they move away from the ion repeller towards the accelerating plates and thus begin their journey from the ion source to the detector.

∏ In some cases, electron impact mass spectrometers are used to produce negative ions. What alterations would have to be made to the source components and their electrical connections for negative ion mass spectrometry?

None of the components would have to be changed, but the polarity of the electrical connection between the ion repeller and the first accelerating plate would have to be reversed. This would allow the negatively charged repeller to cause the negatively charged ions to move out of the source.

SAQ 2.1b

Underline the correct answer from the options given in the sentences below.

(*i*)   Ionising electrons are emitted from the trap/filament/repeller plate.

(*ii*) The collimating magnets affect the ionising electrons/ions/electrons expelled from molecules on ionisation.

(*iii*) In positive ion mass spectrometry, the repeller plate is held at a positive/negative potential with respect to the first accelerating plate.

(*iv*) The trap is held at a positive/negative potential with respect to the filament.

∏   Choose the correct answers from the following list (more than one answer is correct).

The mass spectrometer operates under a vacuum in order to:

(*i*)   ensure that the filament does not burn out.

(*ii*) help to vaporise many of the samples to be analysed.

(*iii*) make electrons collide with sample molecules.

(*iv*) prevent ions, once formed, being lost by collision with atmospheric gases.

(*v*)   remove sample from the instrument after analysis.

(*vi*) remove electrons after the bombardment process.

(*i*)   This is correct. In some ways the filament in a mass spectrom-
        eter behaves like the filament of a light bulb and thus both are
        rapidly oxidised in the presence of air.

(*ii*) This is also correct. Most organic molecules are liquids or
        solids and, particularly in the case of the latter, may have only
        a very low vapour pressure at STP. Thus low pressures aid va-
        porisation and in addition the ion source may also have to be
        heated. This is, in fact, quite common and most instruments
        can be heated to about 200 °C.

(*iii*) This is not correct. Once the vaporised sample molecules are
        in the ion source, the movement and energy of the electrons is
        controlled by the electrical potential between the filament and
        the trap and the magnetic field generated by the collimating
        magnets.

(*iv*) This is true. If the pressure in the ion source was high, the
        ions would have a high probability of colliding with molecules
        of atmospheric gases (G). This usually leads to ion-molecules
        reactions of the type:

$$M^+ + G \rightarrow M + G^+$$

If this happened extensively, we would lose the ions we wish to
analyse and produce only G^+ ions (in air, this would mostly be
N_2^+ and O_2^+ – not very useful!)

Another way of expressing this need for operation under a vacuum
is to say that under these conditions, the ions have a long mean free
path.

(*v*)   This is true, as well. In practice the electron bombardment
        process is a very inefficient means of producing positive ions.
        It has been estimated that probably less than 1% of the sample
        entering the instrument is actually converted into positively

charged ions. Thus, much unused sample has to be removed before a new one can be introduced. As the sample is in the vapour phase, the vacuum is essential, if this is to be an efficient process.

(*vi*) This is incorrect. As stated under (*iii*), the behaviour of the electrons is controlled by electric and magnetic fields. In this case the positive potential on the electron trap leads to their being attracted to it.

Just as they are about to leave the ion source, ions enter the region between the *accelerating plates*. As we see from Fig. 2.1b, each of these plates has a hole in it, so that the ions can pass through them and an electrical potential is applied across these plates (this is a separate potential to that between the repeller and the first accelerating plate). The value of this potential is high (usually between 4000 and 8000 volts, depending on the particular instrument being used) and its function is to increase the kinetic energy of the ions.

The gain in kinetic energy is proportional to the potential difference (V) through which the ions have passed and the mathematical relationship between the two is:

$$\text{kinetic energy (ke)} \; = \; \frac{1}{2}mv^2 \; = \; zV \qquad (2.1b)$$

where z is the charge on the ion. We will use this equation again a little later on.

After acceleration, the ions enter the analyser region and in the case of the instrument we are discussing, analysis is achieved by means of a magnetic field. You will see from Fig. 2.1b that the ions are shown as following a curved flight path while under the influence of the magnetic field.

Any charged particle follows such a curved flight path when in a magnetic field. This can be thought of as the reverse of an experiment you will probably have performed in the physics laboratory in which a current is passed through a loop of wire, thus generating a magnetic field. In the magnetic analyser of a mass spectrometer we have the reverse of this situation. That is we have a current, which is composed of ions rather than electrons, but it is still a current, and we have a magnetic field. So, if we pass the ion current through the field, we might expect the ions to travel in a circular path. In the mass spectrometer we do not allow the ions to travel in a complete circle, but only to traverse the circumference of a sector of the circle. For this reason it is common to refer to such a mass spectrometer as a *magnetic sector* instrument.

Different manufacturers of mass spectrometers may well choose to use magnets that operate over different sectors of the circle. The most common are those employing 60° and 90° sectors.

Let's now examine the forces that affect an ion of mass m and charge z as it traverses the magnetic field of strength B.

If a particle of mass m (which may be our ion) travels with velocity v around the circumference of a circle of radius r, then it experiences a centrifugal force, the magnitude of which is given by the formula, mv^2/r.

In order to keep the ion on its circular path and not have it fly off at a tangent we need to impose a force directed towards the centre of the circle to counterbalance the centrifugal force. This we do in a mass spectrometer by imposing a magnetic force, the magnitude of which is given by the formula, Bzv.

Thus we can write

$$mv^2/r \;=\; Bzv \qquad\qquad (2.1c)$$

when the two forces are exactly balanced.

This balance is required to ensure the ions keep following the circumference of the circle.

Let me now remind you of Eq. 2.1b derived earlier in our discussion of the acceleration process.

$$\frac{1}{2}mv^2 \;=\; zV \qquad\qquad (2.1b)$$

Remember, what a mass spectrometer is required to do is measure the m/z values of ions. Let us now see how we can use Eq. 2.1b and 2.1c to show how this is done. We can see that both contain terms in m and z, and the ratio, m/z is what we wish to measure. The equations also contain one other common term – that is v, the velocity of the ions.

Now, the velocity of an ion as it traverses the magnetic field is governed solely by the kinetic energy it acquires during acceleration, so the value of v in Eq. 2.1b is the same as that in Eq. 2.1c.

We can carry out alegebraic manipulations that allow us to equate expressions for v derived from the two equations and derive a new equation, which is

$$m/z = B^2 r^2 / 2V \qquad (2.1d)$$

∏   See if you can manipulate Eq. 2.1b and Eq. 2.1c to obtain expressions for v (or v^2) and then equate them to obtain the expression for m/z (Eq. 2.1d). Do this before reading on!

$$\frac{1}{2}mv^2 = zV \qquad (2.1b)$$

this can be rearranged to:

$$v^2 = 2zV/m$$

$$Bzv = mv^2/r \qquad (2.1c)$$

cancel v on each side

$$Bz = mv/r$$

this can be rearranged to:

$$v = Bzr/m$$

squaring both sides gives:

$$v^2 = B^2 z^2 r^2 / m^2$$

We can now equate the two expressions in v^2

$$v^2 = \frac{2zV}{m} = \frac{B^2 z^2 r^2}{m^2}$$

cancelling gives:

$$V = \frac{B^2 z r^2}{2m}$$

or

$$m/z = B^2 r^2 / 2V \qquad (2.1d)$$

Don't worry too much if you didn't get through this bit of maths on your own, but do make sure you follow all the steps given in the answer.

Providing Eq. 2.1d is obeyed, an ion with that particular m/z value will traverse the magnetic field and reach the detector.

The *detector* most commonly employed is called an electron multiplier. This is shown in Fig. 2.1c.

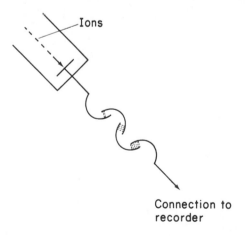

Fig. 2.1c.   *Essentials of an electron multiplier*

It consists of a series of electrodes (known as dynodes) arranged so that they are close to each other. It is enclosed in a vacuum jacket and linked to the end of the mass spectrometer. When ions hit the first dynode a shower of electrons is released which strike the second dynode. This causes the second dynode, in turn, to release a larger shower of electrons to the third and so on. This cascading effect continues through the whole series of dynodes (usually about ten, although Fig. 2.1c shows only four for simplicity). The result of this sequence is that the small electrical current generated when the ions hit the first dynode is greatly amplified (by factors of up to 10^6) and becomes large enough to be passed to a recording device. We will discuss recording devices later.

SAQ 2.1c

As far as you can, sequence the following components of the single focussing magnetic sector mass spectrometer in the order that an ion would experience their effect, after its formation to its detection.

Magnet;
Accelerating plates;
Vacuum system;
Electron multiplier;
Repeller plate.

Let us now return to Eq. 2.1d

$$m/z = B^2 r^2 / 2V \qquad (2.1d)$$

What this equation is telling us is that at any particular fixed values of magnetic field strength, B, and accelerating voltage V, only ions of one particular mass/charge ratio (m/z) will follow the required circular path of radius r to reach the detector. All other ions with different m/z values will be trying to follow circles of different radii, as shown in Fig. 2.1d

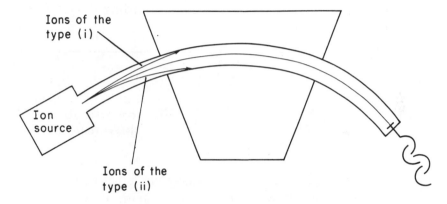

Fig. 2.1d. *Flight path of ions in a mass spectrometer*

SAQ 2.1d

In Fig 2.1d are shown the trajectories of two types of ions that do not follow the correct flight path to reach the detector. These are shown as:

ions of type (i) which are following a circular flight path with a radius larger than that required, and

ions of type (ii) which are following a circular flight path with a radius smaller than that required.

Which of the two types of ions is the heavier?

SAQ 2.1d

Eq. 2.1d also tells us how we can operate a mass spectrometer in order to make ions of different mass/charge ratio follow the correct trajectory to reach the detector.

∏   In operating a mass spectrometer we have to make ions of varying m/z follow the fixed circular path of radius, r. There are two ways this can be achieved. Can you work out what they are?

If m/z is varying and r is fixed, then we can vary either one of the two parameters in Eq. 2.1d. That is we can:

(a) vary V while holding B constant;

or

(b) vary B while holding V constant.

If (a) is used this is called a *voltage scanning* mass spectrometer and if (b) is used this is called *magnetic scanning*. It is cheaper and easier to construct a mass spectrometer with a variable accelerating voltage and a permanent magnet than to construct one with an electromagnet capable of exerting a variable field. For this reason, the first low-cost commercial mass spectrometers were voltage scanning. However, there are advantages to be gained by employing magnetic scanning, one of the most important of which is concerned with the sensitivity of the instrument. *Thus all currently commercially available mass spectrometers of this type which are used for the analysis of organic compounds are magnetic scanning.*

2.2. INLET SYSTEMS

There are several alternative methods of introducing a sample into a mass spectrometer.

∏ Why do you think we need a range of alternative methods?

Samples to be analysed come in a variety of different forms: solids, liquids and gases, single compounds and mixtures. Each brings with it its own problems of handling.

∏ If the mass spectrometer has an electron impact ion source (see Section 2.1) there are three major requirements for the inlet system. Try to decide what these might be, but don't worry if you can't think of three without a little prompting.

The three requirements are:

(*a*) that the sample must be in the vapour phase prior to ionisation;

(*b*) that the sample does not suffer thermal decomposition during the vaporisation process;

(*c*) that during admission of the sample, the pressure inside the mass spectrometer is kept as low as possible.

Thus, the type of inlet system used will depend upon the nature of the sample and its thermal stability. Both gases and liquids with a high vapour pressure at room temperature are usually admitted to the spectrometer via the *cold inlet*. This is shown in Fig. 2.2a and consists of a point at which a sample container is attached, which in turn is connected via a series of vacuum taps to a reservoir for storage of the sample, the ion source and an auxilliary vacuum system separate from that used to evacuate the spectrometer.

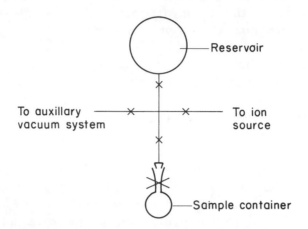

Fig. 2.2a.   *Cold inlet system (× represents a vacuum tap)*

By the appropriate manipulations, sample can be transferred to the reservoir and from there, to the ion source. In the line connecting the ion source to the inlet system is placed a fritted glass disc.

∏   Can you think what this fritted glass disc is for?

It acts as a flow restrictor, allowing some of the sample to slowly leak into the source. This has the effect of producing a steady supply of analyte into the ion source, while also keeping the pressure low. Remember the source pressure is maintained at about 10^{-6} torr. A cold inlet system is usually constructed out of both glass and metal (stainless steel) components.

When the sample is a rather involatile liquid, a similar inlet system in design to that shown in Fig. 2.2a is used, but in this case, the whole system, including the transfer lines, is heated. The sample container is also usually heated.

∏   When handling hot samples, it is preferable to construct the whole inlet system out of glass rather than metal. Why is this so?

The problem in passing hot organic vapours over metal components is that in some cases, the sample may suffer a metal catalysed decomposition. Obviously this must be avoided. Such an *all glass heated inlet system* is often referred to as an *AGHIS*.

A simpler device that is also used for liquids is the *septum inlet*, which can also cope with rather involatile liquids. This consists of a heated evacuated reservoir which is protected from the external atmosphere by a rubber septum of the type used in glc analysis. Only one valve is required to connect the septum inlet to the auxillary pump.

∏   Suggest how a sample might be admitted to the reservoir.

If you have any knowledge of glc, you will probably have answered this question easily. The answer is that the sample is injected through the septum by means of a hypodermic syringe. The liquid is vaporised in the reservoir and again passes through a leak into the ion source. This system is commonly used for admitting the material employed as a calibrant in computer-linked mass spectrometry.

∏   Would any of the above methods be suitable for admitting organic solids?

The answer is really – no. If a solid has a reasonably large vapour pressure at about 100–150 °C, you may consider using a heated inlet system. But, the problem that is likely to arise is that if the heating system has any shortcomings, cold spots may develop in the transfer lines. This will lead to ready condensation of sample vapour, thereby blocking the lines.

Solids are therefore admitted to the ion source using the *direct insertion probe*. This is shown in Fig. 2.2b and consists of a glass sample holder fitted onto a retractable metal rod. This probe unit is inserted directly into the ion source via a vacuum lock.

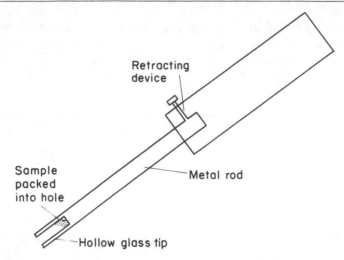

Fig. 2.2b. *Direct insertion probe for solids (not to scale)*

The probe tip is then extended until it is only a short distance away from the electron beam. Both the ion source and the probe tip itself are capable of being heated and at a pressure of about 10^{-6} Torr, heating between 50 °C and 200 °C is sufficient to volatilise most organic solids.

Most modern mass spectrometers will also have inlet systems that are connected to either gas or liquid chromatographs enabling mixture analysis. These are special devices that will be discussed in Part 10 dealing with gas chromatography – mass spectrometry, (gc/ms) and liquid chromatography – mass spectrometry (lc/ms).

SAQ 2.2a

Typically, a mass spectrometer will have a number of inlet systems. What systems are required to meet the following requirements for mass spectrometric analysis?

(*i*) A laboratory performing daily gas analysis with occasional requirements for the analysis of solids. The instrument is not linked to a computer. $\longrightarrow$

**SAQ 2.2a
(cont.)**

(*ii*) A laboratory exclusively dedicated to mixture analysis using a computerised mass spectrometer.

(*iii*) A laboratory performing analysis of single substances that are invariably solids or liquids, using a computerised spectrometer.

Summary

This Part has shown you how the most commonly encountered type of mass spectrometer operates. It has also dealt with the physical principles on which this type of instrument is based. Inlet systems have also been discussed.

Objectives

When you have completed this Part you should be able to:

- describe the major components of a single focussing magnetic sector mass spectrometer with an electron impact ion source and explain their function;

- represent the components of a single focussing magnetic sector mass spectrometer in diagrammatic form;

- describe in detail the components and operation of the electron impact ion source;

- explain why the mass spectrometer is operated under high vacuum;

- explain the function of the magnetic analyser;

- describe the passage of ions through the instrument in terms of the appropriate equations and derive the relationship, $m/z = B^2r/2V$;

- describe the electron multiplier detector;

- explain the mode of operation of this type of mass spectrometer;

- describe the inlet systems used for admitting solids, liquids and gases to the mass spectrometer;

- recognise any limitations of each inlet system;

- decide which is the inlet system of choice for any type of sample.

3. Ionisation and Ion Sources

In this part of the unit you will encounter a detailed treatment of the phenomenon of ionisation. We will begin with a discussion of electron impact ionisation which encompasses the advantages and disadvantages of the method. Alternative methods of ionisation are also discussed. These are chemical ionisation, field ionisation, field desorption and fast atom bombardment. In each case, advantages and disadvantages are highlighted.

3.1. IONISATION BY ELECTRON IMPACT

We have already dealt, briefly, with ionisation of molecules in Part 1. Let us now look at this aspect in more detail. Firstly, we will concentrate on the method of ionisation we have already discussed; electron impact (EI). We will look at its advantages and short comings and then go on to examine alternative methods of ionisation.

As we discussed earlier, when electrons pass through or very near to a molecule, they can bring about its ionisation. Usually this process results in formation of positive ions, but electron attachment and hence negative ion formation is also possible. The probability of electron capture is about 100 times less than that of electron removal and, furthermore, when a moving electron is taken up by a molecule, the translational energy of the electron must be taken up by the molecular ion (M^-). This translational energy is usually converted into vibrational energy.

∏   What might you expect to happen to such a negative molecular ion with a high degree of vibrational energy?

The molecular ion would be expected to fragment very readily. Thus negative fragment ions are observed, but it is relatively uncommon to observe negative molecular ions. Furthermore, the fragment ions are often not very informative. For example, if the molecule contains a nitro group, the spectrum is dominated by the ion, NO_2^- and there are few other ions giving any structural information.

SAQ 3.1a   Can you think of other groups or atoms commonly encountered in organic chemistry which would also be likely to dominate the negative ion spectra of molecules of which they are part? On what basis did you make your choice?

From now on we will largely confine our remarks to positive ion formation by electron bombardment. Fig. 3.1a shows a typical ionisation efficiency curve for positive ion formation, in which the energy of the ionising electrons (eV) is plotted against the number of ions produced (the ionisation efficiency).

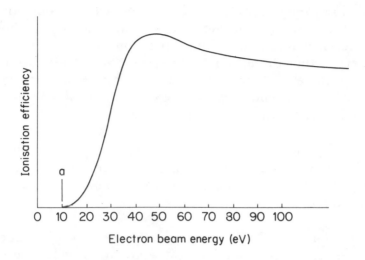

Fig. 3.1a. *A typical ionisation efficiency curve*

The point marked (*a*) on the curve corresponds to the first appearance of ions as the electron beam energy is increased.

∏   What is the beam energy value at point (*a*) known as?

It is known as the *ionisation potential* and this applies to the process in Eq. 3.1a

$$M \rightarrow M^{+\cdot} + e \qquad (3.1a)$$

Thus, in principle, mass spectrometry can be used to measure ionisation potentials. In practice, there are experimental difficulties in making such measurements and thus the results are not terribly accurate. If the ionisation efficiency curve for a fragment ion is examined, then a similar parameter can be measured. This is known as the *appearance potential* and corresponds to the overall energy for the two processes given in Eq. 3.1a and Eq. 3.1b

$$M \rightarrow M^{+\cdot} + e \qquad (3.1a)$$

$$M^{+\cdot} \rightarrow F_1^+ + N_1^\cdot \quad \text{or} \quad M^{+\cdot} \rightarrow F_2^{+\cdot} + N_2 \qquad (3.1b)$$

Information gained from ionisation and appearance potential measurements can be used to determine thermodynamic quantities such
as the heats of formation of ions and bond dissociation energies.
However, further discussion of this aspect is outside the scope of
this programme and the interested reader is referred to the text-
books by Williams and Howe and Rose and Johnstone listed in the
bibliography for a more detailed coverage.

If we look again at Fig. 3.1a, we can see that the number of ions
formed increases dramatically as the beam energy is increased from
the ionisation potential to about 50 eV. It then reaches a virtually
constant value.

∏   Explain why the shape of the curve helps you to understand
       earlier comments that mass spectra are usually recorded us-
       ing 50–70 eV bombarding electrons with preference being
       given to the higher value.

In order for the spectrometer to operate at maximum sensitivity,
we require the maximum yield of ions from any given amount of
sample. The curve shows we would get this at approximately 50 eV.
However, at that value we are very close to the steeply increasing
part of the curve and thus any small unintentional experimental
decrease in beam energy would lead to a dramatic reduction in
ionisation efficiency. If we increase the operational beam energy to
70 eV, we sacrifice only a small amount in ionisation efficiency, but
move well into the plateau region, where small changes in beam
energy have little effect on ionisation efficiency. For this reason
electron impact mass spectra are usually recorded at 70 eV.

∏   There is a further reason for recording mass spectra at ener-
       gies well above the ionisation potential. This was discussed
       in Part 1. Can you recall what it was?

A spectrum recorded with a beam energy set at or about the value
of the ionisation potential will show the presence of only molecular
ions. Fragmentation only becomes important at higher beam energies. The spectra of C_6H_5COOH recorded at 9, 12, 15, 20, 30 and
70 eV are shown in Fig. 3.1b.

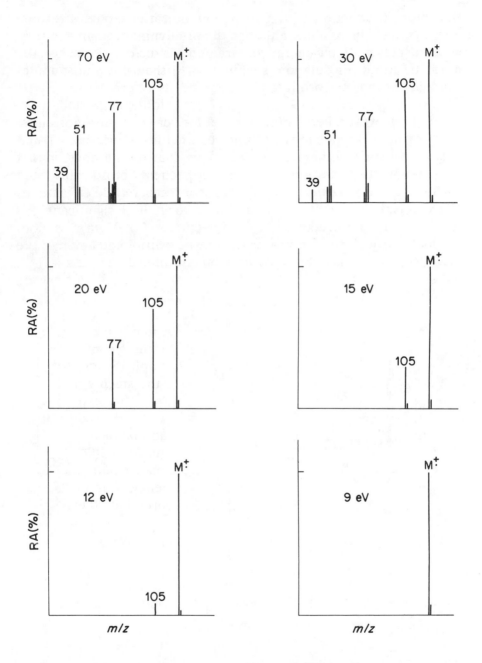

Fig. 3.1b. *Mass spectrum of* C_6H_5COOH *using different beam energies*

Maximum production of fragment ions is not observed until one employs the higher values, although the predominant fragmentation, which occurs with low energy requirements is seen at 12 eV. Usually most structural information is gained when the maximum number of fragment ions are produced.

It should, however, be pointed out that for very complex molecules, many of the low mass ions will not be structurally informative and may in fact conceal the presence of the most important ions, so that it is often worth recording a low voltage spectrum. This is illustrated by the spectra of *genipin* (Fig. 3.1c) recorded at (*a*) 70 eV and (*b*) 10 eV. It is the case that m/z 78 and m/z 96 are both significant ions in elucidating the structure of this molecule. At 70 eV, m/z 96 does not look particularly significant and may be overlooked amongst the plethora of other ions between m/z 60 and m/z 100.

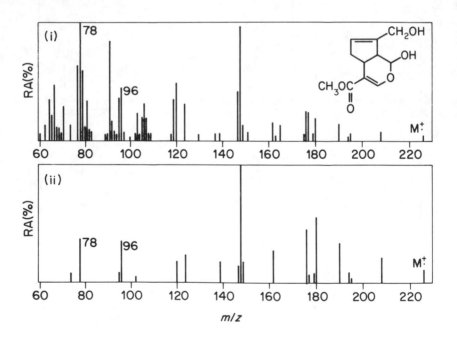

Fig. 3.1c.   *Mass spectrum of genipin at (i) 70 eV and (ii) 10 eV*

The low voltage spectrum clearly reduces this complexity without losing the important ions.

Electron impact ionisation is the most widely used method of ionisation in mass spectrometry. This is because for most molecules it produces both molecular and fragment ions. Thus, it allows determination of both relative molecular mass and molecular structure for such molecules. You will learn about interpreting such spectra later.

In some cases, however, there are problems with electron impact spectra. This is illustrated with some examples.

The mass spectra of $C_2H_5CH(CH_3)OH$ is shown in Fig. 3.1d and that of a C_9H_{20} isomer in Fig. 3.1e.

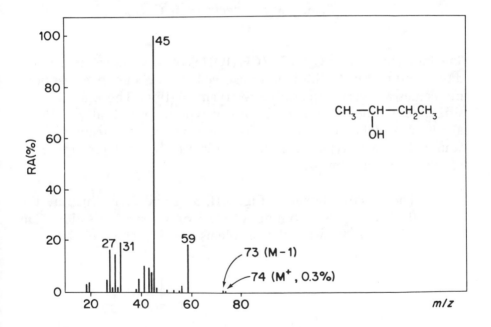

Fig. 3.1d. *Mass spectrum of $C_2H_5CH(CH_3)OH$*

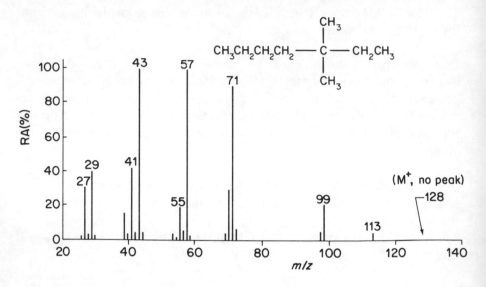

Fig. 3.1e. *Mass spectrum of* C_9H_{20}

The molecular ion of $C_2H_5CH(CH_3)OH$ is of very low relative abundance and that of C_9H_{20} is not observed. This occurs because both molecules are very susceptible to fragmentation. There are a considerable number of molecules of this type for which all or almost all the molecular ions formed fragment before they leave the ion source. For these types of molecule it is difficult to determine their relative molecular mass.

Π   The spectra shown in Fig. 3.1f, 3.1g and 3.1h illustrate the three other shortcomings of electron impact ionisation. Can you work out what the problems are in each case?

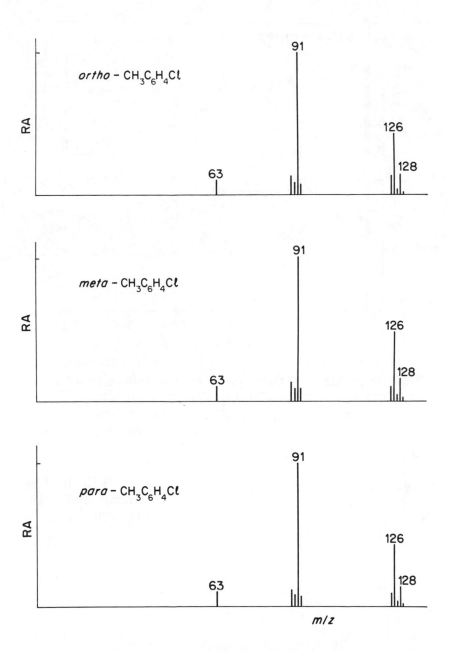

Fig. 3.1f. *The mass spectra of* ortho, meta *and*
para *isomers of chloro methylbenzene*

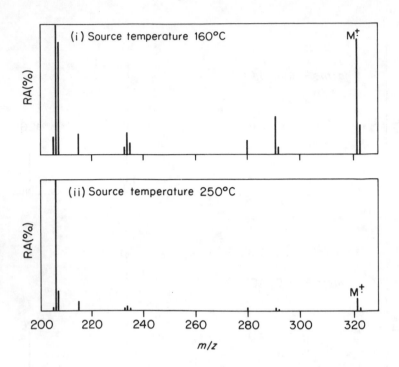

Fig. 3.1g.   *Mass spectra of the peptide derivative*
$C_6H_5CH_2OCONHCH(CHMe_2)CONHCH_2COOMe$ *recorded at*
(i) 160 °C and (ii) 250 °C

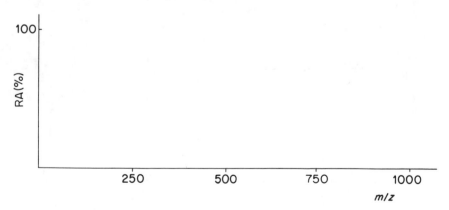

Fig. 3.1h.   *Mass spectrum of polystyrene*

The spectra of the three isomers of $CH_3C_6H_4Cl$ are all virtually identical and thus it is not possible to distinguish between them.

The mass spectrum of the peptide derivative is markedly dependent on the temperature used for vaporisation. This arises as a consequence of one of two reasons. In some cases thermally unstable molecules dissociate prior to ionisation and thus the observed mass spectrum is that of the decomposition product(s). In other cases relatively high temperatures (250 °C or above) are required to obtain volatility and this excess thermal energy, when added to the energy imparted by the ionisation process makes the molecule very prone to fragmentation. This may make it difficult to observe molecular ions and many fragment ions. You may have thought Fig. 3.1h contained a printer's error as no ions are shown. This is not so, the high molecular weight polymer, polystyrene is involatile at temperatures up to 250 °C and thus does not give a mass spectrum.

Thus to summarise the drawbacks of electron impact ionisation:

(*i*)   it may be difficult to measure relative molecular masses for some molecules;

(*ii*) it is difficult to distinguish between isomers;

(*iii*) some compounds may undergo thermal decomposition prior to ionisation or be very prone to fragmentation after ionisation because of the temperature required for vaporisation;

(*iv*) others may simply be too involatile to give a spectrum.

When these problems are encountered, recourse must be made to alternative methods of ionisation. The alternative methods that are in common use when analysis by electron impact is inappropriate are:

(*a*) Chemical Ionisation (CI);
(*b*) Field Ionisation (FI);
(*c*) Field Desorption (FD);
(*d*) Fast Atom Bombardment (FAB).

Each of these will now be discussed in some detail.

3.2.   CHEMICAL IONISATION

Part of Section 2.1 contained a discussion of why the mass spectrometer with an electron impact ion source operates under very low pressure (about 10^{-6} torr).

SAQ 3.2a   In that part of Section 2.1, a number of reasons were advanced for using such low pressures. Can you list them?

I want you to concentrate on just one of the points in the response to the above SAQ, which is that ions can undergo reactions with neutral molecules and this may lead to the production of new ions. This is the basis of the chemical ionisation method. Let us illustrate this by looking at the mass spectrum obtained when methane at a pressure of between 0.1 and 1.0 torr is ionised by electron impact. Initially the molecular ion is formed in the usual way.

$$CH_4 + e \rightarrow CH_4^{+} + 2e$$

But because of the high pressure of methane, there is a significant probability of the molecular ion colliding with another methane molecule. When this happens the most likely reaction is

$$CH_4^{+} + CH_4 \rightarrow CH_3^{\cdot} + CH_5^{+}$$

This is perhaps a somewhat suprising process as the CH_5^{+} ion is not one that is often encountered in solution chemistry. If you want to provoke some discussion with fellow students or chemists with whom you work, ask for opinions on the structure of CH_5^{+} and what orbitals might be used in its bonding.

∏   We won't go into the structure of CH_5^{+} here, but I will ask you if you can say what general reaction type is represented by the reaction of CH_4^{+} with CH_4.

The answer is an acid–base reaction, CH_4^{+} is acting as a proton donor, an acid and CH_4 is acting as a proton acceptor, a base.

Now, if a small amount of the sample to be analysed is introduced into the ion source in the vapour phase, it is species such as CH_5^{+} that act as the means of ionisation of the sample. This is therefore known as the *chemical ionisation* (CI) method. CH_5^{+} leads to ion formation of the analyte (M) usually by means of the protonation reaction shown below

$$CH_5^{+} + M \rightarrow CH_4 + (MH)^{+}$$

The ion $(MH)^{+}$, which will have an m/z value one amu greater than that of the molecular ion, is known as the *quasi-molecular ion*.

A protonation reaction of this type produces quasi-molecular ions with considerably less internal energy than is the case for molecular ions ($\overset{+}{M}$) produced by electron impact.

SAQ 3.2b   What is the consequence of this difference in internal energies likely to be?

SAQ 3.2b

This reduced propensity for fragmentation means that the relative abundance of the quasi-molecular ion in the chemical ionisation mass spectrum is usually larger than that of the corresponding molecular ion in the electron impact (EI) mass spectrum. The comparison of EI and CI mass spectra of proline, given in Fig. 3.2a illustrate this point.

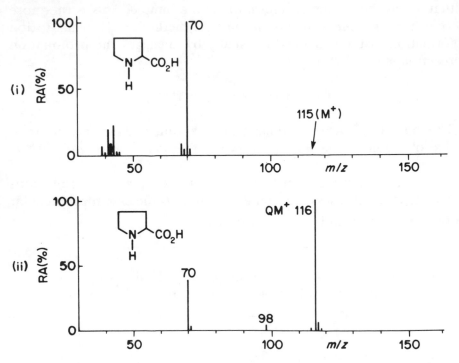

Fig. 3.2a.   *Mass spectrum of proline (i) by electron impact, (ii) by chemical ionisation*

The Cl mass spectrum should help in the determination of relative molecular masses for 'difficult' compounds.

One point that may be worrying you over this method of ionisation is that if the methane, which is known as the *reactant gas*, is ionised by electron impact, what is to stop the sample, M, also forming M^+ ions by an electron impact process? The answer is that the experimental conditions are designed to minimise the electron impact ionisation of M, although it does not eliminate it entirely.

Π   How do you think electron impact ionisation of M is minimised.

Although some sample will be ionised by electron impact, in order to ensure that a statistically unsignificant proportion of sample, M, is ionised by this method compared to that ionised by chemical ionisation, we must do all we can to encourage the latter. This is achieved by using a ratio of reactant gas to sample of at least 1000 to 1, so that there is always a much higher probability of a sample molecule colliding with a CH_5^+ ion than with an electron.

Π   The statement given earlier that Cl mass spectrometry should help in establishing relative molecular masses for 'difficult' compounds has one major *proviso*. Can you think what that might be?

In describing the Cl process, we have shown a quasi-molecular ion to be formed by protonation, thus having a mass to charge ratio one unit greater than the relative molecular mass of the compound. What we are assuming in generalising the technique is that all molecules form $(MH)^+$ ions, so that we can infer that the highest mass ion in the Cl spectrum is always one unit higher than the relative molecular mass of the sample. Although this is a very widespread occurrence, there are exceptions. These arise for two reasons. The first is that the reactant gas may undergo other ion-molecule reactions leading to other reactant ions and these may react differently with the sample. In the case of methane, the CH_4^{+} ion may undergo fragmentation before it collides with a CH_4 molecule.

$$CH_4^{+} \rightarrow CH_3^+ + H^{\cdot}$$

The fragment ion CH_3^+ may itself undergo an ion-molecule reaction.

$$CH_3^+ + CH_4 \rightarrow C_2H_5^+ + H_2$$

The formation of $C_2H_5^+$ does occur to a minor, but significant extent and ethylation of sample molecules is also sometimes observed.

$$M + C_2H_5^+ \rightarrow (M + C_2H_5)^+$$

This gives $(M + C_2H_5)^+$ ions which have a mass 29 units higher than the relative molecular mass of the molecular ion. Fig. 3.2b shows the spectrum of a molecule in which this is occurring.

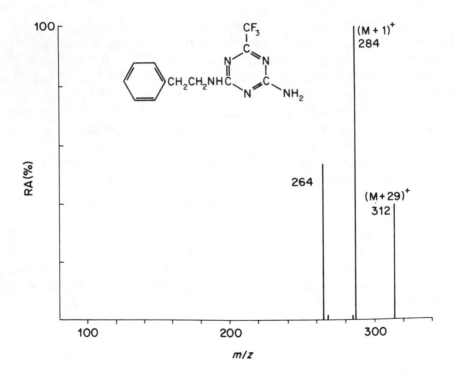

Fig. 3.2b.   *CI mass spectrum showing $(MH)^+$ and*
$(M + C_2H_5)^+$ ions

In some cases CH_5^+ does not protonate the sample. The reaction of CH_5^+ with M to give $(M + H)^+$ depends on the proton affinity of M being higher than that of CH_4. If it is not, then collision between them may still lead to a reaction, but not to simple proton transfer. The most important examples of this are the higher alkanes which have a lower proton affinity than methane and so the ion-molecule reaction proceeds as shown below.

$$CH_5^+ + M \rightarrow (M - H)^+ + CH_4 + H_2$$

These types of exceptions lower the value of the technique for establishing relative molecular masses, but they do not totally invalidate it, since recognising that an ion such as $(M + 29)^+$ or $(M - 1)^+$ may be observed is half the battle when interpreting the spectrum. The insoluble problem that makes a technique quite valueless is not knowing what to expect!

So far our discussion has been exclusively directed towards use of one reactant gas, methane. However, others are used and the most common are ammonia (NH_3) and iso-butane ($(CH_3)_3CH$).

∏   Without going into the full sequence of ion molecule reactions for each of these reactant gases, would you like to suggest what might be the major reactant ion each generates? In attempting this question, use your knowledge of solution chemistry in thinking about what ions are likely to be formed.

The major reactant ion from ammonia is NH_4^+ and that from iso-butane is $(CH_3)_3\overset{+}{C}$. The formation of NH_4^+ from NH_3 directly parallels the behaviour of CH_4 and, of course, NH_4^+ is a well known stable cation. The behaviour of iso-butane does not directly parallel that of CH_4. If it did you might expect to form $C_4H_{11}^+$ and I don't blame you if you wrote this down. Instead, iso-butane forms $(CH_3)_3\overset{+}{C}$, which is quite stable as it is a tertiary carbocation.

∏   Suggest how you might expect NH_4^+ and $(CH_3)_3\overset{+}{C}$ to react in ion molecule reactions.

NH_4^+ and $(CH_3)_3\overset{+}{C}$ both protonate samples to give $(M + H)^+$ ions. You may be surprised to find $(CH_3)_3\overset{+}{C}$ behaves in this way, but of course, the stable molecule, butene is formed in the process.

$$(CH_3)_3\overset{+}{C} + M \rightarrow (MH)^+ + CH_2=C(CH_3)_2$$

The use of these alternative reactant gases can be important in CI mass spectrometry as they may lead to different amounts of fragmentation of the same quasi-molecular ion. This is our first mention of fragmentation in CI mass spectrometry, although the spectrum of proline (Fig. 3.2a) did show fragment ions.

SAQ 3.2c

> The major fragment ion in both the EI and CI spectra of proline occurs at m/z 70. This represents the loss of 45 mass units from $M^{\ddagger}$ in the EI spectrum and the loss of 46 mass units from $(MH)^+$ in the CI spectrum. Suggest a formula for the neutrals produced in each spectrum.

∏   On the basis of the loss of the 46 mass unit neutral in the CI spectrum, suggest a structure for the quasi-molecular ion.

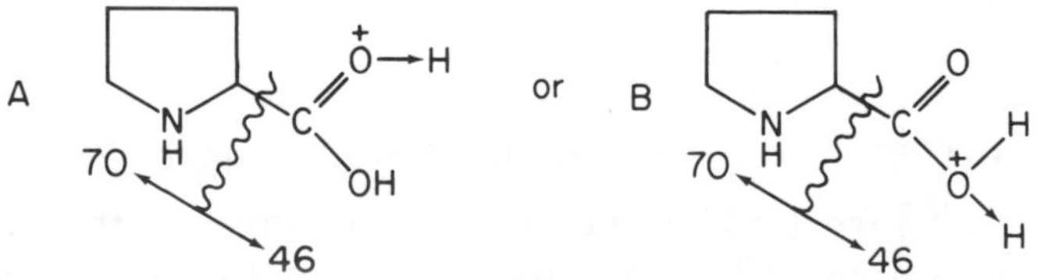

Both A and B are likely structures. In drawing these structures, I have indicated the mass of both the fragment ion and the neutral. As the formation of the quasi-molecular ion is an acid–base reaction, it is reasonable to write the product as involving electron pair donation by a lone pair on a hetero-atom. For either of the structures shown above we can readily understand the loss of a 46 mass unit neutral.

Π   There is another small fragment ion in the CI spectrum at *m/z* 98. How does this arise and does it tell us anything about the structure of the quasi-molecular ion?

m/z 98 corresponds to the loss of H_2O from the quasi-molecular ion. This can easily be explained in terms of structure B.

It doesn't necessarily exclude structure A as a hydrogen transfer may occur, but this is, perhaps, less likely.

These two fragmentations, loss of CO_2H_2 and H_2O do seem to suggest that protonation does not occur at the nitrogen atom, the other atom with an available lone pair of electrons.

SAQ 3.2d   Are quasi-molecular ions odd- or even-electron ions? Can you rationalise the fragmentations of the quasi-molecular ion of proline in terms of its electron configuration?

SAQ 3.2d

The loss of even electron neutrals is a common feature of CI mass spectrometry. As a further example, the behaviour of methylphenylketone is shown below.

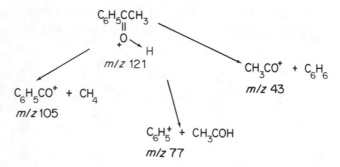

This is the fragmentation pattern of methylphenylketone obtained with CH_4 as reactant gas and it is interesting to compare it with that obtained using iso-butane. In this case no fragment ions, whatsoever, are observed – only the quasi-molecular ion is formed. This is the aspect referred to earlier, that the amount of fragmentation depends on the reactant gas. This arises because the amount of energy transferred to a sample molecule on protonation depends on both the proton affinity of the sample and the acidity (the willingness to release a proton) of the reactant ion.

Acidity increases in the order $NH_4^+ < C_4H_9^+ < CH_5^+$ and thus the internal energy of quasi-molecular ion formed by these reactant ions follows the same order. As the amount of fragmentation depends on the internal energy of the ion, this explains the results given for methylphenyl ketone.

It is also worth pointing out that CI mass spectrometry can distinguish between isomers in some cases. This is illustrated by the fact that the quasi-molecular ions of 5-β-3-keto steroids readily lose H_2O, whereas the 5-α-isomers hardly show this fragmentation.

In general, the amount of fragmentation of quasi-molecular ions is small or non-existent because these ions are formed with low internal energies compared to molecular ions formed by EI. The CI technique and others that also show few fragment ions are termed *soft ionisation methods*.

SAQ 3.2e   We have now spent a considerable time discussing CI mass spectrometry. Before we move on to the next ionisation method, list the advantages and limitations of the CI method, paying particular attention to those problems we outlined for EI at the end of Section 3.1.

3.3. FIELD IONISATION

Field ionisation (FI) and field desorption (FD), which is discussed in Section 3.4, are two very closely related methods of ionisation which rely on essentially the same principle. We will deal with FI first and follow this with a discussion of FD, which corresponds to the historical sequence of their development.

A schematic diagram of an FI ion source is shown in Fig. 3.3a.

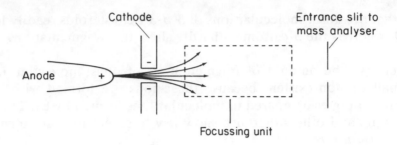

Fig. 3.3a. *Schematic diagram of a field ionisation ion source*

The most important components are the electrodes, an anode and a cathode, which are held at a potential difference of up to 10 kV and are sufficiently close together to develop a potential gradient of up to 10^8 V cm^{-1}.

SAQ 3.3a	Using the figures given above, calculate the distance between the electrodes.

When a sample molecule in the vapour phase impinges on the anode or comes very near to it, the potential gradient it experiences leads to an electron being transferred to the metal anode giving rise to the formation of a positive ion.

Π   What might you expect to happen to the positive ion when it is formed?

It is attracted towards the cathode and as this has a hole in it, some ions pass through.

SAQ 3.3b   These ions will have passed through a high potential difference. What effect will this have on them?

Most mass spectrometers are designed to analyse ions travelling at velocities attained by acceleration through potential differences of 4000 to 8000 volts. The ions leaving the cathode of the FI source will have passed through a potential difference of about 10 000 volts.

SAQ 3.3c   What must be done to these ions so that the mass spectrometer can analyse them?

The focussing device beyond the cathode has two functions. Firstly, it must retard the ions until they reach the correct velocity for analysis and secondly, it must focus the diverging beam of ions coming through the cathode into a coherent beam for mass analysis. We won't discuss the details of how these two feats are achieved.

The anode of the FI source is often termed the *emitter*, because molecules impinge upon it or come very close to it and are emitted as ions. The emitter is usually a sharp blade, a sharp tip or a wire. The efficiency of ion emission can be increased by a factor of about 10^3 by the technique of 'whiskering' the emitter. This will be described in more detail in the next section concerning the field desorption ion source.

Field ionisation usually produces molecular ions, $M^{\ddot{+}}$, but in some cases, quasi-molecular ions $(MH)^+$, are observed. Some samples produce both.

The quasi-molecular ions arise as a consequence of a high concentration of sample molecules existing near the anode, which gives rise to ion-molecule reactions. The formation of $M^{\ddot{+}}$ ions in some cases and $(MH)^+$ in others obviously introduces a degree of uncertainty into the determination of relative molecular masses of uncharacterised substances. This lowers the value of the technique somewhat. However, unlike CI, no higher adduct ions such as $(M + C_2H_5)^+$ are formed.

The FI method produces molecular ions with much lower internal energies than those produced by EI.

SAQ 3.3d	What is the consequence of this as far as the mass spectrum is concerned?

SAQ 3.3d

Although there is less fragmentation than in EI mass spectra, FI spectra usually show some fragment ions and these can be structurally informative. This is well illustrated by a comparison of the EI and FI spectra of xanthosine (Fig. 3.3b). The FI spectrum of xanthosine shows a much enhanced M^{+} peak, but the structurally significant ions at m/z 152 and 133 are also present.

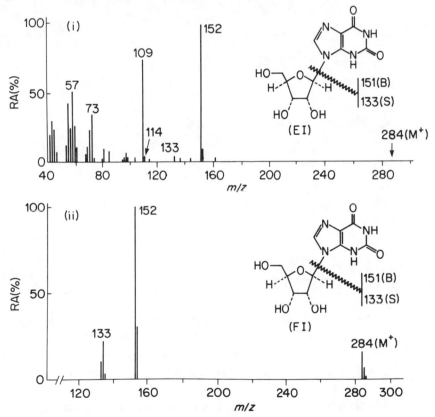

Fig. 3.3b. *Mass spectrum of xanthsosine, (i) using EI and (ii) using FI*

SAQ 3.3e   List any advantages and disadvantages of FI over EI or CI and any limitations of the technique.

3.4. FIELD DESORPTION

The problems that none of the previous methods of ionisation have been able to overcome are how to deal with thermally unstable or involatile compounds. Field desorption (FD) is the first of two techniques we shall discuss which addresses these problems.

∏   If you had to begin designing an ion source to deal with these types of compounds, what fundamental principle would you attempt to use? I don't want you to think in detail about ion sources, just about physical principles.

If vaporising the sample is a problem, it might be overcome if we could produce ions directly from the solid state. This is what field desorption does. It operates on the same principle as FI in that it uses an emitter held at a high temperature with respect to a cathode. But, the sample is placed on the surface of the emitter in the solid state and ions are desorbed directly from the solid towards the cathode.

The anode is first prepared by covering it with a carbon coating in the form of microneedles or 'whiskers', about 0.001 cm in length. This enlarges the surface area for coating with the sample and also increases ionisation efficiency. Whiskering is something of an art, which needs a skilled operator to get good results. The sample is applied in the form of a solution and the solvent subsequently removed by evaporation.

∏   Coating the sample onto the emitter must obviously be done in the laboratory, outside the ion source of the mass spectrometer. How do you think the emitter might be introduced into the source?

If the emitter is attached to a direct insertion probe similar to that described in Section 2.2, this can be inserted into the ion source *via* a vacuum lock.

Field desorption is the 'softest' of the ionisation techniques to be discussed in this section. Generally it only produces molecular ions and/or quasi-molecular ions with virtually no fragmentation. An example of this is shown in Fig. 3.4a, which compares the EI, CI, and FD spectra of the anti-inflammatry compound, dihydrocortisone ($M_r = 364$) which has the structure.

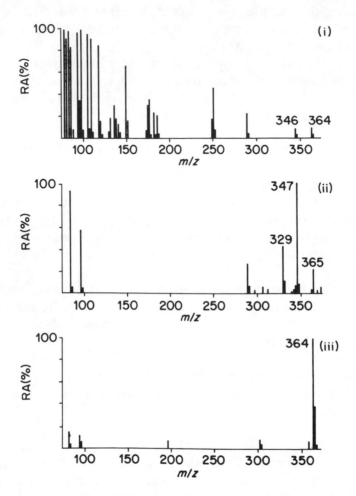

Fig. 3.4a. *Mass spectra of dihydrocortisone obtained by (i) EI; (ii) CI and (iii) FD ionisation*

SAQ 3.4a   What is the consequence of not producing fragment ions?

SAQ 3.4a

There is a further drawback with FD which is related to the persistance of the spectrum. With an EI ion source, the small amount of sample introduced (usually, about 1 mg via an AGHIS or septum inlet or 1 μg on a probe) is usually sufficient to produce many hundreds of mass spectra and as long as sample remains in the instrument, the mass spectrum is produced. This process may go on for an hour or more. Although, in practice, one wouldn't usually need that length of time, it is not uncommon to spend 15 min or so adjusting the instrument to give the best conditions for that particular sample (this is referred to as tuning the instrument) and recording the spectrum.

In the case of FD, some samples will not give persistant spectra. As soon as the emitter is switched on ions are formed, but the ion current very quickly decays to nothing. This gives hardly any time for tuning before recording the spectrum. Thus the spectrum may not be recorded under the most favourable instrumental conditions. Nevertheless, FD has proved to be a very useful technique for obtaining spectra of thermally unstable or involatile compounds. As an example of this, the FD mass spectrum of polystyrene shows ions up to m/z 11 000.

SAQ 3.4b	List the advantages and shortcomings of the FD technique.

SAQ 3.4b

3.5.   FAST ATOM BOMBARDMENT

Fast atom bombardment is an ionisation method that has only recently come into common use in organic mass spectrometry, although it has been used in surface chemistry for a number of years. It is based on the following idea. If a beam of fast moving neutral atoms are directed onto a metal plate coated with a sample, then much of the high kinetic energy of the atoms is transferred to the sample molecules on impact. This energy can be dissipated in various ways, some of which lead to ionisation of the sample.

The bombarding atoms are usually rare gases, either xenon or argon. In order for them to achieve a high kinetic energy, atoms of the gas are first ionised and these ions are then passed through an electric field.

SAQ 3.5a   What will be the result of passing ions such as Xe^{+} through the electric field?

SAQ 3.5a

After acceleration, the fast moving ions pass into a chamber containing further gas atoms and collision of ions and atoms leads to charge exchange.

$$Xe^+(\text{fast}) + Xe(\text{thermal}) \rightarrow Xe(\text{fast}) + Xe^+(\text{thermal})$$

The fast atoms formed in this process retain most of the original kinetic energy of the fast ions and carry on in the original direction.

∏   Any remaining fast ions and the ions with thermal energies can be removed before sample bombardment. How might this be done?

If a deflector plate with a negative potential is employed, the ions will be directed towards it.

As previously mentioned, when the fast atoms bombard the sample, both positive and negative ions are formed. If the plate on which the sample is placed is held at a suitable potential with respect to an ion exit plate, either positive or negative ions can be directed towards the analyser. A schematic representation of the FAB source is shown in Fig. 3.5a.

The sample is usually applied to the plate in the form of a solution in an inert involatile liquid (often called the *matrix* material) such as glycerol.

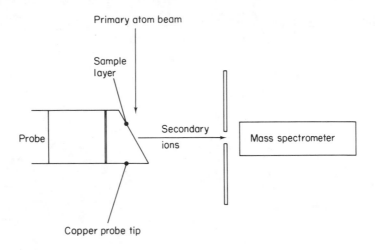

Fig. 3.5a. *Schematic representation of a FAB source*

This aids ionisation and although at the time of writing (1985), the detailed mechanism of FAB ionisation is not fully understood, it is believed to be important to form a monolayer of sample at the surface of the matrix material to obtain efficient ionisation. It is also important to dissolve the sample in the matrix; a suspension of sample in the matrix does not give good spectra.

∏   There is a disadvantage in using a matrix material. Can you think what that might be?

The disadvantage is that the matrix also forms ions on bombardment in addition to those formed by the sample. This obviously complicates the spectrum, but, once more, as this is an expected occurrence it is not too difficult to make allowance for it.

Once more the solution of sample in the matrix are applied to the plate in the laboratory. The plate is part of a probe and is inserted into the ion source through a vacuum lock.

The spectra produced usually provide relatively abundant molecular or quasi-molecular ions and also show some structurally important fragment ions. This is illustrated by the spectrum of β-chaconine ($M_r = 705$) shown in Fig. 3.5b.

Although this method is still undergoing development, it does look very promising. So much so, that all commercial mass spectrometer manufacturers now offer FAB sources for their instruments. An especially important point, compared to FD, is that ion beams usually persist for reasonable times of about 10–15 minutes.

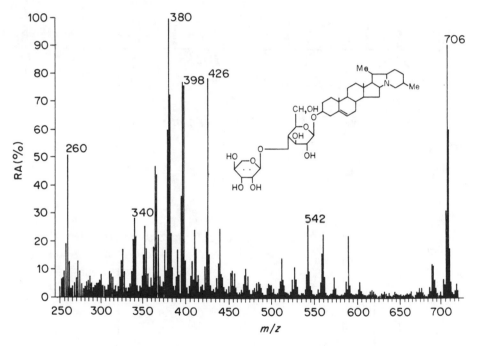

Fig. 3.5b. *Mass spectrum of β-chaconine obtained using a FAB source*

Both fast atom bombardment and field desorption have been used extensively for obtaining mass spectra of salts. The exact form of the spectrum obtained is very dependent on the nature of the cation and anion. Two examples involving both positive and negative ion FAB spectra are shown in Fig. 3.5c and Fig. 3.5d.

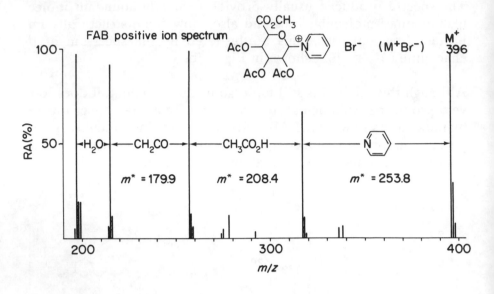

Fig. 3.5c. *FAB positive ion spectrum*

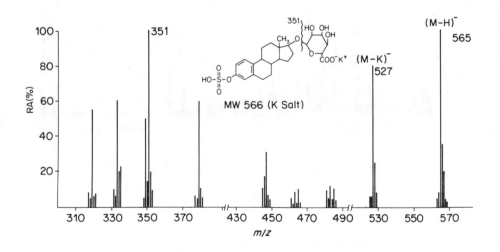

Fig. 3.5d. *Negative ion FAB, β-Estradiol-3-sulfate-17-glucuronide (K salt)*

SAQ 3.5b	List the advantages and disadvantages of FAB.

3.6. SPARK SOURCE MASS SPECTROMETRY

There is one other type of ion source you should know about. This is the *spark source*. It is another method of examining involatile compounds, but is only of use in analysing inorganic materials.

A powdered sample is mixed with graphite and the mixture pressed into rods. Two rods are mounted close to each other in the source and a large electrical potential applied to them. An electric discharge (a spark) passes between the rods (acting as electrodes) and ions of the electrode material are formed in the spark.

The ions formed are virtually always atomic ions such as Na^+, Mg^{2+}, Cl^+ etc, and the technique is used for elemental analysis of inorganic compounds. By using an analyser of Mattauch-Herzog geometry and a photoplate detector, (see Part 4) large number of elements can be detected simultaneously. The method is also used for quantitative elemental analysis, as the intensities of ions are proportional to the amounts of each element present. The method is used for analysis of archaelogical samples and the determination of trace metal quantities in materials as diverse as semi-conductors and biological samples.

∏   Do you know another analytical method that can be used for
       the same task?

The answer is atomic spectroscopy, particularly atomic emission
spectroscopy.

SAQ 3.6a	Why is spark source mass spectrometry of virtu-ally no use to the organic chemist?

3.7.   REVISION

After completing this long, but important section on ion sources,
revise the major points by answering the following SAQs. Do not
proceed to the next section until you are happy about the answers.

SAQ 3.7a	*Electron Impact Ionisation*
	(*i*) EI ionisation can produce both positive and negative ions. Why is it more difficult to produce negative molecular ions than their positive counterparts?
	(*ii*) Draw an ionisation efficiency curve and explain why this leads mass spectrometrists to use 70 eV bombarding electrons. ⟶

SAQ 3.7a (cont.)

(*iii*) What might be the advantage of recording a low voltage spectrum?

(*iv*) List the major shortcomings of the EI method.

SAQ 3.7b

Chemical Ionisation

(*i*) Ion-molecule reactions of methane lead to two predominant species being formed. Write equations for their formation.

(*ii*) What properties govern the ease of proton transfer from a reactant ion to a sample molecule? ⟶

SAQ 3.7b
(cont.)

(*iii*) The CI mass spectra of methyl esters of carboxylic acids show fragment ions formed by loss of CH_3OH and the elements of CH_3COOH. Rationalise this behaviour in terms of the electron configuration and possible structures of the pseudo-molecular ion.

(*iv*) Compare the advantages and disadvantages of the EI and CI methods.

SAQ 3.7c

Field Ionisation and Field Desorption

(*i*) These are two techniques based on the same principle. Describe this and state how FD differs from FI.

(*ii*) Explain what is meant by 'whiskering'.

(*iii*) The EI, FD and FI spectra of D-glucose, ($M_r = 180$) are shown in Fig. 3.7a. Use these to highlight the comparative features of the techniques. $\longrightarrow$

SAQ 3.7c
(cont.)

(*iv*) Explain the origin of the ions at *m/z* 163, 145 and 127 in the FI spectrum.

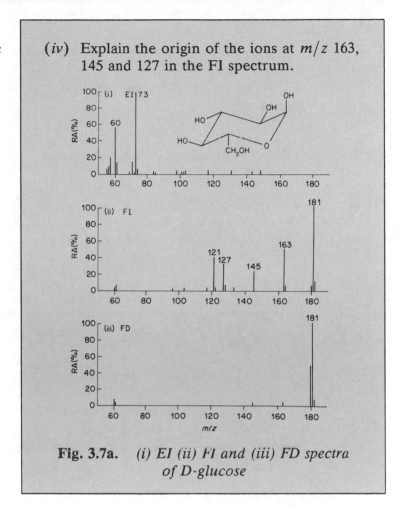

Fig. 3.7a. *(i) EI (ii) FI and (iii) FD spectra of D-glucose*

SAQ 3.7c

SAQ 3.7d   *Fast Atom Bombardment*

(*i*)   How are fast atoms generated?

(*ii*) What is meant by the term 'matrix material'?

SAQ 3.7e   On the basis of the discussion of ion sources, describe the features you would seek in a perfect ionisation method.

Summary

This Part has covered all the methods of ionisation commonly encountered in organic mass spectrometry. These are electron impact, chemical ionisation, field ionisation, field desorption and fast atom bombardment. You have also encountered one suitable for dealing with inorganic samples. For each method you have examined spectra and considered advantages and disadvantages.

Objectives

After studying this Part you should be able to:

● describe the principles behind electron impact ionisation, chemical ionisation, field ionisation, field desorption and fast atom bombardment;

● discuss the advantages and disadvantages of each method of ionisation;

● describe the main operational features of each method of ionisation.

4. Mass Analysis

In this part of the unit you will deal with the commonly-used methods of mass analysis. You will also be introduced to the need for accurate mass measurement and the instrumental features required to carry this out. The important features of scan rate and multiple ion monitoring will also be covered.

4.1. MASS ANALYSERS

When we were discussing the mass spectrometer in Section 2.1, we encountered the most common type of mass analyser – a magnet.

SAQ 4.1a

> By way of revision, can you describe how a magnetic analyser functions and write down the equation governing the passage of an ion (mass, m; charge, z; kinetic energy, $\frac{1}{2}mv^2$) through a magnetic field of strength, B?

SAQ 4.1a

SAQ 4.1b | Can you use the relationship derived in SAQ 4.1a, together with the relationship between accelerating voltage and kinetic energy of the ion, to derive an equation relating m/z to field strength and accelerating voltage?

4.1.1. Resolving Power of a Mass Analyser

Such magnetic analysers are perfectly adequate for measuring m/z values to the nearest whole number and they can do this quite satisfactorily for ions with m/z values up to about 5000. The ability of analysers to measure m/z satisfactorily is quantified in the form of a parameter known as the *resolving power*.

As an example of the calculation of resolving power, let us consider what resolving power is needed to identify ions differing by one relative atomic mass unit in the region of 5000: for example ions of m/z 5000 and 5001.

Resolving power is defined as the mass to be measured divided by the difference in masses to be identified (m/Δm). In this example the resolving power required is, thus,

$$\frac{5000}{5001 - 5000} = 5000$$

SAQ 4.1c

Let's suppose we are using an instrument with a resolving power (rp) of 5000, but that the mass region we are interested in is 500. How accurately could we measure such masses?

We might just pause here and ask what we mean by resolution. It is the ability to distinguish two ions in the mass spectrum which are usually displayed as peaks on a trace or on an oscilloscope.

∏   Drawn out below in Fig. 4.1a are pairs of peaks. Which would you say are clearly resolved.

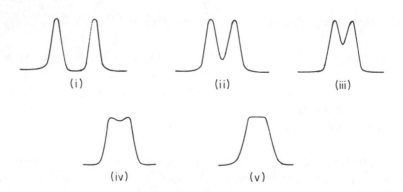

Fig. 4.1a.   *Schematic representation of two neighbouring peaks*

The way the series of peaks is displayed in Fig. 4.1a, they become less clearly resolved in moving from (*i*) to (*v*). One could easily distinguish the peaks in (*i*) and (*ii*) and reasonably easily do so in (*iii*). It might be difficult to do so with certainly in (*iv*) and it is not even possible to see two separate peaks in (*v*). Whether different people consider peaks to be sufficiently well resolved can obviously be a subjective assessment. Some people might be happy with the resolution in (*iv*), while others might require the peaks to be completely separated as in (*i*). In order to obtain a standard definition of resolution and hence, resolving power, mass spectrometrists only accept a pair of peaks as being adequately resolved if the height of the overlapping portion between them is 10% or less of the peak height, Fig. 4.1b. This is known as the *10% valley definition.*

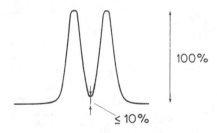

Fig. 4.1b. *Representation of the 10% valley definition of resolution*

Now let's return to the measurement of m/z, but as we shall assume in this discussion that all ions are singly charged, we can talk about the measurement of mass. Consider the molecular ion of methane, CH_4^{+}.

SAQ 4.1d	What is the mass of the methane molecular ion?

SAQ 4.1e

Assuming we had the right kind of mass spectrometer for the job, would you expect the accurately measured mass of CH_4^+ to be a whole number? Explain your answer.

SAQ 4.1f

Calculate the mass of CH_3OH^+ to an accuracy of five decimal places using the A_r values given in the response to SAQ 4.1e

The value calculated in SAQ 4.1f is, of course, also the relative molecular mass of the methanol molecule, because the mass of the electron lost on ionisation is negligible. Thus determination of the accurate mass of the molecular ion gives the accurate M_r of the corresponding molecule.

This information can be used in the examination of uncharacterised samples to determine molecular formulae, as each molecular formula has its own unique accurate mass value, when measured to 5 decimal places. When masses are measured to this accuracy, this is called *high resolution* mass spectrometry, whereas measuring mass to the nearest whole number is called *low resolution* mass spectrometry.

SAQ 4.1g

As an example of the identification of molecular formulae by means of accurate mass measurement of molecular ions, imagine we are presented with a colourless liquid and a colourless gas to identify by mass spectrometry. These two compounds have the same nominal mass of 32, however, the accurate mass of the molecular ion of the liquid is found to be 32.0263 ± 0.0001 and that of the gas is 31.9898 ± 0.0001. Use this information and the list of accurate A_r values in the response to SAQ. 4.1e to identify the molecules.

SAQ 4.1h   If we had a mixture of methanol and oxygen in the mass spectrometer, what resolving power would be needed to distinguish between the molecular ions?

SAQ 4.1i   Calculate the resolving power needed to distinguish between the following pairs of ions:

(*i*) $C_8H_{16}^{+\cdot}$ and $C_7H_{12}O^{+\cdot}$

(*ii*) $C_{24}H_{50}^{+\cdot}$ and $C_{23}H_{46}O^{+\cdot}$

(*iii*) $C_{40}H_{82}^{+\cdot}$ and $C_{39}H_{78}O^{+\cdot}$

Thus, $C_8H_{16}^{+\cdot}$ and $C_7H_{12}O^{+\cdot}$ can be resolved by a magnetic analyser as the required resolving power is less than the value of 5000 which can be achieved with such an instrument. However, the other two pairs of ions require resolving powers much higher than can be achieved with this type of instrument.

4.2. DOUBLE FOCUSSING MASS SPECTROMETER

You might, now be wondering what limits the resolving power of a mass spectrometer and what can be done to increase it. Let's find the answers to these questions.

I hope you will recall the equations used to describe the behaviour of an ion in the mass spectrometer. These are:

$$zV = \frac{1}{2}mv^2 \qquad (2.1b)$$

$$mv^2/r = Bzv \qquad (2.1c)$$

Combination of these leads to the equation

$$m/z = B^2r^2/2V \qquad (2.1d)$$

SAQ 4.2a   Examination of these three equations reveals there are three properties of an ion which are vital in mass spectrometry. What are these three properties?

SAQ 4.2b   In combining Eq. 2.1b and 2.1c to generate Eq. 2.1d, what assumption do we make?

If the velocity of the ions remains constant, so does their kinetic energy. The principle behind the analysis of ions by the mass spectrometer is that all ions of the same m/z have the same kinetic energy and hence the same velocity. The magnet separates them according to their mass, charge and velocity and is therefore sometimes referred to as a means of *velocity focussing*.

The resolving power of the mass spectrometer is limited by the fact that, in practice, not all ions of the same m/z have exactly the same kinetic energy after acceleration. This arises as a consequence of the Boltzmann distribution of thermal energies of ions and field inhomogeneity in the ion source. The second of these reasons means that some ions experience slightly different potentials from other ions of the same m/z, depending on exactly whereabouts in the source they are formed. Thus all ions are formed with a spread of kinetic energies which can be represented graphically as shown in Fig. 4.2a.

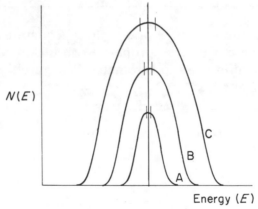

Fig. 4.2a.   *Plot of number of ions of energy E (N(E)) against energy*

This shows three curves with kinetic energy spreads of different magnitudes.

∏   Assuming you wanted to determine the kinetic energy at the peak maximum for each curve, in which curves are accurate measures of kinetic energy most easily and least easily achieved?

The curve (A) has the sharpest maximum and thus in this case the kinetic energy can be measured with most certainty. The curve (C) has a relatively flat top and thus a measure of the position of the peak maximum would involve the least certainty.

If we recall the principle on which the magnetic analyser functions, which is that all ions of the same m/z have the same kinetic energy (ie the same velocity), we will see that any variation in kinetic energy will be assumed by the magnet to be a variation in mass. Hence Fig. 4.2a could be redrawn with the *abcissa* as a mass scale rather than an energy scale, Fig. 4.2b.

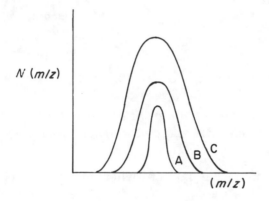

Fig. 4.2b. *Plots of the number of ions of mass to charge ratio m/z (N(m/z)) against m/z*

Thus, the ion with the greatest spread of kinetic energies will have the largest uncertainty in the measurement of its mass. It is this that limits the resolving power of the mass spectrometer.

∏   On the basis of the above discussion, what must we do if we want to make mass measurement more accurate?

We must reduce the spread of kinetic energies for each ion to a minimum, thereby making the peaks as 'sharp' and well defined as possible.

The peaks are made sharper by placing an extra focussing device between the ion source and the magnetic analyser, which reduces the kinetic energy spread. This takes the form of a pair of curved metal plates, called the electrostatic analyser or electric sector, with an electrical potential maintained across them. Such an arrangement is shown in Fig. 4.2c.

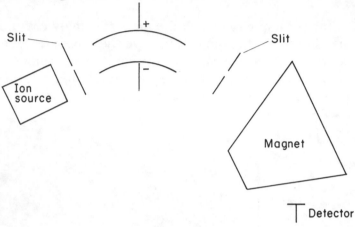

Fig. 4.2c.   *Schematic diagram of mass spectrometer using both electrostatic and magnetic focussing*

Let us now examine what happens when an ion passes between such a pair of plates. The ions will enter the plates with a kinetic energy arising from acceleration and given by Eq. 2.1b.

$$ke \;=\; \frac{1}{2}mv^2 \;=\; zV \qquad\qquad (2.1b)$$

When the ions enter the region between the plates they experience the potential difference, E, between them.

∏   Assuming we are analysing positive ions and the bottom plate is negatively charged with respect to the top one, what will happen to the ions?

The ions are deflected towards the bottom plate. That is, the ions experience a force acting at right angles to the direction of their flight. The magnitude of the force (F) acting on an ion of charge z, is given by Eq. 4.2a.

$$F \;=\; zE \qquad\qquad (4.2a)$$

Thus the electric force will tend to deflect the ions from their original flight path into a circular path as shown in Fig. 4.2d.

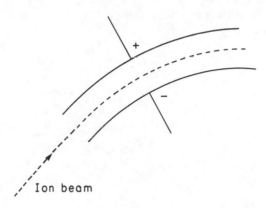

Fig. 4.2d. *Passage of an ion through an electrostatic analyser*

Provided that this electric force is balanced by the centrifugal force the ions attain as a consequence of their velocity, v, they follow a circular path which is the circumference of a circle of radius R.

∏ Write an equation balancing the electric force and the centrifugal force for an ion of mass m and charge z.

The equation is:

$$zE = mv^2/R \qquad (4.2b)$$

∏ Combine Eq. 2.1b with Eq. 4.2b to obtain a new equation relating R to V and E.

$$zV = \frac{1}{2}mv^2 \qquad (2.1b)$$

$$zE = mv^2/R \qquad (4.2b)$$

$$mv^2 = 2zV = zER$$

$$R = 2V/E \qquad (4.2c)$$

Thus ions accelerated through a potential of *V* and passing through the field *E* all follow the curve of radius *R* *irrespective of their m/z values* provided they have the same kinetic energies. Another way of putting this is to say that ions passing through the field *E* will follow curves of different radii depending on their kinetic energies as shown in Fig. 4.2e.

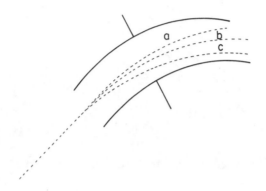

Fig 4.2e.   *Flight path of ions, of different kinetic energy, in the electrostatic analyser*

∏   In Fig. 4.2e which of the flight paths *a*, *b* or *c* will the ions of highest kinetic energy follow?

The ion following flight path (*a*) has the highest kinetic energy. This ion has the highest centrifugal force, which will override the electric force and thus the ion follows the circular path of largest radius.

Thus ions of different kinetic energies follow different paths when between the curved plates of the electrostatic analyser. If you refer back to Fig. 4.2c you will notice there is a slit placed between the electrostatic analyser and the magnetic analyser and only ions that have followed the circle of radius *R* will be correctly positioned to pass through to the magnetic analyser. In this way we can reduce the spread of kinetic energies for any ion entering the magnetic field. This is shown in Fig. 4.2f.

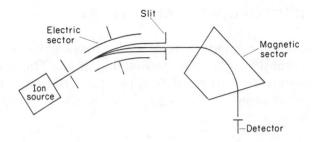

Fig. 4.2f. *Double focussing mass spectrometer*

In this way the resolving power of the instrument is improved. A mass spectrometer of this type is called a *double focussing* instrument. In such a mass spectrometer, the electric sector acts as an energy analyser and the magnetic sector as a mass analyser. With such instruments, resolving powers of up to 100 000 can be obtained.

The mass spectrometer shown schematically in Fig. 4.2f is said to have Nier–Johnson geometry after the scientists who first used this design.

Π   What would you expect to happen to positive ions if the polarity of the magnet was reversed?

If the magnet polarity was reversed, positive ions would attempt to curve in the opposite direction while in the magnetic field. Instruments of this type are used and are said to have Mattauch–Herzog geometry (Fig. 4.2g). These instruments normally employ photographic detectors and are usually used in conjunction with spark sources.

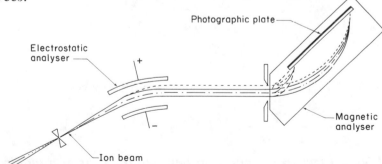

Fig. 4.2g. *Mass spectrometer of Mattauch–Herzog geometry*

4.3.   QUADRUPOLE MASS ANALYSER

There is one other common type of mass analyser that works on an entirely different principle to those detailed in 4.2. This is called the *quadrupole analyser*. This consists of two pairs of precisely parallel rods arranged as shown in Fig. 4.3a

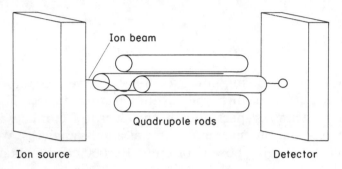

Fig. 4.3a.   *Arrangement of rods in a quadrupole mass analyser*

Ions are formed using the sources described in Section 2.3 and accelerated into the analyser in the usual way, but using a very small accelerating voltage. The rods are arranged as shown in Fig. 4.3b, which also shows the arrangement of electrical connections between them. A voltage made up of a dc component, U, and a radio frequency (rf) component, $V \cos \omega t$, is applied to adjacent rods.

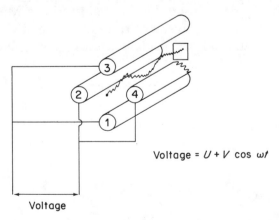

Voltage = $U + V \cos \omega t$

Voltage

Fig. 4.3b.   *Electrical connection between the rods of a quadrupole mass analyser*

Opposite rods are electrically connected. Let's now consider how these potentials affect ions, but let's do it in stages.

∏   Let us examine just one pair of rods, say 2 and 3 in Fig. 4.3b and firstly assume they are connected only by the dc voltage. If rod 3 is at a positive potential with respect to rod 2, what will be the effect on positive ions passing near them?

The positive ions will be drawn towards the negative rod, 2.

∏   Which will be deflected most easily by the rod voltage, ions of low or high kinetic energy?

Ions of low kinetic energy are deflected most easily (Fig. 4.3c)

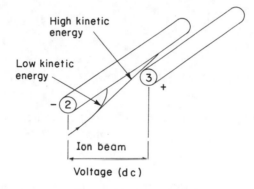

Fig. 4.3c.   *Deflection of ions between rods at a fixed potential difference*

Consider the effect of the variable rf component on the potential between these rods. The rf component is effectively an ac power supply which alternates at 10^8 Hz as compared with the 50 Hz of normal household supply.

∏   Let us now consider that only the rf field is connected to rods 2 and 3. What will be the effect of this rf field on ions passing close to the rods?

The positive ions would feel an attraction to rod 2 when the rf polarity is the same as is shown in Fig. 4.3c and an attraction to rod 3 when the polarity is reversed.

This change in polarity is occurring at a very high rate and if it supplements the dc fixed potential the ions follow an eratic path as shown in Fig. 4.3d.

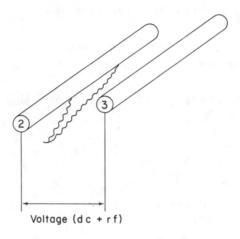

Voltage (d c + r f)

Fig. 4.3d. *Deflection of ions between rods at a variable potential difference*

Now, the same dc voltage and rf field are also applied to rods 1 and 4 simultaneously to their application to rods 2 and 3. The dc connections are such that 4 is at a negative potential with respect to 1, but the rf field applied to 1 and 4 is 180° out of phase with that applied to 2 and 3. It is difficult to envisage in a simple way what this does to the path of ions, but effectively they follow an oscillating trajectory between the rods. At any one pair of dc voltage and rf values only ions of one particular kinetic energy pass directly between the rods and out the other end as shown in Fig. 4.3b. All other ions collide with the rods.

Now, if both the dc voltage and the rf field are varied together, while keeping the ratio between them constant, ions of different mass to charge ratio pass between the rods and reach the detector. In this way, mass scanning is accomplished. Because of this filtering action, this type of analyser is often called a *quadrupole mass filter*. This type of analyser has about the same maximum resolving power as a single focussing magnetic sector instrument.

SAQ 4.3a	Can you recall what the maximum value is for such an instrument?

The quadrupole analyser has some advantages over a magnetic sector instrument. These are:

(*a*) it is relatively cheap to build;

(*b*) it is lighter and physically smaller;

(*c*) it is more robust;

(*d*) computer control of rod voltages can be effected much more accurately than the corresponding control of a magnetic field in a magnetic sector instrument;

(*e*) the mass scale is linear (with a magnetic sector analyser, $m/z \propto B^2$);

(*f*) it is a very fast scanning instrument.

4.4. SCAN RATE

We have not yet mentioned scan rates, that is, the time taken to record the spectrum – but this is a good point to do so. For some applications of mass spectrometry, particularly those in which the mass spectrometer operates together with either a gas or liquid chromatograph, scan rate or scan time is very important. Scan times are con-

ventionally measured in seconds per decade of mass. (s decade^{-1}).
A decade of mass covers a range of mass which changes by a factor
of ten (eg m/z 20 to 200 or 50 to 500).

∏   If a spectrometer is scanned at 10 s decade^{-1}, how long would
       it take to scan from m/z 1 to 1000?

It would take 30 s; 10 s each for m/z 1 to 10, m/z 10 to 100 and m/z
100 to 1000. You may think this is a most perculiar way of measuring scan time, but it originates with magnetic scanning instruments
where the magnet follows an exponential scan law.

The fastest magnetic scanning instruments employ special laminated
magnets, but cannot usually operate faster than 0.1 s decade^{-1}. If
you try to make the magnet go faster, funny things start happening
to the scan law because of eddy currents in the magnet. Most scan
no faster than 1 s decade^{-1} and all require time to reset themselves
before the next scan. The *cycle time* (time for scanning and reset-
ting) is therefore important, especially in gc-ms and lc-ms where
continuous repetative scanning is used extensively. The quadrupole
mass analyser can scan its full mass range in a few milliseconds and
its reset time is even less.

The combination of advantages (d) and (f) given in Section 4.3
make the quadrupole mass spectrometer the instrument of choice
for the method of analysis known as *selected ion monitoring* or
multiple ion monitoring. In this technique one does not scan the
whole mass spectrum, but selects just a small number of characteristic ions and rapidly switches between the instrument parameters
necessary to record these selected ions. This method is usually applied to mixtures of compounds and is used for sensitive quantitative
analysis. Obviously, the rapidity and accuracy with which one can
switch between these selected masses with a quadrupole instrument
is invaluable.

4.5.   REVISION EXERCISES

This concludes our discussion of mass analysers. There follows a set
of SAQs which act a revision test of this section. Please answer these

questions now and if you have difficulty with any of them, revise the appropriate section before proceeding.

SAQ 4.5a Calculate the resolving power needed to distinguish between $C_{27}H_{56}^{+\cdot}$ and $C_{26}H_{52}O^{+\cdot}$ ($^{12}C = 12.00000$; $^{1}H = 1.00783$; $^{16}O = 15.99492$).

SAQ 4.5b What is the maximum permitted overlap between a pair of peaks which are said to be resolved?

SAQ 4.5c What is it that limits the resolving power of a mass spectrometer?

SAQ 4.5d

Show that ions of the same kinetic energy, when accelerated through a potential of V and passed through an electric sector with a potential of E, all follow a curved flight path of radius R while in the electric sector, independent of their m/z values.

SAQ 4.5e

Fill in the missing words or phrases in the following description of a quadrupole analyser.

A quadrupole analyser consists of .... parallel rods with adjacent .... linked to a .... consisting of a .... component and an .... component. The m/z range is scanned by altering both the .... and the ...., while the ratio between them is ....

SAQ 4.5f

Suppose you are asked to recommend the design of mass spectrometer to be included in an unmanned space probe to Mars, what type of ion source and analyser would you suggest for the instrument, given that the following points are of importance.

(i) The spectrometer will be analysing for relatively simple organic molecules of low M_r (up to 150), but needs positive identification.

(ii) The instrument must be small and not too heavy to fit into the probe, but it must be robust to withstand take-off and landing.

(iii) It has to be reliable – there are no service engineers on Mars.

(iv) It has to be computer controlled – this is an unmanned flight.

Summary

This Part has covered the two most important methods of mass analysis, using magnetic or quadrupole analysers, and has highlighted the advantages and disadvantages of each. There has also been discussion of the importance of accurate mass measurements and of the double focussing system required to carry out such analysis. Other important features such as resolving power, scan rates and multiple ion monitoring have also been described.

Objectives

After studying this Part you should be able to:

- explain what is meant by resolving power of a mass spectrometer, why it is important and how it is calculated;

- explain what limits the resolving power in a single focussing mass spectrometer;

- describe the double focussing instrument that is used for accurate mass measurement;

- describe the quadrupole analyser.

5. Metastable Ions

In this section you will encounter an entirely new type of ion that we have not discussed before, the metastable ion. It is something that often leads to confusion among students and it is hoped that this section will show you that it is really quite simple and very important. Thus, we will discuss the origins of metastable ions and how to make use of their presence in understanding mass spectral fragmentation.

5.1. THE ORIGIN OF METASTABLE IONS

The photographic trace of a mass spectrum, Fig. 5.1a, shows a number of sharp peaks at integer m/z values and also some much weaker, broader peaks occurring at non-integer m/z values (marked m^*). These are metastable ions and although they complicate the spectrum somewhat, they can be very useful.

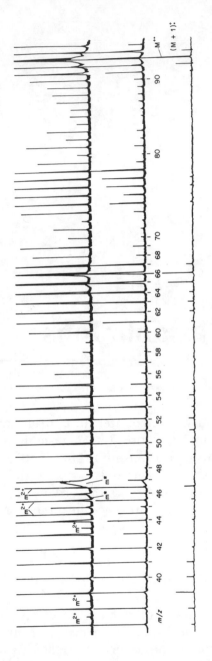

Fig. 5.1a.   *Photographic trace of a mass spectrum*

SAQ 5.1a

If you recall our general discussion in Section 1.3 concerning the energetics of the formation of molecular and fragment ions, you will remember that molecular ions with less than a certain threshold energy do not have enough energy to decompose. These are analysed and recorded as M^{+} ions. Those with internal energies over the threshold decompose in the ion source and are analysed and recorded as fragment ions.

Can you represent this on a plot of internal energy distribution? You were shown this earlier.

In fact this description is a little simplistic. It assumed there were only two types of molecular ions: those with sufficient energy to decompose to fragment ions very quickly, before leaving the ion source, and those with insufficient energy to decompose before reaching the detector.

∏   There must be a finite probability of creating some ions of
        energy which is intermediate between these two extremes.
        What do you think would happen to these ions?

These molecular ions of intermediate energy, would have lifetimes
sufficiently long for them to leave the ion source, but they would
then decompose during flight, before detection.

Ions spend about 10^{-6} s in the ion source and take about 10^{-5} s
to travel from the source to the detector. Thus, any molecular ions
that have high internal energies and hence high rate constants for
decomposition ($k > 10^6$ s^{-1}) will give rise to fragment ions and any
molecular ions of low internal energies and low rate constants ($k
< 10^5$ s^{-1}) will be recorded as M$\overset{+}{\cdot}$. Those ions with $10^6 > k$ s$^{-1} >
10^{-5}$ will decompose in flight and these give rise to metastable ions.

In the above discussion we referred only to molecular ions decom-
posing to fragment ions, but the same arguments also apply to de-
composing fragment ions, as well.

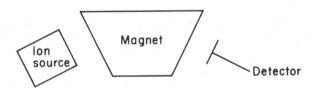

Fig. 5.1b.   *Sketch of single focussing mass spectrometer*

∏   Fig. 5.1b is a simple diagram of a single focussing mass spec-
        trometer. Can you indicate on it where metastable ions are
        formed.

Metastable ions are formed in any part of the mass spectrometer
between the exit from the ion source and the detector. In fact we

shall only concern ourselves with the ions that are formed in the small region between the exit from the ion source and the magnet. This is known as the field free region because it is situated between the regions of the accelerating field and the magnetic field.

Let us now consider what happens to an ion that decomposes in the field free region, the fragmentation being given by

$$M_1^+ \rightarrow M_2^+ + (M_1 - M_2)$$

The ion M_1^+ is accelerated out of the ion source and attains a velocity v_1, so that it has a kinetic energy of $\frac{1}{2}m_1v_1^2$.

SAQ 5.1b   Can you recall the relationship between accelerating potential, V and kinetic energy of the ion being accelerated?

The ion M_1^+ has a momentum m_1v_1 and if it decomposes after acceleration, then the law of conservation of momentum tells us that the momentum of the ion will be shared between the products of decomposition, which are the ion M_2^+ and the neutral $(M_1 - M_2)$. Here, we are going to make the assumption that the ion, M_2^+ and the neutral $(M_1 - M_2)$ move apart with equal velocities, v_2.

∏   Can you now write down an equation which describes the conservation of momentum?

The equation is

$$m_1v_1 = m_2v_2 + (m_1 - m_2)v_2 \qquad (5.1a)$$

∏   Use Eq. 5.1a to show $v_1 = v_2$

$$m_1 v_1 = m_2 v_2 + (m_1 - m_2) v_2$$

$$m_1 v_1 = m_2 v_2 + m_1 v_2 - m_2 v_2$$

$$m_1 v_1 = m_1 v_2$$

$$\therefore \quad v_1 = v_2$$

From now on we will call the velocity, $v (= v_1 = v_2)$.

Remember from the above discussion, the ion M_1^+ is accelerated such that

$$zV = \frac{1}{2} m_1 v^2 \qquad (2.1b)$$

This ion decomposes in the field free region and thus the ion M_2^+ is analysed by the magnet.

The equation governing the behaviour of M_2^+ in the magnetic field is

$$m_2 v^2 / r = Bzv \qquad (2.1c)$$

Using Eq. 2.1b and 2.1c, let us eliminate all terms in v and thereby derive a new equation relating m_1, m_2 and z to B, r and V.

Recalling

$$\frac{1}{2} m_1 v^2 = zV \qquad (2.1b)$$

and

$$m_2 v^2 / r = BzV \qquad (2.1c)$$

From Eq. 2.1c

$$v = Bzr / m_2$$

$$\therefore \qquad v^2 = B^2 z^2 r^2 / m_2^2$$

From Eq. 2.1b

$$v^2 = 2zV/m_1$$

$$B^2z^2r^2/m_2^2 = 2zV/m_1$$

$$m_2^2/m_1z = B^2r^2/2V \qquad (5.2b)$$

This is the same as the Eq. 5.1c

$$m^*/z = B^2r^2/2V \qquad (5.1c)$$

where

$$m^* = m_2^2/m_1 \qquad (5.1d)$$

This formula (Eq. 5.1d) predicts that the mass of the metastable ion (m^*) will always be less than m_2. If you are not mathematically minded, you might not be too convinced of this. Let's take two examples. Suppose an ion of m/z 60 decomposes by loss of a hydrogen atom to m/z 59. What would be the m^* value corresponding to this? From Eq. 5.1d.

$$m^* = 59^2/60 = 58.02$$

Next suppose an ion of m/z 600 decomposes to m/z 599, here m^* = $599^2/600$ = 598.00.

In both cases m^* is less than m_2.

You will notice that the masses of metastable ions (m^*) have only been quoted to two decimal places. This is because the energy focussing provided by the magnet is negligible and as the metastable ions are produced with a range of energies, the metastable ion peaks are broad compared with the normal ion peaks (see Fig. 5.1a). Thus their positions cannot be measured with such precision as those of the fragment ions produced in the source. In fact, they are normally only quoted to one decimal place.

How does this energy distribution arise? Well, fragmentation is rather like a shrapnel burst. The fragments fly off in all directions upon decomposition, but they are generally carried forwards by the high velocity imparted from the accelerating voltage. Those projected forwards will get a little ahead, and those backwards a little behind the average metastable ions, so that the peak broadens out.

As a simple example of the points we have just covered, the partial mass spectrum for the fragmentation, $M_1^+ \rightarrow M_2^+ + (M_1 - M_2)$ is shown in Fig. 5.1c.

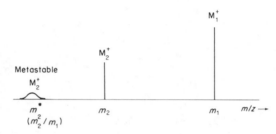

Fig. 5.1c.   *Partial mass spectrum for the fragmentation*
$$M_1^+ \rightarrow M_2^+ + (M_1 - M_2)$$

In concluding this section, it must be emphasised that there is nothing special about metastable ions. They are simply fragment ions formed after an ion leaves the source and therefore they have less kinetic energy than the corresponding ions formed in the source.

5.2.   THE USE OF METASTABLE IONS

Metastable ions can be useful in helping to establish fragmentation routes, since if one sees all three ions, M_1^+, M_2^+ and $(M^*)^+$ in the mass spectrum, then at least a proportion of M_2^+ ions (those formed in metastable decompositions) *must* be formed directly from M_1^+ and it is quite likely that, even if there is an alternative route for formation of M_2^+ ions, a considerable proportion of the M_2^+ ions formed in the source will also be formed directly from M_1^+.

As an example, given below is the mass spectrum of benzene Fig. 5.2a.

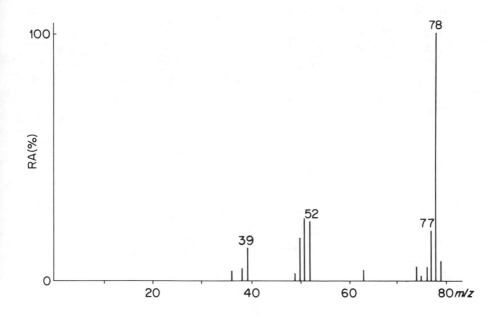

Fig. 5.2a. *Mass spectrum of benzene*

Metastable ions are observed in the spectrum at m/z 76.0, 74.1, 34.7 and 19.5.

SAQ 5.2a   All the metastable ions observed in the mass spectrum of benzene, arise from the molecular ion (m/z 78). Using the formula

$$m^* = m_2^2/m_1$$

assign the fragment ions (M_2^+). As an example the fragmentation

$$C_6H_6^{+\cdot} \rightarrow C_6H_5^+ + H^\cdot$$

$$m/z\ 78 \qquad m/z\ 77$$

shows a metastable ion at $77^2/78 = 76.0\ m/z$

SAQ 5.2a

From the observed metastable ions and the calculations performed in SAQ 5.2a, a fragmentation pattern for the molecular ion of benzene can be established. This is shown below.

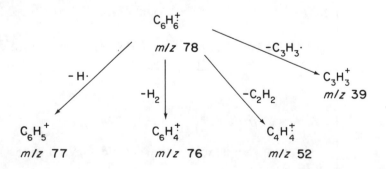

SAQ 5.2b

Other metastable ions are observed in the benzene spectrum:

M^*	M_1^+	$\rightarrow$	M_2^+
50.0	52	$\rightarrow$	51
48.1	52	$\rightarrow$	50
33.8	77	$\rightarrow$	51
32.9	76	$\rightarrow$	50

Use these metastable ions and those used in SAQ5.2a for the $C_6H_6^+$ ion, to draw up a complete fragmentation pattern for benzene.

In the double focussing instrument there are two field free regions. These are between the ion source and the electric sector (first field free region), and between the electric and magnetic sector (second field free region). Metastable ions observed in the mass spectrum of compounds analysed by such an instrument are formed in the second field free region. Metastable ions formed in the first free region are deflected by the electric sector and do not reach the detector. They can, however, be recorded using special techniques. The interested reader is referred to Rose and Johnstone (1982) or Williams and Howe (1972).

In conclusion, it is important to point out that metastable ions provide useful *positive* evidence for specific fragmentation routes, but they cannot be used in the opposite sense. Thus, the fact that no metastable ion is observed corresponding to a specific fragmentation does *not* mean that process is not occurring. It may well be occurring without giving rise to observable metastable ions. There are a variety of reasons why metastable ions are not observed including the energetics of the process concerned and the sensitivity of the instrument. So remember you can draw positive conclusions from the presence of metastable ions, but you cannot draw any conclusions from their absence.

The following SAQ is a revision exercise.

SAQ 5.2c	Fig. 5.1a is actually the mass spectrum of aniline, $C_6H_5NH_2$. The two metastable ions shown at m/z 46.8 and 45.9 correspond to fragmentations of the molecular ion $(M^{\ddagger})$ and the fragment ion $(M - H)^+$ respectively. What are the daughter ions formed in these two fragmentation processes?
	Can you suggest a formula for the neutral species lost?

SAQ 5.2c

Summary

After studying this Part you should be aware of how metastable ions are formed and how to use them as an aid to elucidating mass spectral fragmentation pathways.

Objectives

When you have finished this Part, you should be able to:

● explain why metastable ions are formed;

● explain in which region of the mass spectrometer such ions are formed;

● derive the equation $m^* = m_2^2/m_1$;

● use this equation to assign metastable ions to parent and daughter ions;

● use metastable data to give information about fragmentation patterns.

6. Ion Detection and Recording

We have now reached the final section of the mass spectrometer, the detector. We shall concentrate on the electron multiplier detector.

Display and recording devices are also covered in this section and special emphasis is laid on the most important method of dealing with mass spectral data, the on-line computer.

6.1. ION DETECTION, DISPLAY AND RECORDING

We have already discussed one type of detector, which is that most commonly employed.

SAQ 6.1a

Can you remember what the most commonly employed type of detector is called and how it works?

SAQ 6.1a

There are other methods of electrical detection, but these are relatively rarely used and are outside the scope of this text. The interested reader is referred to Rose and Johnstone (1982) for a description of these methods.

The amplified signal from the electron multiplier is usually relayed to either an oscilloscope, a chart recorder or a computer. The oscilloscope is useful for displaying the peak arising from a single ion in the mass spectrum when carrying out accurate mass measurements without a computer or when adjusting the instrumental parameters to give maximum sensitivity (tuning the instrument). The oscilloscope is also used in gaining a preliminary scan of the whole spectrum of the sample being analysed. This allows the experienced operator to tell whether there is a sufficiently strong spectrum for recording in a more permanent way or whether more sample should be admitted to the source.

A common method of taking a permanent record of the spectrum is to use a chart recorder. The chart recorder usually employs photosensitive paper with the image of the spectrum being induced by allowing a light beam to traverse the paper.

∏   Have you any idea why this photographic chart recorder is used rather than the more obvious, pen recorder? Don't worry too much if you can't think of the answer.

The reason is that the photographic chart recorder has a much faster response than a pen recorder. The latter has a considerable lag between the pen reaching the maximum of a peak and returning to the base-line to begin recording another. The fast response is required because mass spectra are usually scanned at relatively rapid rates. A typical spectrometer scan speed when using a chart recorder is 10 s decade^{-1}.

How long would it take to scan from m/z 10 to m/z 1000 at 10 s decade^{-1}?

A typical chart paper spectrum was shown in Fig. 1.4a. *This is reproduced again here*, Fig. 6.1a. As you can see there are three traces.

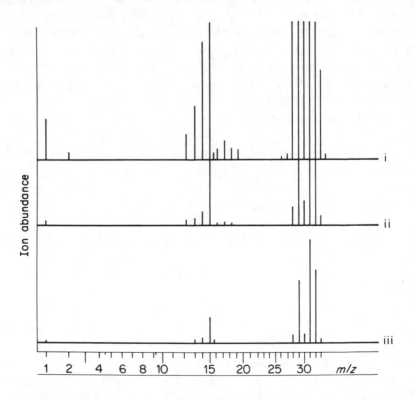

Fig. 6.1a.   *Mass spectrum of CH_3OH*

SAQ 6.1c
> Can you recall how these traces differ and why we need more than one?

SAQ 6.1d
> How are spectra usually presented?

Spectra can also be recorded using a photographic plate as a detector. This is not very common but is used with mass spectrometers of Mattauch–Herzog geometry and spark sources.

6.2. THE USE OF COMPUTERS

The most important method of recording mass spectra now used is that employing an on-line computer. We will briefly outline how this is done. The whole system is often known as a *Data System* and consists of a computer, an interface to the mass spectrometer, a VDU and a printer–plotter.

∏ If you know anything about computers and computer interfacing, you may be able to answer this question, but don't worry if you can't. If you wish to feed the electrical signal from the electron multiplier into a computer, there is one vital thing that you will have to do to that signal. Do you know what this is?

To make a continually varying electrical current or voltage (an analogue signal) into a form which a computer can handle, it has to digitised, as computers are digital devices. This is shown in Fig. 6.2a.

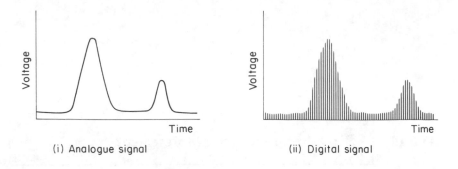

Fig. 6.2a.   *Analogue and digital forms of a mass spectrometer output*

This digitisation is achieved by means of an Analogue-to-Digital Converter (ADC). The data system contains a very accurate crystal oscillator clock which defines precise regular time intervals at which the analogue voltage is sampled by the ADC. This results in a regular series of voltage pulses being fed into the computer.

∏   In Fig. 6.2a (*ii*), which obviously only represents a small portion of a mass spectrum, there are some 60 voltage readings. For a whole spectrum, there will obviously be many hundreds or thousands of such readings. Can you suggest how we might reduce the amount of information being fed to the computer without reducing its usefulness or accuracy?

There are two obvious ways we could approach this problem. One is to allow longer time intervals between sampling the voltage. But, if we do this we will not define the shape of the peak so well and we will see shortly that defining the shape of the peak is important.

The second thing that could be done is to reject voltage readings below a certain value as these generally arise from the base-line noise of the signal. We want information about the peaks not the base-line noise, so if we include a threshold device, which triggers the computer to take in data only when the voltage exceeds a certain value, we will dramatically reduce the information fed to the computer. This is shown in Fig. 6.2b.

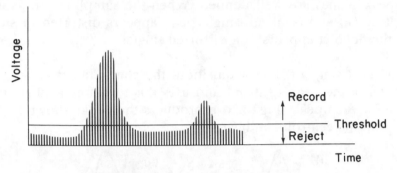

Fig. 6.2b. *Digital signal showing a threshold setting*

To return to the point about sampling rates, Fig. 6.2c shows a single peak and a doublet in analogue form and digital form using different sampling rates.

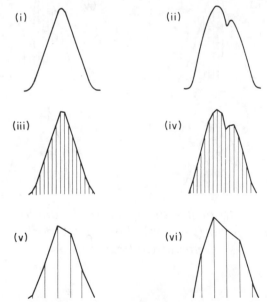

Fig. 6.2c. *Peaks from a mass spectrum in analogue form (i and ii) and in digital form iii to vi, using different sampling rates*

Π   Can you explain what problems are arising in the bottom pair of representations (*v* and *vi*) using a low sampling rate?

When the sampling rate is high, as shown in Fig. 6.2c (*iii*) and (*iv*), the peak shapes are well defined. When the sampling rate is low, Fig. 6.2c (*v*) and (*vi*), the single peak appears distorted in shape and the doublet appears as a distorted singlet.

In the next stage of data acquisition, the computer calculates the position of the peak centroid and uses this to define a single time value for each peak. Fig. 6.2d reproduces the peaks shown in Fig. 6.2c with the centroids indicated.

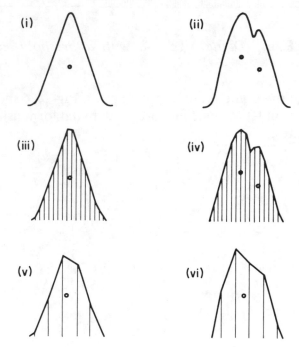

Fig. 6.2d.   *Peaks from a mass spectrum in analogue form (i and ii) and in digital form (iii to vi) showing the position of each peak centroid (indicated by ○)*

∏   For a symmetrically shaped peak, where would you expect to find the centroid with respect to the complete peak profile?

The centroid should be directly below the peak maximum, if the peak is symmetrical. Fig. 6.2d (*iii*) and (*iv*) shows this to be the case for peaks recorded at a high sampling rate, but not with a low sampling rate, especially if doublets are not resolved (*v* and *vi*).

The computer uses this centroid position to define the peak. By effectively dropping a perpendicular from the centroid to the time axis, the peak is described by one time value. It is best if this always corresponds to the time for the peak maximum, because each peak will then be described in the same way.

∏   If, in a separate operation, the computer also sums the voltage readings for all measurements describing a peak, what information will it gain?

It will gain a parameter that is effectively a measure of the peak area or intensity and the computer can ratio the values for all peaks to give the relative intensities.

The computer can therefore store two pieces of information, defining each peak. These are the time value for the centroid and the peak area. This total information for all peaks produced is effectively the mass spectrum. The computer is, of course, faced with one more problem – it needs to convert the time value for each peak to a mass value. In order to do this, it effectively establishes the time against mass relationship for the spectrometer, which is done by recording the spectrum of a reference compound. This is shown in Fig. 6.2e where the curve shown in Fig. 6.2e (*ii*) represents the required relationship.

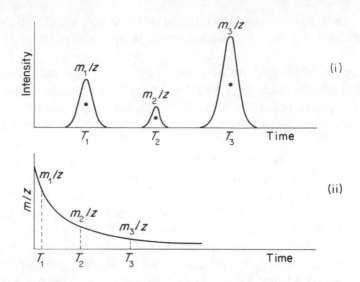

Fig. 6.2e. *(i) Mass spectrum in the form intensity against time.
(ii) Curve showing the m/z to time conversion, the scan law*

This is shown here as a graph, although the computer does not actually produce such a curve. The relationship is called the *scan law* of the mass spectrometer and is stored in the computer's memory.

∏   Can you now suggest how the mass spectrum of an unknown compound is obtained?

The computer has the scan law of the instrument stored in its memory. If an unknown sample is now examined using exactly the same instrumental settings, a series of peak centroid times and peak areas will be obtained for the sample ions. The computer can effectively use the time values to read off the appropriate mass values from the scan law curve. The computer thus holds the mass spectrum in terms of masses and relative intensities and can print this data in tabular form or plot it as a bar diagram. The process we have just described is known as *data acquisition*.

Computers can also be used to control the operation of the mass spectrometer and for processing data after it has been acquired. In the instrument control function, the computer can be used to optimise conditions for obtaining a spectrum by altering source and analyser settings. In the data processing function it can be used for converting accurate mass measurements into molecular formulae; for comparing one spectrum with another, particularly using a library of standard spectra, and in a variety of ways that are important in combined gc-ms and hplc-ms. Some of these will be discussed in more detail in Part 10.

SAQ 6.2a	Briefly list the steps involved in the acquisition of mass spectral data.

Summary

With this Part we have completed our coverage of instrumental aspects of mass spectrometry and you should now be aware both how ions are detected and how spectra are displayed and recorded. With the increased availability of mass spectrometers with on-line computers, it is important that you are familiar with the basic introduction to this aspect in this section.

Objectives

When you have studied Part 6 you should be able to:

● describe how an electron multiplier detector functions (revision of Section 2.1);

● describe why spectra are sometimes displayed on an oscilloscope;

● describe how spectra are recorded on a chart recorder and why photo-sensitive chart paper is used;

● explain how a computer is used to acquire mass spectral data;

● briefly list the uses of a computer in mass spectrometry.

7.   Detection and Use of Isotopes in Mass Spectrometry

7.1.   INTRODUCTION

Almost all the elements exist as a number of naturally occurring isotopes, which are either stable or long-lived enough for them to be of significance in analytical chemistry. Such elements possess at least one isotope arising from one or more additional neutrons in the nucleus, and their mass spectra consist of clusters of ions each derived from one of the isotopes. Recall (Section 1.4) that the mass spectrometer produces positive ions from atoms or molecules in the vapour phase, separates them according to their mass-to-charge (m/z) ratios and then records their intensities. Intensities are usually measured relative to the most intense ion in the spectrum which is given an arbitrary value of 100%.

∏   Fig. 7.1a shows the mass spectrum of cyclohexene, determined with a low resolution mass spectrometer and plotted out on ordinary chart paper just as it would appear from the instrument. Which is the base peak of the mass spectrum, and what are the intensities of the molecular ion and the second most intense fragment ion in the spectrum, and their m/z values?

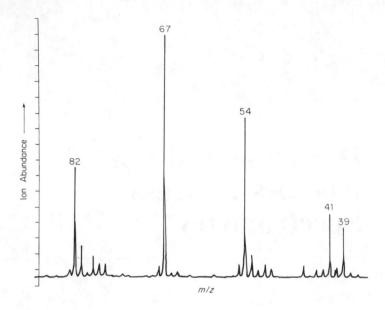

Fig. 7.1a. *Mass spectrum of cyclohexene*

The base peak is by definition the most intense ion in the spectrum, m/z 67. This is measured and found to be 57 mm. The molecular ion is the highest significant mass observed in the spectrum, in this case m/z 82. Its height is 26 mm and therefore its relative abundance (RA) or intensity (I%) is 26/57 × 100 or 46% to the nearest whole %. The second most intense ion in the spectrum of cyclohexene is m/z 54, and its intensity turns out to be 66%, using the same method as before. It is usual in presenting mass spectral data to tabulate them as in Fig. 7.1b.

m/z	I(%)
82	46
[67]	100
54	66

Fig. 7.1b. *Tabulated mass spectrum of cyclohexene*

Base peaks are usually distinguished in some way eg by using square brackets as we have done, or by underlining them or using a bold or italic type. Data are also frequently presented in rows rather than columns eg

$$\text{Cyclohexene} \quad m/z \quad 82 \quad [67] \quad 54$$
$$\text{I}(\%) \quad 46 \quad 100 \quad 66$$

which has the merit of taking up less space though it is more diffi-cult to read. We do not mind which method you use but do stick to the same format. Did you notice in measuring the peaks in Fig. 7.1a that it was difficult to be very accurate? This is why in most mass spectrometric analyses intensities are only quoted to the near-est %. Added to which the variation in operating parameters such as temperature and pressure in the mass spectrometer cause changes of several percent in relative abundance when compounds are re-determined by the same instrument, let alone a different model in say Japan! You will not impress anyone by quoting the eight decimal places off your calculator.

7.2. APPLICATION TO ELEMENTS OTHER THAN HALOGENS

From a study of the m/z values at high resolution, and the rela-tive abundances (RA) of the atomic or molecular species present in the cluster determined at low resolution, we can learn a great deal about the number and kind of elements present in an unknown sam-ple. In the years between the two world wars physicists used early mass spectrometers based on A.J. Dempster's 1918 design, or F.W. Aston's 1919 mass spectrograph (so called because it had a photo-graphic plate detector) to measure the atomic weight of the naturally occurring isotopes to accuracies of around 10 ppm and their RA's to around 1 in 10 000. Today these data are collected in tables to be found in many reference books such as the well known *Hand-book of Chemistry and Physics* published by the Chemical Rubber Company Press (the 'Rubber Book'), and specialist texts on mass spectrometry eg J.H. Beynon's *Mass Spectrometry and its Applica-tion to Organic Chemistry*, Elsevier, 1960, Appendix 3.

Sometimes you will meet a mass spectrum which shows an unfamiliar or unusual cluster of ions. When these are clearly the molecular ions or closely related species you will then have to scour the reference data to find out which element or elements give rise to the observed patterns. This sounds a formidable task bearing in mind the 90 odd elements available, but in practice you will often have a good idea as to which elements are likely to be present, from the origin of the sample. This will narrow down your search considerably.

In some cases though there will be no useful information supplied with the sample and you can only rely on your basic knowledge of the relative industrial importance and natural occurrence of the elements and eliminate the most unlikely first. We once received an extract of a contaminated batch of flour. It contained all sorts of esters of high molecular weight fatty acids, which we painstakingly identified hoping to find a 'smelly' one among them. After wrestling with the problem for two weeks we noticed a persistent and strange cluster of ions centred around m/z 256. It was S_8 – flowers of sulphur. Did you know that sulphur will dissolve in some organic solvents as S_8 and what's more, chromatograph by gc? As you will find later, sulphur has a ^{34}S isotope which is easily picked up by mass spectrometry, provided you are expecting it! The flour had been transported in a tanker which had previously been used for sulphur powder and not properly cleaned in between. In the baking process H_2S and organic sulphides were being formed and as I am sure you know these taste and smell horrible even in minute amounts (and H_2S is extremely toxic).

Well, S_8 is a difficult example to start with, so I would now like you to start with a simpler problem but one also related to pollution work.

∏   Fig. 7.2a shows the M^+ cluster of a compound containing an unknown heavy metal. From other analytical evidence the element concerned may be mercury, Hg; cadmium, Cd; or lead, Pb. Which do you think it is? Refer to the data in Fig. 7.2b to help you.

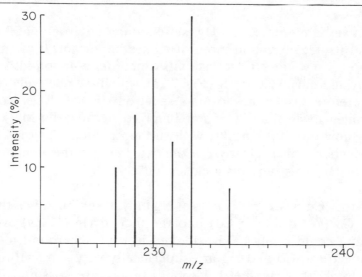

Fig. 7.2a. *M⁺ cluster of a heavy metal compound*

Element	At No	Mass No	Relative Abundance (%)*
Cd	48	106	1.2
		108	0.9
		110	12.4
		111	12.8
		112	24.1
		113	12.3
		114	28.9
		116	7.6
Hg	80	196	0.2
		198	10.0
		199	16.8
		200	23.1
		201	13.2
		202	29.8
		204	6.9
Pb	82	204	1.4
		206	25.2
		207	21.7
		208	51.7

* to 3 figure accuracy

Fig. 7.2b. *Isotopes of Cd, Hg and Pb*

The answer is mercury, Hg. Hg is a common environmental pollutant and its detection and monitoring are important. It often turns up in soil and water samples as dialkylmercury compounds such as dimethyl mercury, $(CH_3)_2Hg$ (Fig. 7.2a), which are relatively volatile and therefore susceptible to mass spectrometric analysis. You can tell straight away that *Pb* is not involved because its M^{+} cluster would show only four peaks, with the most intense at the *highest* mass number. This is clearly *not* the case here, so the element must be Cd or Hg. This requires a closer look.

Notice that Cd has *eight* isotopes, Hg only *seven* but since the first two in Cd (^{106}Cd and ^{108}Cd) and the first in Hg (^{196}Hg) are very small and could easily be overlooked in the background to a real mass spectrum, Cd and Hg are superficially very similar. Looking more closely then, notice that some isotopes are missing, 109 and 115 in Cd and 203 in Hg and the relative abundances vary too. In Cd clusters it is the fifth and seventh isotopes (^{112}Cd and ^{114}Cd) which are most intense, whereas in a Hg cluster it is the fourth and sixth (^{200}Hg and ^{202}Hg). Even so we can't be quite certain because these principal isotopes span a similar three m/z range. As a final check then, you must measure the actual peak intensity of these key ions. For Hg they should have the ratio $1:1.29$, while for Cd the ratio is $1:1.20$, which shows a significant distinction.

From this exercise you should have gathered that mass spectrometry is a good technique for identifying unknown elements especially when only minute samples are available. But will it work every time?

∏   What sort of elements could you *not* recognise by the technique you have just used?

I hope you remembered that I began this section by saying that *almost all* the elements exist as a number of naturally occurring isotopes. This implied that there are some *monoisotopic* elements about ie those which have only one isotope. Unless these give ionised *atoms* when their compounds are examined and can be recognised by their mass alone, which is uncommon, there is no way of knowing which of these elements are contained in a *single* peak without further information.

SAQ 7.2a

Can you think of an example of a monoisotopic element? (If you have access to a 'Rubber Book' have a glance through the Table of the Relative Atomic Masses of the Elements, but don't spend too much time on this).

Of the monoisotopic elements it is the non-metals fluorine, phosphorus, and iodine which are most commonly met with in mass spectrometry, as they occur in covalent volatile compounds. But be warned, all the metals and arsenic listed in the response to SAQ 7.2a do have volatile compounds such as halides, alkyl, aryl and organometallic derivatives whose spectra you may meet one day! However, the great majority of mass spectra arise from organic compounds and it is to these we next turn our attention.

∏   Benzene (C_6H_6) gives an intense M^{+} at m/z 78, which is the base peak of the spectrum. A small peak appears at (M + 1), m/z 79. What is this peak due to? Calculate its RA to two significant figures. Use the data in Fig. 7.2c to help you.

Isotopes	Relative Atomic Mass (A_r)	%	A_r	%	A_r	%
H	1	100	2	0.016		
C	12	100	13	1.08		
N	14	100	15	0.36		
O	16	100	17	0.04	18	0.20
F	19	100				
Si	28	100	29	5.07	30	3.31
P	31	100				
S	32	100	33	0.78	34	4.39
Cl	35	100			37	32.4
Br	79	100			81	97.5
I	127	100				

Fig. 7.2c.   *Natural abundances of the isotopes of some common atoms*

The (M + 1) peak in benzene is due to the presence of ^{13}C and ^{2}H (deuterium). Reference to Fig. 7.2c shows that as ^{2}H is very small (0.016% relative to ^{1}H as 100%) its contribution can be neglected to two significant figures, as it is only 6 × 0.016% or 0.096%. The RA of ^{13}C is 6.6%. If you said 1.1% remember that when the benzene molecule is made there is a 1.1% chance that a ^{13}C isotope will be selected *each* time a carbon atom is built into the structure, so the total probability that a ^{13}C is present in the completed molecule is *six* times the RA of ^{13}C, 6 × 1.1% or 6.6%.

This shows us that even minor isotopes give rise to significant peaks in mass spectra. This is particularly so in organic molecules where many atoms of the same type appear. Isotopes with such low natural abundance as ^{2}H, ^{13}C, ^{15}N and ^{18}O can give rise to significant (M + 1) and (M + 2) peaks simply because it is possible to have so many of them in the one molecule. As an aside, do not run away with the idea that this is only of interest to organic chemists. Remember that mass spectra can be obtained for any compound which is volatile enough to give a significant vapour pressure at 10^{-6} to 10^{-7} torr. The common factor is covalent bonding and not all covalent compounds contain carbon!

In General Molecular Ions are Rarely Single Peaks.

Notice that in Fig. 7.2c the abundances of the isotopes are presented as %'s of the abundance of the most common isotope ie they are *normalised*. The normalisation of mass spectra has been explained previously (Section 7.1). Most mass spectra you will meet in this Unit and in your work will be plotted in terms of the base peak (most abundant peak) as 100% and all other ions as %'s of this. This makes for consistency and comparability of your data with those, say, of a Japanese chemist working in Yokohama ten years ago. Normalised values are universally adopted in the world of mass spectrometry. Physicists however still use data in the form shown in Fig. 7.2a ie *un-normalised* and it is important that you can recognize and inter-convert the two.

Now try SAQ 7.2b and 7.2c to make sure you can do this.

SAQ 7.2b	Taking the data for lead in Fig. 7.2b, plot the normalised spectrum of the Pb^{+} ion cluster.

SAQ 7.2c
A normalised spectrum of a molecular ion cluster is shown in Fig. 7.2d. Calculate the isotopic abundances of the element concerned. Given that the species has the formula $(CH_3)_2X$, what is the element X?

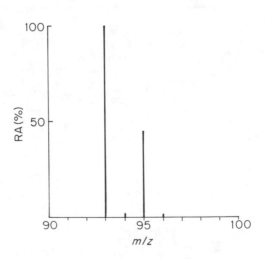

Fig 7.2d. *Mass spectrum of $M^+_{\cdot}$ of $(CH_3)_2X$*

You may have noticed in Fig. 7.2d small peaks at m/z 94 and 96, and wondered what these could be due to. Of course, if carbon is present in the compound, reference to Fig. 7.2c tells you that there will be some M† present which will contain ^{13}C instead of ^{12}C. The amount of ^{13}C in m/z 94 is given by the relative intensity of the peak multiplied by the number n of carbon atoms present multiplied by the RA of ^{13}C ie $(I \times n \times RA)\%$. This is a general result for any minor isotope relative to the most abundant one.

$$\text{Isotope intensity} = (I \times n \times RA)\% \qquad (7.2a)$$

So in Fig. 7.2d we would expect ^{13}C at m/z 94 to be

$$1.0 \times 2 \times 1.08\% \quad \text{or} \quad 2.2\%.$$

∏ Remembering formulae is always difficult. Can you think of a mnemonic which would help you to 'fix' this one?

Well, we thought of 'I never Rang Ann'. If you have a better one let us know! What matters is that you can easily recall Eq. 7.2a and use it to check (M + 1) and any higher mass isotope peaks for the correct intensity. This gives us a method of confirming that the supposed molecular formula is consistent with the isotopic intensities.

∏ What do you predict the m/z 96 ion intensity in Fig. 7.2d to be?

Applying 'I never Rang Ann', $n = 2$ and RA $= 1.08\%$ as before, but $I = 0.445$, so $I \times n \times RA = 0.445 \times 2 \times 1.08 = 0.96\%$. There is also H present in $(CH_3)_2Cu$ so according to Fig. 7.2c there will also be some deuterium present and this too will contribute to both m/z 94 and 96.

∏ We have ignored deuterium in calculating the RA's of the m/z 94 and 96 ions in Fig. 7.2d. Are we justified in this?

We are. You should have applied Eq. 7.2a to calculate the RA at m/z 94 due to ^{2}H. $I = 1$, $n = 6$, and RA for ^{2}H is 0.016%. Hence the intensity of the ion at m/z 94 due to ^{2}H is predicted to be $1.0 \times 6 \times 0.0116\%$ or 0.096%. This is negligible on the scale of

Fig. 7.2d compared to the 2.2% ^{13}C ion so we can safely ignore the deuterium. The deuterium containing ion at m/z 96 will be even less. Nevertheless, there are such low abundance species present in the M‡ cluster.

As the number of C and H atoms increases in larger organic molecules, account does need to be taken of their contribution to the (M + 1) ions. This can be done in a fairly straightforward way, when there are no elements present, other than C, H and N, having (M + 1) isotopes, if we apply the formula in Eq. 7.2b.

$$\frac{(M + 1)}{M} = (1.1\% \times nC)\% + (0.016 \times nH)\% + (0.36 \times nN)\%$$

$$(7.2b)$$

Here nC, nH and nN are the number of C, H and N atoms respectively.

Our 'I never Rang Ann' formula is actually a particular case of this more general one. Let's apply it to $(CH_3)_2Cu$ to calculate the intensity of the (M + 1) peak. Since this compound does not contain an element having a (M + 1) isotope other than C and H, it should work and give the same result to the same degree of accuracy. Eq. 7.2b consists of C, H and N terms. There is no N present, in $(CH_3)_2Cu$ so the final term in Eq. 7.2b is zero. We have already seen that ^{2}H is negligible so the second term can be ignored, too. This leaves (1.1% × nC) or (1.1 × 2) = 2.2%, the same result that Eq. 7.2a gave us. So it does work, as it should.

Eq. 7.2b can be used for calculating (M + 1) intensities of ions which contain elements such as O, F, P, Cl, Br and I as well as C, H and N because the former group have either very low or no (M + 1) isotopes (see Fig. 7.2c). Essentially then, Eq. 7.2b allows calculation of the relative abundance of the (M + 1) isotope peak in the great majority of carbon compounds, because these only contain C, H, N, O, the odd halogen, and the occasional P or S. Only the elements Si ((M + 1) = 5.07%) and S ((M + 1) = 0.78%) from Fig. 7.2c are not included. However, you could easily incorporate Si or S instead of N. If you were dealing with a class of compounds which contained either Si or S but not N, then you would replace the 'nitrogen' term in Eq. 7.2b by either (5.07 × nSi)% or (0.78% × nS atoms).

The relative abundance of (M + 2) peaks in organic compounds due to the presence of two ^{13}C and/or ^{18}O atoms can be calculated from a simple formula too. However, there is a snag. By reference to Fig. 7.2c, you will see that there are a number of common elements which have significant (M + 2) isotopes, Si and S again, and also ^{37}Cl and ^{81}Br, so the list of atoms which may be present for this calculation to work is reduced to C, H, N, O, F, P and I. The formula is given in Eq. 7.2c.

$$\frac{(M + 2)}{M} = \frac{(1.1 + nC)^2}{200}\% + (0.2 \times nO)\% \qquad (7.2c)$$

You will see later how we calculate the relative intensities of ion clusters which contain elements with significant (M + 2) isotopes, such as ^{37}Cl and ^{81}Br.

SAQ 7.2d

(*i*) Calculate the abundance of the (M + 1) peaks of CO, N_2, and $CH_2=CH_2$ ($M^{+} = 28$).

(*ii*) Calculate the abundance of the (M + 1) peak of naphthalene, $C_{10}H_8$ ($M^{+} = 128$). Could you expect to distinguish naphthalene from cyclohexanecarboxylic acid, $C_7H_{12}O_2$ (also M_r 128) by the height of its (M + 1) peak alone?

(*iii*) Calculate the abundances of the (M + 1) and (M + 2) peaks in the mass spectrum of cholesterol, $C_{27}H_{45}OH$.

(*iv*) Calculate the abundance of the (M + 2) peak in the mass spectrum of trimethyl phosphate, $(CH_3O)_3PO$.

(*v*) At what carbon number, approximately, would you expect the (M + 2) peak of an organic compound containing C, H, N, F,

$\longrightarrow$

SAQ 7.2d
(cont.)

P or I and 4 oxygen atoms to appear above a general background level of 2% (relative to $M^+ = 100\%$)?

You should now be able to see why we made the statement earlier that ions in a mass spectrum are rarely represented by single peaks. They normally appear as clusters. This leads us to a problem. I wonder if you can see what it is? How do we define the molecular ion when in practice even simple molecules give rise to a molecular ion for *each* isotopic species present?

Π   What would you propose as a working definition of a 'molecular ion'?

The answer is that the molecular ion is that *single species in which all the elements present are found in their most abundant forms.* Any other isotopic species in the cluster are then defined as (M − 1), (M + 1), (M + 2) etc relative to the most abundant isotope species. Congratulations if you thought it through and reached this definition. The point is that the M^+ defined in this way is *not* necessarily the most abundant ion in the cluster in terms of total intensity, though it usually is. Working out which is the ion of greatest RA can get quite complex when several multi-isotope elements are present. Selecting the most abundant isotope for each element and adding up their relative atomic masses is straightforward – well fairly!

SAQ 7.2e   What are the molecular ions of:

(*i*)   1-chloronaphthalene, $C_{10}H_7Cl$;
(*ii*)   diethylmercury, $(C_2H_5)_2Hg$;
(*iii*) tetraethyl lead, $(C_2H_5)_4Pb$ and
(*iv*)   ferrocene, dicyclopentadienyliron $(C_5H_5)_2Fe$?

SAQ 7.2e

7.3. HALOGEN ISOTOPES

7.3.1. Introduction

In Section 7.2 we pointed out that fluorine and iodine were among the few monoisotopic elements while in Fig. 7.2c you saw that chlorine and bromine had 32.4% and 97.5% (M + 2) isotopes, respectively. Since Cl and Br are fairly common elements, it is a good bet that when you see an ion in a mass spectrum, particularly of course a molecular ion, which is a cluster of intense peaks each differing by 2 amu, that either or both of these elements are present. They are relatively easy to spot!

Before we go on to consider how to interpret ion clusters due to Cl_x, Br_y, or $Cl_x + Br_y$, where x and y are small integers, let us briefly consider how we would spot the presence of F or I in a sample.

∏ How would you identify the presence of F or I by a mass spectrometric and a non-mass spectrometric method?

There are standard inorganic qualitative tests for these elements eg sodium fusion, but these would need fairly large amounts of sample. High resolution mass spectrometry would be the best approach, provided you had access to a double focussing instrument, because both F and I are *mass deficient* atoms. That is, their accurate atomic weights are less than a whole number, unlike the common elements H, and N which are *mass proficient*. The accurate A_r of fluorine is 18.998495 and of iodine 126.904470 (see Fig. 7.3a).

Isotope	Atomic Weight
^{1}H	1.007825
^{14}N	14.003074
^{16}O	15.994915
^{19}F	18.998495
^{28}Si	27.976929
^{31}P	30.973765
^{32}S	31.972073
^{35}Cl	34.968851
^{79}Br	78.918329
^{127}I	126.904470

Fig 7.3a. *Precise relative atomic mass (A_r) of the elements of Fig. 7.2c (based upon $^{12}C = 12.0000$)*

This means that compounds containing F and I will usually have precise atomic masses, relative to $^{12}C = 12.0000$, which are less than the nominal mass and this will alert you to their possible presence. It is for this reason also that polyfluoro compounds such as polyfluorokerosene, PFK, $CF_3(CF_2)_nCF_3$ are used as calibrants in high resolution work. Ions from these polymeric mixtures turn up in regular series such as CF_3^+, $CF_3CF_2^+$, $CF_3(CF_2)_2^+$; m/z 69, 119, 169 respectively. They are significantly mass deficient:

Ion	Accurate Mass ($^{12}C = 12.0000$)
CF_3^+	68.99549
$CF_3CF_2^+$	118.99248
$CF_3(CF_2)_2^+$	168.98947

This deficiency takes them well out of the range of accurate masses expected for most organic compounds, which are mass proficient. They fulfil one of the desirable characteristics of a calibrant – lack of interference with the sample peaks.

If only low resolution mass spectra are available, the presence of F or I can be spotted by two features:

(*a*)   mass increments of 18, 36, 54, 72 etc for each F, and 126, 252, 378, 504 etc for each I present, respectively;

(*b*)   successive losses of 127(I˙) or 128 (HI) amu from the M⁺ of an iodocompound.

∏   Why are the increments for F, 18 amu, and I, 126 amu, when the A_r of F is 19 and I 127?

Well of course, in order to substitute a F or I into an organic molecule, a H must be replaced. Hence the nett increase is 1 amu less than the A_r. As an example of this, look at Fig. 7.3b and Fig. 7.3c. Fig. 7.3b shows the mass spectrum of 1,3-dinitrobenzene and Fig. 7.3c is of a derivative of this compound.

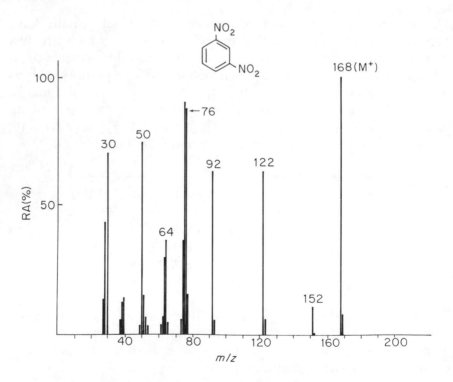

Fig. 7.3b.   *Mass spectrum of 1,3-dinitrobenzene*

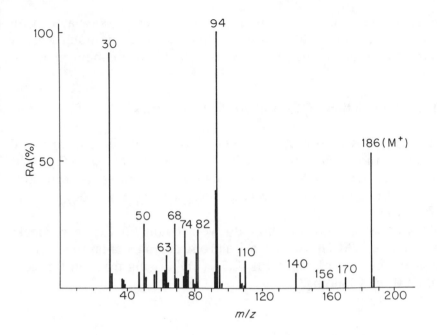

Fig. 7.3c. *Mass spectrum of a derivative of 1,3-dinitrobenzene*

You will see that the spectra are rather different but concentrating on the molecular ion regions the additional atom or group has increased M_r to 186 from 168, an increment of 18 amu. Hence Fig. 7.3c must be the spectrum of dinitrofluorobenzene (in fact it is the 2,4-dinitro isomer):

You might have expected that the mass spectrum would have shown signs of the loss of F from either the M^+ or one of the fragment ions seen in the spectrum, ie ions 19 amu to lower mass, but there are no such ions. The presence of F is only obvious because of the mass increase of 18 amu which is unique to F. There is no other common group which causes such a mass shift. In case you are wondering why F atoms are not readily lost when the positive ions fragment the answer lies in the very high bond energy associated with C—F bonds, about 423 kJ mol^{-1}. There are usually much weaker bonds in an organic molecular ion which would fragment first.

7.3.2. Calculation of Cl and Br Isotope Ratios

When one Cl is present in a molecule, any ion derived from it still containing the Cl will show a $^{35}Cl : ^{37}Cl$ ratio of 100 : 32.4 (Fig. 7.2c, see also Fig. 7.3d (*i*)). Similarly a single Br containing ion will show a $^{79}Br : ^{81}Br$ ratio of 100 : 97.5 (Fig. 7.2c, see also Fig. 7.3e).

This is quite obvious, but how do we account for the three peaks seen in Fig. 7.3d (*ii*) which arise from an ion containing two Cl atoms or the four peaks in Fig. 7.3d (*iii*), arising from an ion containing three Cl atoms?

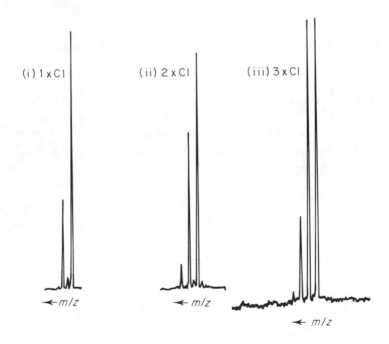

Fig. 7.3d. *Patterns produced by elements with abundant stable isotopes. Examples of ions with one (i), two (ii) and three (iii) Cl atoms respectively*

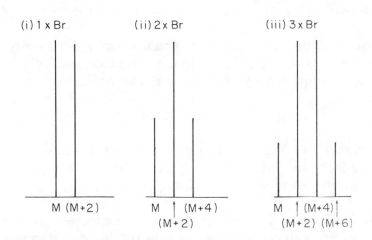

Fig. 7.3e. *Patterns produced by elements with abundant stable iso-topes. Examples of ions with one (i), two (ii) and three (iii) Br atoms, respectively*

Notice first of all there are $(n + 1)$ isotope peaks, where n is the number of halogen atoms present. The intensities of the isotopic species can be found by applying the formula in Eq. 7.3a.

$$(a + b)^n \qquad (7.3a)$$

Where a = relative abundance of the light isotope
b = relative abundance of the heavy isotope
n = number of halogen atoms present.

Abundances a and b can be to any desired accuracy but for our purposes we will use whole number ratios.

eg $^{35}Cl = 3$ (almost) and $^{37}Cl = 1$

$^{79}Br = 1$ and $^{81}Br = 1$ (almost)

Let's first take the case where $n = 1$ for both Cl and Br. Eq. 7.3a gives the relative abundances of the isotope peaks as 3 : 1 and 1 : 1, as it should of course. Now consider $n = 2$. For Cl_2, Eq. 7.3a becomes

$$(a + b)^2 = (a + b)(a + b) = a^2 + 2ab + b^2 \quad (7.3b)$$

If you now substitute 3 everywhere you have an a and 1 everywhere you have a b, work out the products, and replace the addition signs by proportionality signs you obtain the following ratios:

$$3^2 + 2 \times 3 \times 1 + 1^2, \quad \text{which becomes } 9:6:1$$

∏   Now check Fig. 7.3d (ii) to see if our calculation is correct (NB the ion intensities in Figure 7.3d (ii) read right to left and have not been normalised).

I hope you found that the ratios were as expected, give or take 5% or so. Remember that mass spectra result from scanning over an assembly of ions formed in the mass spectrometer source over a few microseconds, and their reproducibility is not that good. Do not expect relative abundances determined under these conditions to be very accurate. There are many links in the chain, eg source, analyser, multiplier, amplifiers, data systems, and lastly plotters all of which are liable to random and systematic errors. There are special mass spectrometers called *isotope ratio* instruments which are used to determine isotopic abundances very accurately eg for ^{13}C dating. They do not 'scan' as qualitative mass spectrometers do and avoid some of the sources of error. We will not deal with them here.

SAQ 7.3a   Use Eq. 7.3b to calculate the ratios of the isotope peaks expected for a Br_2-containing ion. Compare your prediction with Fig. 7.3e (ii).

Let's continue now with the case $n = 3$.

$$(a + b)^3 = (a + b)(a^2 + 2ab + b^2)$$

$$= a^3 + 3a^2b + 3ab^2 + b^3 \qquad (7.3c)$$

For Cl, $a^3 = 27$, $3a^2b = 27$, $3ab^2 = 9$, $b^3 = 1$, so the relative abundances for the $M^{\ddagger}$, $(M + 2)^{\ddagger}$, $(M + 4)^{\ddagger}$ and $(M + 6)^{\ddagger}$ ions are $27 : 27 : 9 : 1$ approximately.

SAQ 7.3b | Use Eq. 7.3c to calculate the ratios of the isotope peaks expected for a Br_3-containing ion. Compare your prediction with Fig. 7.3e (*iii*).

Lastly, for $n = 4$, we find that the expanded formula $(a + b)^4$ has the terms:

$$[a^4 + 4a^3b + 6a^2b^2 + 4ab^3 + b^4] \qquad (7.3d)$$

∏   What are the relative abundances of the $M^{\ddagger}$, $(M + 2)^{\ddagger}$, $(M + 4)^{\ddagger}$, $(M + 6)^{\ddagger}$ and $(M + 8)^{\ddagger}$ species in the molecular ion clusters of molecules containing Cl_4 and Br_4 atoms, respectively?

For Cl_4, the relative abundances are

$$3^4 : 4 \times 3^3 \times 1 : 6 \times 3^2 \times 1^2 : 4 \times 3 \times 1^2 : 1^4$$

or $81 : 108 : 54 : 12 : 1$.

For Br_4, the relative abundances are

$$1^4 : 4 \times 1^3 \times 1 : 6 \times 1^2 \times 1^2 : 4 \times 1 \times 1^3 : 1^4$$

or $1 : 4 : 6 : 4 : 1$,

which is much easier to remember!

SAQ 7.3c	(*i*) Expand the expression $(a + b)^5$ and use it to calculate the relative abundances of the ions in the Cl_5 and Br_5 clusters.
	(*ii*) Expand the expression $(a + b)^6$ and use it to calculate the relative abundances of the ions in the Cl_6 and Br_6 clusters. (Optional)

You may have found the algebra and arithmetic involved in these exercises and SAQ 7.3c quite difficult. If so, don't worry, you won't have to do more than recognise the typical clusters in future. In fact, there are some features of them worth noting. For example, did you notice that when n is $\geqslant 3$, the $(M + 2)^{+}$ ion in the Cl_n cluster is as or more abundant than the M^{+} ion? As n increases, this trend is more marked. In all the Cl_n clusters, the ion of highest mass (ie where all the chlorines are ^{37}Cl) becomes less abundant, being negligible for $n > 5$. This means that it is easily missed, and the cluster taken for that due to Cl_{n-1} rather than Cl_n. It is important to look carefully at clusters suspected of being due to chlorines to identify the peak of highest mass. This will help you decide what n is, since the mass of this ion is $(M + 2n)^{+}$.

All the bromine clusters have symmetrical abundances, using the approximation $^{79}Br = {}^{81}Br = 1$ relatively. For example, Br_2 shows the ratio $1:2:1$; Br_3, $1:3:3:1$; and Br_4, $1:4:6:4:1$. If you are familiar with nmr spectra these intensity ratios may ring a bell, but don't be like a student of mine who once described a Br_3 compound as having a M^{+} consisting of a quartet with a coupling constant of 2 amu! The resemblance is purely coincidental, but quite distinctive nevertheless.

Let us now move on to consider how to calculate the relative abundances of clusters of ions due to the presence of *both* chlorine and bromine. This is quite common in practice.

∏   Can you think of a simple way of extending Eq. 7.3a to cover the case of ions containing *both* Cl and Br?

The extended formula uses the product of $(a + b)$ and $(c + d)$, raising each to the power appropriate to the number of each type of isotope present, that is:

$$(a + b)^{n} \times (c + d)^{m} \tag{7.3e}$$

where a is the relative abundance of ^{35}Cl, b is that of ^{37}Cl, c that of ^{79}Br, d that of ^{81}Br, and n is the number of chlorine and m the number of bromine atoms present in the molecule. If you thought this out for yourself, well done.

Taking the simplest case first, a molecule containing just one chlorine and one bromine, let's use Eq. 7.3e to calculate the expected
relative abundances. Eq. 7.3e becomes $(a + b)(c + d)$ when
$n = m = 1$, and expands to give

$$ac + ad + bc + bd$$

Here though we need to remember that ad and bc will in practice
have the same mass, $(M + 2)^{+\cdot}$, so we should group them together
giving:

$$ac + (ad + bc) + bd \qquad (7.3f)$$

Substituting $a = 3$, $b = 1$, $c = 1$ and $d = 1$ into Eq. 7.3f we find
that the $M^{+\cdot} : (M + 2)^{+\cdot} : (M + 4)^{+\cdot}$ are predicted as $3:4:1$ to the
same accuracy we used previously.

SAQ 7.3d	Fig. 7.3f shows the un-normalised mass spectrum of 4-bromochlorobenzene. Are the molecular ions in the predicted $3:4:1$ ratio? What other ions containing halogens are present in the spectrum?

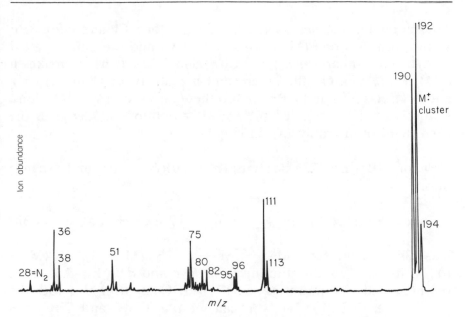

Fig 7.3f. *Mass spectrum of 4-bromochlorobenzene*

Next, let's consider clusters containing Cl_2 Br and $ClBr_2$. Will they look alike, or can they be easily distinguished by mass spectrometry? Eq. 7.3e becomes $(a + b)^2(c + d)$ for Cl_2Br and expands into

$$a^2c + (2abc + a^2d) + (b^2c + 2abd) + b^2d \qquad (7.3g)$$

In Eq. 7.3g, the two contributors to the $(M + 2)^+$ and $(M + 4)^+$ have been grouped together. How do we know that

$$2abc + a^2d$$

are the correct $(M + 2)^+$ species, and

$$b^2c + 2abd$$

the correct $(M + 4)^+$ species?

Well, consider each term separately and work out how many $(M + 2)$ isotopes it contains. If there is only one, then it is a $(M + 2)^+$ ion, if two, a $(M + 4)^+$ ion, and if three then a $(M + 6)^+$ ion.

Thus in the term $2abc$, a stands for ^{35}Cl, b for ^{37}Cl, and c for ^{79}Br, so this term has one (M + 2) isotope, ^{37}Cl and therefore is a (M + 2)‡ ion. Similarly a^2d has $^{35}Cl(a)$ and ^{81}Br (d) which makes it a (M + 2)‡ ion. On the other hand b^2c has two ^{37}Cl atoms (b^2), and $2abd$ has ^{37}Cl and ^{81}Br (bd) so these terms are (M + 4)‡ ions. Putting $a = 3$, $b = 1$, $c = 1$ and $d = 1$ into Eq. 7.3g gives the predicted isotope ratios as $3:15:7:1$.

Now for $ClBr_2$ Eq. 7.3e becomes $(a + b)(c + d)^2$ and expands into:

$$ac^2 + (2acd + bc^2) + (ad^2 + 2bcd) + bd^2 \qquad (7.3h)$$

Eq. 7.3h is in the form M‡ + (M + 2)‡ + (M + 4)‡ + (M + 6)‡ as in Eq. 7.3g. Substituting for a b, c and d in Eq. 7.3h gives the predicted isotope ratios as $3:7:5:1$. Comparing this with the previous result $3:15:7:1$, you can see that Cl_2Br and $ClBr_2$ can indeed be distinguished by measuring the relative abundances of the isotopes in the cluster.

∏   Are the relative abundances we have just calculated for Cl_2Br and $ClBr_2$ sufficiently different from those for Cl_3 and Br_3 to enable *all four* clusters to be distinguished by relative abundance measurement?

The relative abundances are as follows.

	M‡	(M + 2)‡	(M + 4)‡	(M + 6)‡
Cl_3	27	27	9	1
Cl_2Br	3	15	7	1
$ClBr_2$	3	7	5	1
Br_3	1	3	3	1

I hope you can see they *are* all distinct.

This approach is taken one step further in SAQ 7.3e. You might have had enough of the algebra by now and take the results as read, but I hope you will have a go at one of them, just to prove you can do it.

SAQ 7.3e

Calculate the relative abundances of the ions in the following clusters:

(i) Cl_3Br;

(ii) Cl_2Br_2;

(iii) $ClBr_3$.

Would mass spectrometry enable these to be distinguished from one another, and also from Cl_4 and Br_4 clusters?

Fig. 7.3g summarises the relative abundances of the ions in the halogen clusters we have been considering in this section.

Composition	M	(M + 2)	(M + 4)	(M + 6)	(M + 8)	(M + 10)	(M + 12)
Cl_2	9	6	1				
ClBr	3	4	1				
Br_2	1	2	1				
Cl_3	27	27	9	1			
Cl_2Br	9	15	7	1			
$ClBr_2$	3	7	5	1			
Br_3	1	3	3	1			
Cl_4	81	108	54	12	1		
Cl_3Br	27	54	36	10	1		
Cl_2Br_2	9	24	22	8	1		
$ClBr_3$	3	10	12	6	1		
Br_4	1	4	6	4	1		
Cl_5	243	405	270	90	15	1	
Br_5	1	5	10	10	5	1	
Cl_6	729	1458	405	540	135	18	1
Br_6	1	6	15	20	15	6	1

Fig. 7.3g. *Relative abundances of ions in clusters containing chlorine and/or bromine*

7.3.3. Simplified Method for Combinations of Cl, Br and other Isotopic Elements

In simple cases, there is a short cut for obtaining an isotopic pattern. Consider Cl_3Br again:

$$\text{For } Cl_3, (a + b)^n = (a + b)^3 = 27:27:9:1 \text{ (Fig. 7.3g).}$$

For Br, the relative abundances are $1:1$.

Combining these $(27:27:9:1)$ $(1:1)$

$$
\begin{array}{l}
(27:27:9:1) \times 1 = 27:27:\ 9:\ 1 \\
(27:27:9:1) \times 1 = \ \ \ \ \ \ 27:27:\ 9:1 \\
\hline
\text{Total} \ \ \ \ \ 27:54:36:10:1
\end{array}
$$

That is, each of the terms relating to chlorine abundances $(27:27:9:1)$ is multiplied by each term relating to the bromine abundances $(1:1)$. After each such multiplication, the resulting terms are moved one column to the right, then each column is totalled, giving the total relative abundances as $27:54:36:10:1$. This is the same as the result calculated in SAQ 7.3e (i) and given in Fig. 7.3g.

Π   Check the relative abundances given in Fig. 7.3g for Cl_2Br_2 by this method.

For Cl_2, relative abundances from Fig. 7.3g are $9:6:1$.

For Br_2, they are $1:2:1$.

Combining these we have $(9:6:1)$ $(1:2:1)$

$$
\begin{array}{l}
(9:6:1) \times 1 = 9:\ 6:\ 1 \\
(9:6:1) \times 2 = \ \ \ \ \ 18:12:2 \\
(9:6:1) \times 1 = \ \ \ \ \ \ \ \ \ 9:6:1 \\
\hline
\text{Total} \ : \ 9:24:22:8:1
\end{array}
$$

which is the result as given in Fig. 7.3g for Cl_2Br_2.

This quick method is also useful for other simple combinations eg HgBr. The isotopes of Hg were given in Fig. 7.2d as 196, 0.2%; 198, 10%; 199, 16.8%; 200, 23.1%; 201, 13.2%; 202, 29.8% and 204, 6.9%. Neglecting ^{196}Hg and converting into ratios by dividing through by 6.9 (the % of the next most abundant isotope, 204) we have $1.4:2.4:3.3:1.9:4.3:1$.

Combining these with $1:1$ for bromine, we have

$$(1.4:2.4:3.3:1.9:4.3:1) (1:1)$$

$$(1.4:2.4:3.3:1.9:4.3:1) \times 1 = 1.4:2.4:3.3:1.9:4.3:1$$
$$(1.4:2.4:3.3:1.9:4.3:1) \times 1 = \quad\quad 1.4:2.4:3.3:1.9:4.3:1$$

$$\text{Total} \quad 1.4:3.8:5.7:5.2:6.2:5.3:1$$

That is, the ion containing $HgBr^+$ would show a cluster of peaks in the ratio $1.4:3.8:5.7:5.2:6.2:5.3:1$. Of course, such calculations can be done to any desired accuracy. We have used two significant figures which is usually enough.

SAQ 7.3f

Calculate the relative abundances, using the quick calculation method, of the major isotope peaks in ions containing:

(*i*) S_2Cl_2;
(*ii*) $CuBr_2$;
(*iii*) $SClBr$;

You will have noticed in the answer to SAQ 7.3f (*iii*) that the cluster of ions resulting from SClBr is little different to that derived from ClBr, namely $3:4:1$, except that SClBr shows a small $(M + 6)^+$ of 1.4%, relative to $M^+ = 100\%$. Although you might think this could easily be overlooked, an experienced mass spectroscopist would spot the difference, especially if a high resolution mass measurement had shown the presence of sulphur.

7.4. AN APPROXIMATE METHOD FOR CALCULATING RELATIVE ABUNDANCES OF CLUSTERS DUE TO MINOR ISOTOPES

The methods given in Section 7.3 will of course allow you to calculate the relative abundances of any isotope whatever its natural abundance. They are not just applicable to Cl and Br. We have used these as examples of the approach because they are very common isotopic elements. We shall now describe a general formula which can be used for isotopes which have a low, but significant natural abundance such as ^{29}Si and ^{34}S.

In Section 7.2, we gave an example of a flour sample contaminated by sulphur, giving rise to S_8^+ in the mass spectrum, and said that once the possibility of sulphur contamination was realised, S_8^+ was easy to spot as a cluster of ions around m/z 256. Let us now return to the job of working out the relative abundances of the more intense ions in this cluster.

The general formula $(a + b)^n$ where $a = 95\%$, $b = 4\%$, and $n = 8$ is obviously going to be painful to evaluate fully. Fortunately, there is no real need to.

∏   Why is it unnecessary, when b is a minor isotope to fully expand the expression $(a + b)^8$ in order to find the correct coefficients for the nine terms in the expansion?

The expansion of $(a + b)^8$ will have terms in b^8, b^7a, b^6a^2, etc. When b is very small compared to a (in the case of $^{32}S:^{34}S$, $a : b = 24.5:1$) terms in b^8, b^7a, b^6a^2 etc will be very much smaller than terms in a^8, a^7b, a^6b^2 etc. Thus, the intensities of ions arising from the combinations $b^8, b^7a, b^6a^2, b^5a^3$, and b^4a^4 can be ignored compared with those from a^8, a^7b, a^6b^2 and a^5b^3.

This is well illustrated by looking at the binomial theorem, according to which:

$$(a + b)^n = a^n + na^{n-1}b + \frac{n(n - 1)a^{n-2}b^2}{2 \times 1} +$$

$$\frac{n(n - 1)(n - 2)a^{n-3}b^3}{3 \times 2 \times 1} +$$

$$\frac{n(n - 1)(n - 2)(n - 3)a^{n-4}b^4}{4 \times 3 \times 2 \times 1} + \ldots + b^n \qquad (7.4a)$$

When $a \gg b$, it can be seen that any terms in this expansion beyond the fourth or fifth become very small relative to the first three. In mass spectrometry applications we can neglect them beyond $(M + 6)^+$ species when the isotopes have a mass difference of two, or beyond $(M + 2)^+$ when the isotopes have a mass difference of one.

Let us now apply Eq. 7.4a to S_8 to see what happens in practice.

$$(95 + 4)^8 = 95^8 : 8 \times 95^7 \times 4 : 28 \times 95^6 \times 4^2 : 56 \times 95^5 \times 4^3 :$$

$$70 \times 95^4 \times 4^4 : \ldots : 4^8$$

Ignoring terms after the fourth and dividing through by 95^5 we have:

$$(95 + 4)^8 = 95^3 : 8 \times 95^2 \times 4 : 28 \times 95 \times 4^2 : 56 \times 4^3$$

$$= 857 : \quad 289 \quad : \quad 43 \quad : \quad 4$$

Normalising, we predict the ratios of the S_8^+ cluster for m/z $256:258:260:262$ to be $100:34:5:0.5$

The mass spectrum of sulphur is shown in Fig. 7.4a and the $S^4{}^+$ to $S^8{}^+$ clusters in more detail in Fig. 7.4b for comparison with this result.

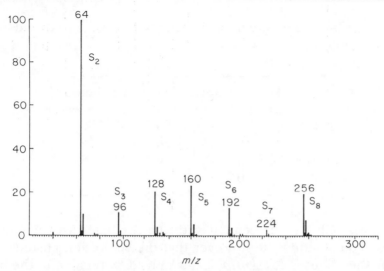

Fig. 7.4a. *Mass spectrum of sulphur determined using direct insertion probe at 70 eV, 200 °C*

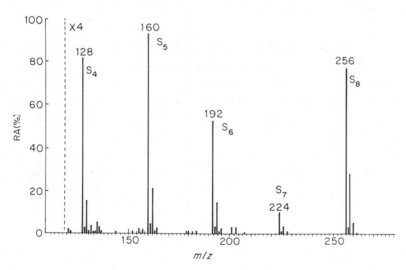

Fig. 7.4b. *Mass spectrum of sulphur, m/z 120–280 amplified four times to show S_4–S_8 clusters*

Π   Compare the relative intensities of the m/z 256, 258, 260 and 262 in Fig. 7.4b with the ratios $100:34:5:0.5$ predicted by Eq. 7.4a.

From Fig. 7.4b., the actual peak heights are measured and normalised as follows:

	Height (mm)	%	Predicted %
m/z 256	45	100	100
m/z 258	16	36	34
m/z 260	3	6.7	5
m/z 262	0.3	0.7	0.5

This shows only fair agreement with the predicted relative abundances. You should note however that the actual ion current carried by these ions is low (m/z 256 is only 20% relative to the base
peak of the spectrum, m/z 64, Fig. 7.4a). This means that we are
asking quite a lot of the data system in this case. It is being asked
to determine accurately from a 3 s decade^{-1} scan the ratios of ions
of abundances down to 1% of the base peak. Perhaps we have done
rather well!

What we did in the S_8 calculation leads to a useful generalisation for
dealing with cluster ions arising from one minor isotope. (Minor in
this context means up to 10% relative abundance, with say four or
more of the atoms concerned present in the molecule). We divided
through by 95^5, ie a^{n-3}. If we do this in the general expression (Eq.
7.4a) and neglect all terms after the fourth, we get:

$$\frac{a^n}{a^{n-3}} + \frac{na^{n-1}}{a^{n-3}} b + \frac{n(n-1)a^{n-2}}{2a^{n-3}} b^2 + \frac{n(n-1)(n-2)a^{n-3}b^3}{6a^{n-3}}$$

or

$$a^3 + na^2b + \frac{n}{2}(n-1)ab^2 + \frac{n}{6}(n-1)(n-2)b^3$$

Hence the relative abundances of the first four ions in *any* such
cluster will be approximately:

$$a^3 : na^2b : \frac{n}{2}(n-1)ab^2 : \frac{n}{6}(n-1)(n-2)b^3 \qquad (7.4b)$$

where a is the relative amount of the more abundant isotope and b that of the less abundant isotope, and n is the number of these isotopes in the molecule.

SAQ 7.4a

Use the expression derived in Eq. 7.4b to calculate the relative abundances of the first four ions in Cl_4 cluster, and compare your result with that found by expanding $(a + b)^4$ in Fig. 7.3g.

SAQ 7.4b

In a similar way to the previous exercise in the text, compare the observed and calculated relative abundances in the S_4^+ and S_5^+ clusters shown in Fig. 7.4b.

SAQ 7.4c

Reference to Fig. 7.2c shows that ^{33}S has a natural abundance of 0.78% relative to ^{32}S = 100%.

(*i*) What is the relative abundance of (M + 1)$^{+}$ in the S^{8+} cluster? (Hint – use 'I never Rang Ann')

(*ii*) What is the contribution made by species containing two ^{33}S to the (M + 2)$^{+}$?

(*iii*) Would (M + 3)$^{+}$ be observable in the S_8^{+} cluster? In attempting (*ii*) and (*iii*) you will find the approximate binomial expansion we have just discussed most helpful (Eq. 7.4b).

SAQ 7.4d

> A common method of analysing polyhydroxy compounds such as glucose is to convert each —OH group into a —$OSi(CH_3)_3$ (trimethylsilyl, TMS) group. This is then used in gc-ms analysis. How intense would you expect the $(M + 1)^+$ ions to be for the penta-TMS derivative of glucose, $C_6H_{12}O_6$, whose formula is $C_{21}H_{47}O_6Si_5$?

SAQ 7.4e

> Fig. 7.4c shows the mass spectrum of the pesticide DDT, which contains a number of atoms of a halogen. What halogen is present, and how many atoms of it are there? How do you think the cluster m/z 235/237/239 has been formed from M^+?

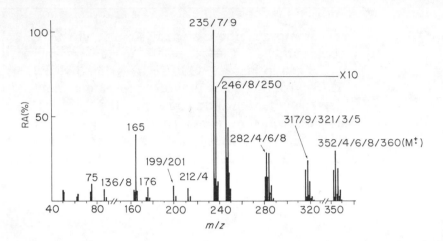

Fig. 7.4c. *Mass spectrum of the pesticide DDT*

Fig. 7.4c. shows two features commonly employed by mass spectrometrists when producing bar diagrams. Firstly, if the molecule has a high M_r, then in order to keep the figure reasonably compact, sections of the mass range that contain no useful ions may be omitted. Thus the m/z scale may be non-continuous. For the DDT mass spectrum note the discontinuities between m/z 80 and m/z 136 and between m/z 320 and m/z 350, both indicated by $//$.

Secondly, when important ions in the spectrum have a very low relative abundance they are sometimes shown with deliberately enhanced abundances. This is especially important when it is necessary to show complex isotope patterns. Usually such peaks are enhanced by a constant factor (eg $\times 5$ or $\times 10$ etc). In Fig. 7.4c, all the ions from m/z 246 onwards are enhanced by $\times 10$.

Summary

Mass spectrometry enables us to detect and quantify the isotopes possessed by most elements. Relative isotopic intensities can be calculated. Compounds of the same M_r but different empirical formula may be distinguished by comparing the RA of their (M + 1) peaks with the theoretical values.

Fluorine and iodine and monoisotopic elements and therefore their compounds cannot be readily distinguished by their isotopic patterns. However, they are mass deficient and therefore can be detected by high resolution mass measurement.

Systems containing two isotopes eg Cl and Br were discussed.

Objectives

Now that you have completed this Part you should be able to:

- recognise that most elements have distinctive isotopic intensity patterns;

- appreciate that mass spectrometry is the best general technique to measure the abundances of the isotopic species, and their masses;

- recognise the patterns produced by common elements such as C, H, O, N, Si, S, Cl, Br, Cu, Cd, Hg, Pb;

- identify unknown elements from their isotopic abundance patterns;

- apply simple formulae to calculate the isotopic intensities of (M + 1) and (M + 2) peaks for the ions produced in the spectra of organic molecules containing C, H, N, O, F, P, Cl, Br, I; and C, H, N, O, F, P, I respectively;

- adapt these formulae to take account of the presence of other elements with (M + 1) isotopes such as ^{29}Si and ^{33}S.

- define what is meant by the molecular ion of a compound, and calculate the relative mass M_r of the molecular ion of any given compound.

- appreciate that fluorine and iodine are mass deficient atoms and may be detected by high resolution mass measurements;

- recognise that fluorine and iodine can also be detected in low resolution mass spectra by noting mass increments of 18 and 126 respectively for each hydrogen replaced;.

- explain why polyfluoro compounds are used for mass calibration in mass spectroscopy;

- apply the formula $(a + b)^n$ to calculate the RAs of clusters containing chlorine or bromine, and recognise the typical patterns produced by Cl_n up to $n = 6$;

- apply the formula $(a + b)^n(c + d)^m$ to calculate the RA of clusters containing chlorine and bromine for all the combinations of Cl and Br up to Cl_4 and Br_4;

- recognise that Br_n clusters are typically symmetrical and the patterns of those up to Br_6;

- apply a simplified method for the calculation of the RA of ions in clusters containing Cl and Br combined with other isotopic elements such as Si, S, Cu and Hg;

- realise that when the RA of an isotope is less than about 10% of the major isotope, an approximate form of the binomial expansion (Eq. 7.4b) can be applied to the calculation of $(a + b)^n$ to give the RA of the first four isotope peaks in the cluster with sufficient accuracy for most analytical applications.

8. Modes of Fragmentation

In this section, we will be looking at the ways in which the energy given to molecular ions in the mass spectrometer is moved around and used to bring about fragmentations, how molecular ions and their daughter ions can be stabilised, and the mechanisms of the fragmentations themselves in terms of how the bonds actually break.

We will first of all briefly consider energy, and introduce the Quasi-Equilibrium Theory which describes how the ionisation energy is distributed in the ionised molecule, leading to the breaking of the weaker bonds first.

Next we consider how positive ions can be stabilised by inductive and mesomeric effects and begin to predict which positive ions are likely to be prominent in a mass spectrum because they are specially stabilised by these effects. If you have studied organic chemistry and know about these effects you will not need to study Section 8.2 too deeply, because it is really there to help those who are not too familiar with the theory of organic chemistry as it applies to positive ions.

Lastly, in Section 8.3 the mechanisms of bond cleavages, which are important in mass spectrometry, are described. Here again, if you have studied modern organic chemistry, these mechanisms will be familiar and you may not need to do more than read over the material briefly and try the exercises and SAQs just to make sure you can still work the old curly arrows! For those not familiar with curly arrow notation this may come a bit hard, but you do not need to become an expert with them in order to interpret mass spectra. It does help, however, to know which types of cleavage occur regularly and which not, so do give it a try.

8.1. ENERGY FACTORS

In the examples presented in the previous sections you will have noticed the loss of various atoms or whole groups of atoms from the molecular ions. In some cases the daughter ions so formed also fragmented, and so on. Recall that in electron impact ionisation the molecular ions are formed with extra energy, up to 60 eV in fact, depending on how much of the ionising electron's energy (70 eV) they acquire in the process. Covalent bond energies in organic compounds lie in the range 10–20 eV (commonly 10–12 eV). So it is obviously possible for $M^{+\cdot}$ species to fragment several times in succession until the remaining positive ion no longer has sufficient excess energy to break a bond. We will now consider *how* such *cascade* processes, Fig. 8.1a, occur.

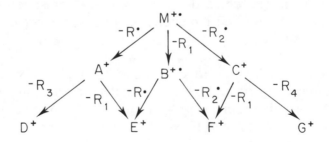

Fig. 8.1a.   *A cascade process*

Π   Apart from the possession of sufficient excess energy, can you think of any other condition that must be satisfied for an ion to fragment? If you need a hint – think about what is happening at the molecular level in fragmentation.

Fragmentation involves the breaking of bonds, and bonds only break when sufficient vibrational energy is concentrated in them to cause the vibrating atoms to move apart beyond a bonding distance. Hence the other condition is that the excess energy given to the ion by the ionising electron must be *rapidly* moved about the molecule and concentrated in the bond which breaks. What do we mean by rapid in this context? Well, vibrations have frequencies of 10^{10}–10^{12} Hz at 0 to 200 °C.

∏ Does this satisfy the condition for rapid energy distribution about the ionised molecule?

Yes, it certainly does! Ions have lifetimes of 10^{-5} to 10^{-6} s in the mass spectrometer source, while a molecular vibration of 10^{10}–10^{12} Hz takes 10^{-10}–10^{-12} s. There is time for around a million vibrations to take place before the molecular ion fragments. The initial excess energy is distributed rapidly around the various bonds of the molecular ion and in theory any of the bonds might break. This idea of what happens to the initial input of energy to the molecular ion is called the *Quasi-Equilibrium Theory* (QET).

∏ Do all the bonds in a molecular ion fragment in practice?

Reference to the mass spectra presented in the previous Parts shows clearly that only few bonds break in such processes, so the answer to this question is, no. Some bonds clearly never break, while others do so readily. We will now consider what lies behind this.

Firstly, it might simply be due to insufficient energy being available at 70 eV, even if all the impacting electron's energy was transferred to the $M^{\ddagger}$ ion. But only 10–20 eV is need to break most organic bonds, and the QET predicts that the necessary energy can be rapidly moved about the molecule through the mechanism of vibrations. No, this won't do.

∏ What happens if the energy of the ionising electrons is increased beyond 70 eV? Do more bonds fragment?

No great changes occur in the mass spectra obtained as the electron beam energy is increased. Eventually the second ionisation energy is reached and M^{2+} are formed. This was explained in Section 3.2. If you were uncertain about this point, read Section 3.2 again.

So, it is not just a question of energy. If it were, it would be possible to break molecules down to their constituent atoms by electron impact. This wouldn't be too useful to analytical chemists. They want, ideally, a few fragments uniquely characteristic of the structure, not a 'soup' of atoms!

A second explanation might be that as soon as the redistribution of energy caused the bond energy of a weak bond in the molecule to be exceeded, that bond breaks. The majority of M^{+} would fragment in such a way, since the QET predicts that the redistribution of smaller amounts of energy will occur more rapidly. The weak bond, or bonds, will fragment, reducing the overall energy, and there will be insufficient vibrational energy left for any stronger bonds to be broken.

This is an attractive idea, which seems to fit in well with observed fragmentation behaviour. For example, organic compounds containing halogen atoms such as Cl, Br and I readily lose halogen atoms. This might be expected as C—Cl, C—Br and C—I bonds are weaker than C—C, C—H and C—N bonds and much weaker than C=C, C=N, and C=O bonds.

In general, *fragments resulting from breaking of weak bonds are prominent in mass spectra* (or further ions formed in a cascade from them – see Fig. 8.1a).

8.2.   STABILISATION FACTORS

Fragments derived in the way described in Section 8.1 are seldom the only ions found in a mass spectrum. Indeed, frequently quite unexpected fragmentations occur eg in the mass spectrum of aniline (Fig. 5.la) the weakest bond is the C—NH_2, yet the $(M - NH_2)^{+}$ ion, m/z 77, formed by loss of $H_2N^{\cdot}$, is very weak. Instead the M^{+} fragments by loss of HCN to m/z 66, as is confirmed by a metastable ion at m/z 46.8. This shows that other factors are at work, to cause the aniline M^+ ion to rearrange itself so as to expel $HC\equiv N$ rather than $H_2N^{\cdot}$.

∏   What do you think these further factors might be?

The answer lies in the formation of a more stable positive ion, or a stable neutral species, or both. Stable neutral species are usually small molecules with multiple bonds such as $HC\equiv N$, $HC=$ $-CH$, $H_2C=CH_2$, $C=O$, $O=C=O$, $S=O$, $O=S=O$, $CH_2=C=O$, or less-commonly small singly bonded molecules such as H_2O, H_2S, CH_3OH, HCl, HBr and HI. The multiple bonds in particular can remove large amounts of energy from the parent ions if they are formed in a high vibrational state.

Positive ions can be stabilised in two ways. If you are familiar with organic reaction mechanisms you will have met these ideas before and can skim through the next few paragraphs. But in order to understand and predict fragmentations you need to understand how positive centres are stabilised by neighbouring groups in the molecule. The neighbouring group must be electron-donating, so as to counteract the positive charge and try to neutralise it.

The first way it can do this is by generally donating electron density towards the positive centre by what is termed an *inductive* effect. Atoms or groups which donate electron density are called $+I$ groups for short ($-I$ groups withdraw electron density). The most common $+I$ group is the alkyl group.

This explains the well-known order of stability of carbocations (C^+ species) which is:

$$
\begin{array}{ccccccc}
& CH_3 & & CH_3 & & H & & H \\
& | & & | & & | & & | \\
CH_3 \!\rightarrow\! C^+ & > & CH_3 \!\rightarrow\! C^+ & > & CH_3 \!\rightarrow\! C^+ & \gg & H \!-\! C^+ \\
| & & | & & | & & | \\
CH_3 & & H & & H & & H
\end{array}
$$

The arrows on the bonds show the direction of the I effect. Clearly the carbocation which has three $+I$, CH_3 groups attached to it will be much more stable that CH_3^+ which has none. The ions

m/z 57 (($CH_3)_3C^+$) and 43 ($CH_3)_2CH^+$) are usually intense in the spectra of compounds containing these end-groups, while m/z 29 ($CH_3CH_2^+$) and 15 (CH_3^+) are weak.

There are two useful rules to remember about I effects. Firstly, atoms which lie to the left of carbon in the Periodic Table (or groups consisting largely of such atoms) are $+I$; those which lie to the right of carbon (or groups consisting largely of such atoms) are $-I$. The $-I$ groups *destabilise* positive ions. The second is that $\pm I$ effects only operate *at the most* over two bonds, the effect falls off rapidly with distance down a carbon chain.

The second way a neighbouring group can help to stabilise a positive centre is by conjugation with it through multiple bonds. Where conjugation (alternate single and double or triple bonds) exists, this is a powerful effect and far outweighs the inductive effect. It is called the *mesomeric* (M) effect. You may have met it before under the name resonance. Resonance *is* involved in explaining how the $\pm M$ effect works but the $\pm M$ effect itself defines how a group will behave when attached to a conjugated system. *If the system is positively charged, a $+M$ group will stabilise it by helping to spread (delocalise) that charge.*

Fig. 8.2a presents a range of organic groups classified according to their I and M effects. There is no need to try to memorise all these, they are for reference when needed. You will see how they work in the examples which follow. Before you look at these, there is one point of difficulty which should be clear from Fig. 8.2a(*ii*).

∏   Can you say what the difficulty is?

The problem is, how do we predict the effect of those atoms and groups which are $+M$, $-I$? Do they stabilise, or destabilise, a positive ion? Or do they both stabilise and destabilise, according to the particular structure involved? This is a fairly large category containing some common groups so we would be in some difficulty if the effect was unpredictable.

(*i*) Inductive Effects of Groups ($\pm$I)

$-$I Groups			$+$I Groups
$-NO_2$	$-CHO$	$-RC=CR_2$	$-CH_3$
$-C{\equiv}N$	$-COR$	$-C{\equiv}CH$	$-CH_2R$
$-COOH$	$-F$	$-C{\equiv}CR$	$-CHR_2$
$-COOR$	$-Cl$	$-SO_2OH$	$-CR_3$
$-OH$	$-Br$	$-SH$	
$-OR$	$-I$	$-SR$	
C_6H_5	$-CH=CH_2$	$-NH_2$	

$-$I groups are better electron attractors than H$-$

$+$I groups are poorer electrons attractors than H$-$

(*ii*) Mesomeric Effects of Groups ($\pm$M)

$+$M, $-$I Groups		$-$M, $-$I Groups		$+$M, $+$I Groups
$-F$	$-SH$	$-NO_2$	$-CONH_2$	$-CH_3$
$-Cl$	$-SR$	$-C{\equiv}N$	$-SO_2R$	$-CH_2R$
$-Br$	$-NH_2$	$-CHO$	$-CF_3$	$-CHR_2$
$-I$	$-NHR$	$-COR$	$-CCl_3$	$-CR_3$
$-OH$	$-NR_2$	$-COOH$		
$-OR$	$-NHCOR$	$-COOR$		
$-OCOR$	$-C_6H_5$			
$-CH=CH_2$	$-CH=CR_2$			

$+$M and $+$I groups supply electron density to conjugated systems

$-$I and $-$M groups withdraw electron density from conjugated systems

Fig. 8.2a

Fortunately, in the vast majority of cases, it isn't unpredictable. The $+$M effect nearly always *wins* over the $-$I effect. In other words,

+ M, −I groups usually work to *stabilise* a positive ion. The reason is that the + M effect operates through the more weakly held π electrons in the double bonds of the ion. These are more readily distorted and the effect is transmitted over any number of conjugated bonds, unlike the I effect which is short range. Now for some examples.

8.2.1. Benzene and its Molecular Ion

Benzene has two Kekule resonance forms:

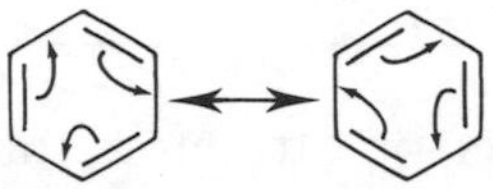

Notice how they are converted by movements of two π electrons as shown by the 'curly' arrows. The benzene molecular ion is formed by loss of a π electron, giving *A*:

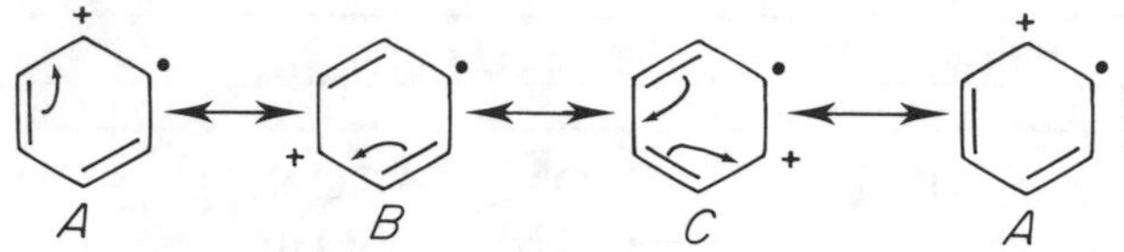

Having reached ion *C* by resonance *via B* we can only return to A, as shown. Thus there appear to be three resonance forms of the benzene M$^{+\cdot}$ ion.

However, we could just as easily have formed *A* by placing the positive charge on the other carbon atom of the ionised double bond eg

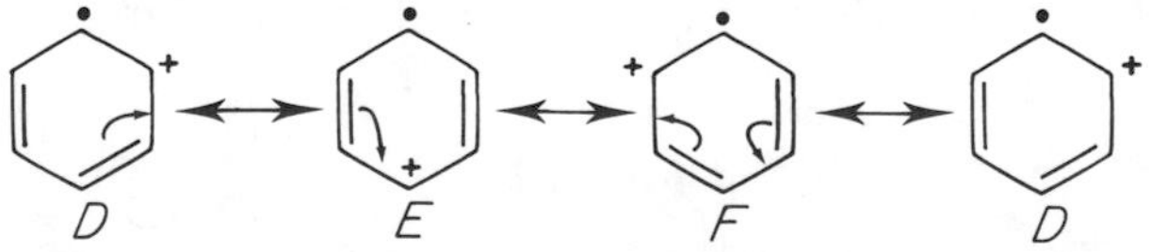

so every possible resonance form of the benzene molecular ion can be derived from *A* and *D*. We can now see why the molecular ions of aromatic hydrocarbons are so very intense and also why they tend to fragment, when they do, by loss of HC≡CH units (Fig. 8.2b). They have a large number of resonance forms and contain

$$\begin{array}{cc} | & | \\ HC{=}CH \end{array} \text{ units.}$$

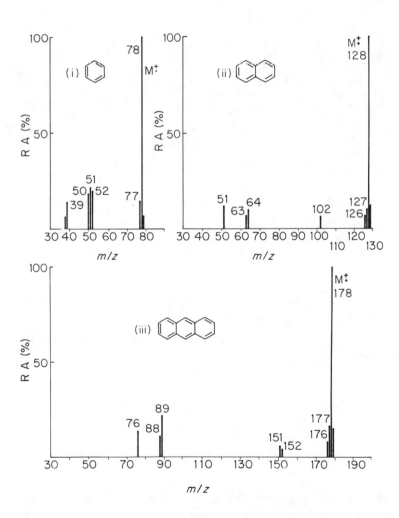

Fig. 8.2b.   *Mass spectra of benzene (i), naphthalene (ii) and anthracene (iii)*

SAQ 8.2a

Fig. 8.2b illustrates the mass spectra of benzene, naphthalene and anthracene. Which ions
are formed from the loss of HC≡CH from (*i*)
benzene, (*ii*) naphthalene and (*iii*) anthracene.

Incidentally, there is an alternative method for the interconversion
of the (benzene)$^{+}$ resonance forms, using single electron shifts. This
introduces the ⌐ or 'fish-hook' notation much used in mass
spectrometry:

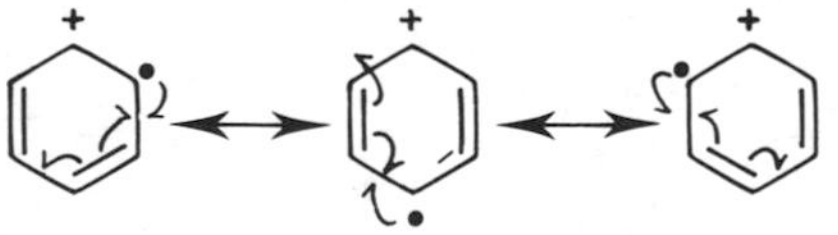

$$A \qquad\qquad B' \equiv B \text{ or } E \qquad C' \equiv C \text{ or } A$$

This shows that the positive charge (+) and the free electron
(.) can be positioned on the benzene ring in any combination.
The ⌐ arrow shows the movement of a single electron from the
position shown by the tail to that of the head of the arrow.

8.2.2. Methoxybenzene (Anisole)

This could be ionised on the benzene ring, or on the oxygen, but as we shall see, this is not important. We start with the electron lost from the π system, giving:

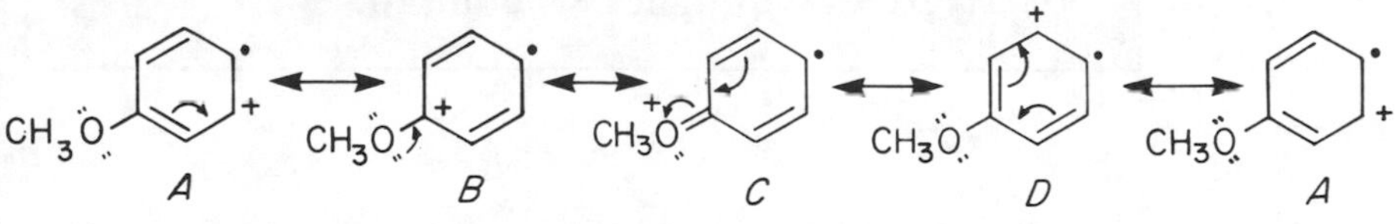

Ions *A* and *B* are similar to those formed from benzene in Section 8.2.1 but *C* is a key ion. It is formed by the use of one of the two lone pairs of p-electrons on the oxygen atom. This is how the oxygen atom exerts its + M effect to help stabilise the methoxybenzene molecular ion.

It is also possible that the molecular ion is formed by loss of one of these relatively easily ionised oxygen p-electrons, giving:

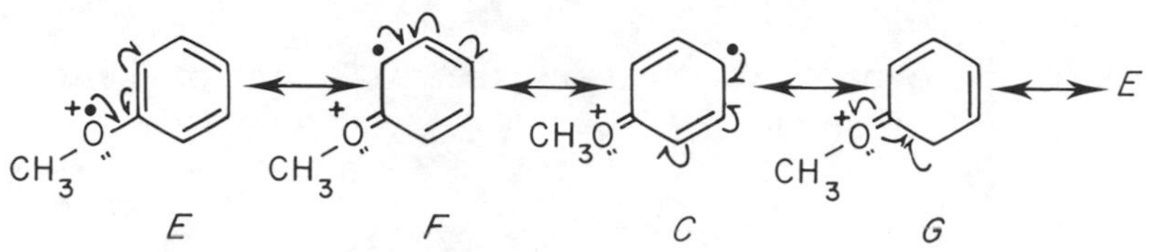

Ion *C* is formed in this sequence too! This shows that it does not matter where the initial loss of the electron occurs. All the possible forms of M⁺ are interconverted by resonance. This result is quite general for the + M groups in Fig. 8.2a (*ii*).

| SAQ 8.2b | Show how the methoxybenzene ion *B* can be converted into ion *G* (2 steps). |

SAQ 8.2c

Show how the $M^{\ddagger}$ ion of phenyl ethanoate

$$\begin{array}{c} O \\ \| \end{array}$$

($C_6H_5OCCH_3$) might be stabilised.

SAQ 8.2d

Would you expect the $M^{\ddagger}$ ion of benzyl ethanoate

$$\begin{array}{c} O \\ \| \end{array}$$

($C_6H_5CH_2OCCH_3$) to be as stable as the

$M^{\ddagger}$ ion of phenyl ethanoate?

8.2.3. 1-Phenylethanone (Acetophenone) $C_6H_5COCH_3$

1-Phenylethanone could be ionised on the benzene ring, or the carbonyl group. Let's start with the benzene ring:

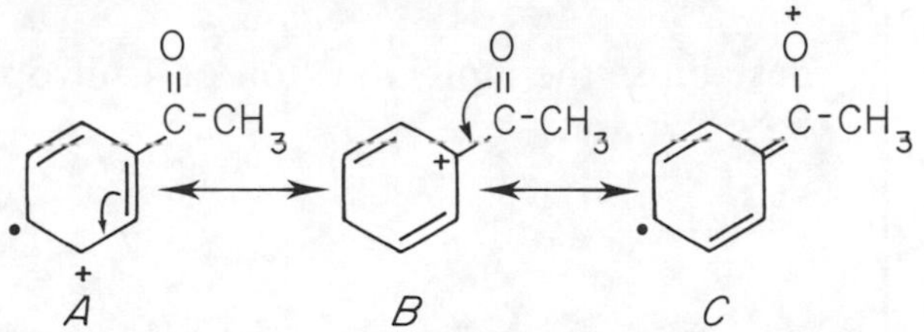

It would appear we can form C by using the π-electrons of the carbonyl group, which *are* conjugated with the benzene ring. Thus A would be more stabilised. But wait a minute. Isn't oxygen more electron attracting than carbon? It would be more likely to attract the electrons of the benzene ring itself, like this:

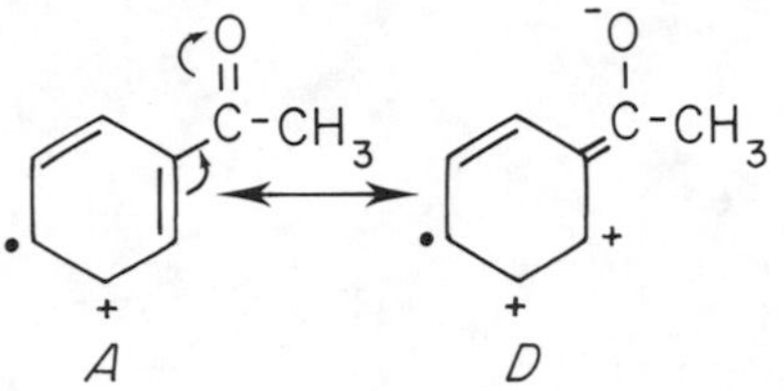

In D, you can see that two positive charges form on adjacent carbon atoms, *a highly unstable arrangement*. We can conclude that $-M$ groups, such as those in Fig. 8.2a (*ii*) will *destabilise* molecular ions like A.

Ionisation in 1-phenylethanone could also take place on the carbonyl oxygen atoms to give E:

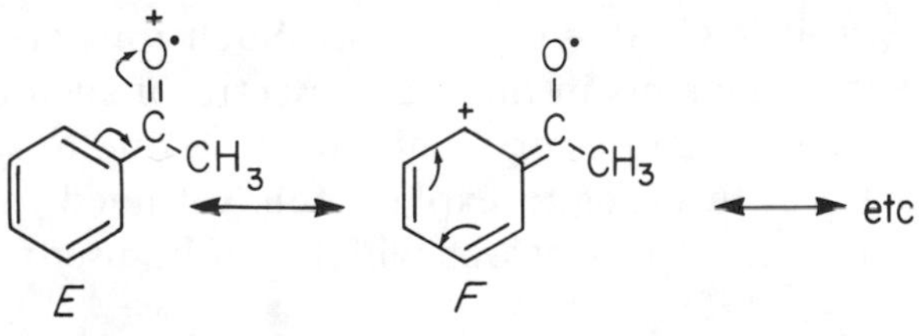

E can be stabilised by the $+M$ effect of the phenyl group as shown, so it is far more likely the M^{+} of this type of compound will exist as E-type ions with the positive charge occurring on the $-M$ group.

SAQ 8.2e

Show how the nitro group $(-\overset{\overset{\textstyle O}{\|}}{\underset{+}{N}}-O^-)$ acts to

destabilise the molecular ion of 4-nitrophenol,

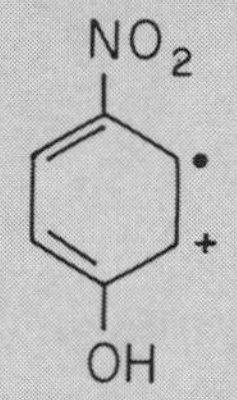

NO$_2$

OH

8.2.4. The Acyl Ion $R-\overset{+}{C}=O$

The acyl ion itself is $CH_3\overset{+}{C}=O$, m/z 43. Such ions are frequently found as intense fragments in the mass spectra of such compounds as aldehydes, ketones and esters. This must be due to some innate stability they have. In order to explain this we need to remember that oxygen has lone-pair electrons which can be used:

$$R-\overset{+}{C}=\overset{..}{\underset{..}{O}} \longleftrightarrow R-C\equiv\overset{+}{\underset{..}{O}}$$

If R is a $+I$ group, this will also help to stabilise the acyl ion.

8.2.5. The Phenyl Ion, $C_6H_5^+$, (m/z 77)

The phenyl ion is frequently found in the spectra of phenyl compounds and is usefully diagnostic, since few other compounds produce ions of m/z 77.

∏　　Try writing resonance structures for the $C_6H_5^+$ ion.

You probably wrote:

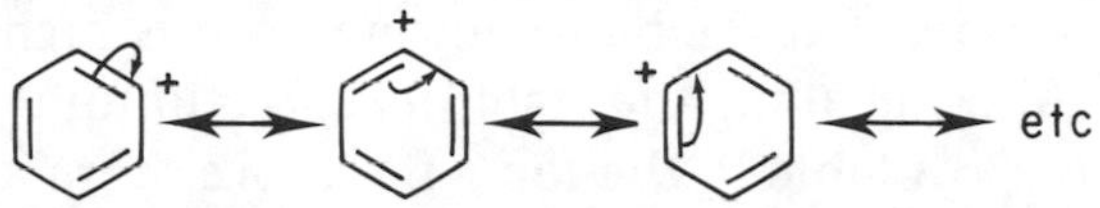

or some such, showing that the positive charge is delocalised round the ring. Nine out of ten students would too! Unfortunately, it's wrong – you cannot delocalise the positive charge in $C_6H_5^+$ because the charge is *in the plane of the ring*. Fig. 8.2c attempts to show this, using phenylethanone as the source of the $C_6H_5^+$ ion.

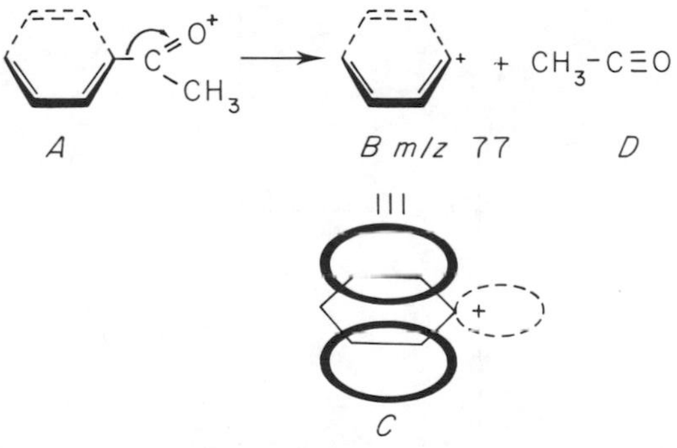

Fig. 8.2c. *Formation and structure of $C_6H_5^+$*

When the bond between the benzene ring and the acyl group breaks, the positive charge occupies the position left by the departing group, as shown in *B*, Fig. 8.2c. This is an empty sp^2 orbital, shown by dashed line in *C*, Fig. 8.2c. Electrons in the π orbitals of the benzene ring shown by the shaded regions cannot interact with this orbital, so the positive ion is *not* especially stabilised. Phenyl cations are still aromatic however, so are reasonably stable compared to alternative, non-aromatic ions which might be formed in the fragmentation of the M^+ ion.

Π   Which do you think would be the most intense fragment ion
         line in the spectrum of phenylethanone, $C_6H_5^+$ or $CH_3C\overset{+}{O}$?

The spectrum of phenylethanone is shown in Fig. 8.2d. It clearly
shows that $C_6H_5^+$ is more stable than $CH_3C\overset{+}{O}$ because the RA's are
in the ratio 70 : 30 approximately. If you chose $CH_3C\overset{+}{O}$ because it has
two resonance forms to help stabilise it (Section 8.2.4) and $C_6H_5^+$
has no resonance forms which delocalise the positive charge, you
have to bear in mind that the benzene ring itself is highly stabilised.
But you were thinking along the right lines to add up the number of
resonance forms available to the ion – the more there are, the more
stable it tends to be.

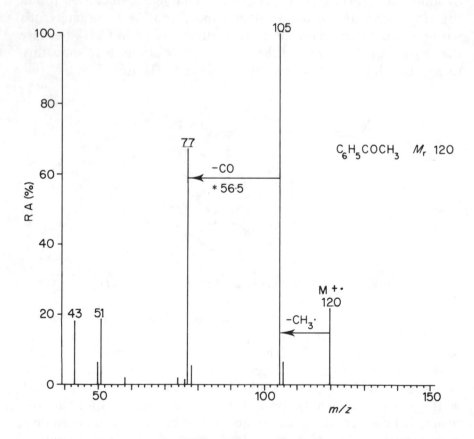

Fig. 8.2d.   *Mass spectrum of phenylethanone (acetophenone)*

Fig. 8.2d shows the base peak at m/z 105. Clearly this is due to loss of a CH_3 group (15 amu.). There must be something specially stable about this ion for it to be so intense relative to the alternative $CH_3\overset{+}{C}O$ ion formed by loss of C_6H_5.

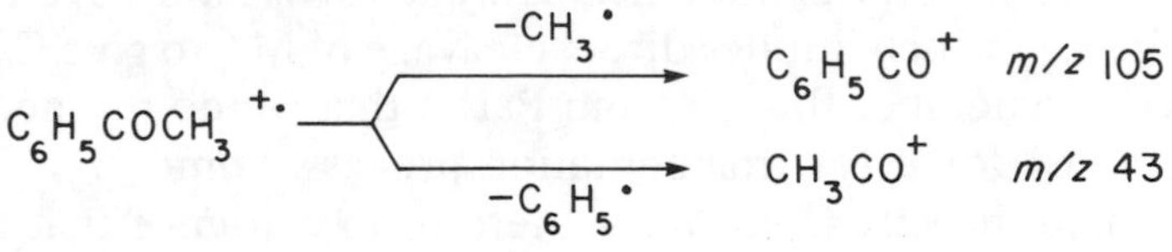

$$C_6H_5COCH_3{}^{+\cdot} \quad \underset{-C_6H_5{}^\bullet}{\overset{-CH_3{}^\bullet}{\longrightarrow}} \quad \begin{array}{ll} C_6H_5CO^+ & m/z\ 105 \\ CH_3CO^+ & m/z\ 43 \end{array}$$

In the next example we will consider why $C_6H_5\overset{+}{C}O$ is so preferred in the fragmentation of phenylethanone.

8.2.6. The Benzoyl Ion, $C_6H_5\overset{+}{C}O$, m/z 105

This ion, like any carbonyl ion, could have two structures A and A^1 in Fig. 8.2e, which are then delocalised using the π electrons of the benzene ring to give B, C, D and E.

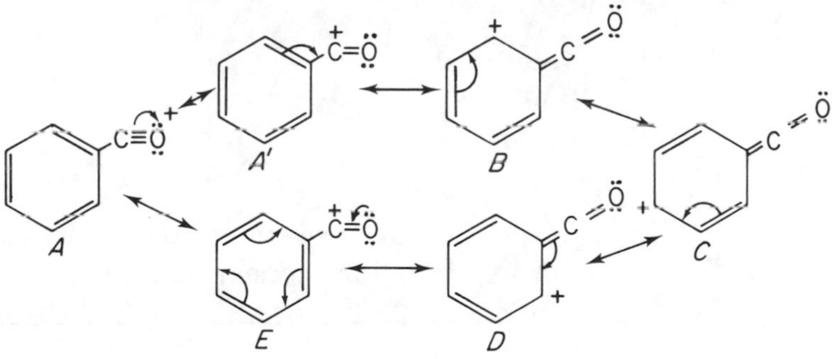

Fig. 8.2e. *Resonance structures of the $C_6H_5\overset{+}{C}O$ ion*

Thus $C_6H_5\overset{+}{C}O$ has six resonance forms to stabilise it. Here the phenyl group is acting as a $+M$ group to stabilise the benzoyl ion and this explains why it is the base peak in the $C_6H_5COCH_3$ spectrum. An interesting feature of the spectrum is a metastable (m*) ion at m/z 56.5 (not shown on Fig. 8.2d as it is too small) which shows that the benzoyl ion decomposes by loss of CO to give $C_6H_5^+$

$$C_6H_5\overset{+}{\underset{}{C}O} \longrightarrow C_6H_5^+ + CO, \; m^* = \frac{77^2}{105} = 56.5$$

This route to m/z 77 is not perhaps what you would have expected from Fig. 8.2c which implies direct cleavage of $M^{\ddot +}$ to give $C_6H_5^+$ and $CH_3CO^{\cdot}$. Remember though from Part 5 that we do not necessarily observe a m^* for *every* fragmentation process. Some m/z 77 could have formed directly from $M^{\ddot +}$ – there is no evidence that they did not!

This is a good example of how the existence of a stabilised ion influences the fragmentation pathways of a molecular ion. Most of the positive ion current is carried by the two stabilised ions, m/z 105 and 77 in the spectrum of phenylethanone.

SAQ 8.2f

In compounds of the type RC_6H_4COX, would the following substituents (ie R) stabilise, destabilise, or have little effect on the benzoyl ion produced:

(*i*)   3-methoxy;
(*ii*)   4-methoxy;
(*iii*) 3-cyano;
(*iv*)   4-nitro?

Try to explain your answers by means of resonance forms of the ions concerned.

8.2.7. Some Basic Rules about Resonance Structures

Now that you have seen some examples of mesomeric effects in action you might find it useful if we summarise the basic rules for writing down the resonance structures and qualitatively assessing their relative importance.

1. Resonance structures are interconvertible by one or a series of short electron shifts: electrons do not *jump* across space.

2. Resonating structures must have the same number of electrons:

   Thou shalt neither create nor destroy electrons!

3. Resonance structures involving even numbers of electrons are more stable than those involving odd numbers of electrons.

4. The more bonds that are involved in the resonance, the greater is the stability of the resulting structure eg benzene, resonance energy 144 kJ mol^{-1}, naphthalene 244 kJ mol^{-1}.

5. The closer the stabilities of resonance structures, the greater is the degree of resonance and the lower the energy of the system eg the Kekulé forms of benzene which are equivalent.

6. Resonance can only occur between structures that correspond to very nearly the same relative positions of *all* the atoms involved. Bond distances and angles should remain the same and be compatible with the overlap of the orbitals being used eg resonance cannot occur between *isomers* of structures.

SAQ 8.2g   (Optional for those who feel they need practice in use of ⌒ and ⌒⃗)

                                                                    ⟶

SAQ 8.2g
(cont.)

Show how the following ions are stabilised:

(i) $C_6H_5\overset{+}{C}H_2$; (ii) $CH_3CH=CH-\overset{+}{C}O$; (iii) $CH_3\overset{..}{S}-CH=CH-\overset{+}{C}H_2$;

(iv) ⬠$^+$　　(v) H✕H⊓⁺ (vi) $\left(Cl-\!\!\left\langle\bigcirc\right\rangle\!\!\right)_2\overset{+}{C}H$

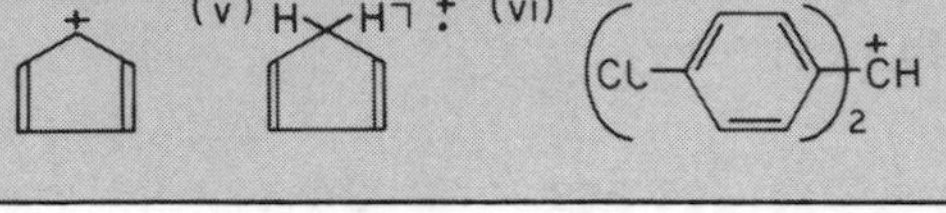

8.3. MECHANISMS OF BOND CLEAVAGES IN MASS SPECTROMETRY

In Section 8.2.5 we showed how the phenyl cation might have been formed from $C_6H_5COCH_3^{+\cdot}$ (Fig. 8.2c). This involved breaking the phenyl to carbonyl bond by placing both the electrons on the carbonyl group:

$$\underset{\diagup}{\overset{\diagdown}{C}}-\overset{R}{\underset{|}{C}}=\overset{+\cdot}{\underset{..}{O}} \rightarrow \overset{\diagdown}{\underset{\diagup}{C}}{}^+ + R-C\equiv O^\cdot \tag{8.3a}$$

You may have recognised this as an example of a *heterolytic* cleavage of a covalent bond:

$$X\!-\!\overset{\curvearrowleft}{Y} \rightarrow \overset{+}{X} + Y^-$$

where Y is more electronegative than X. In such a cleavage, one of the fragments (normally the more electronegative) gains *both* the bonding electrons. The opposite process:

$$Y\overset{\frown}{-}X \rightarrow \overset{+}{Y} + X^-$$

is theoretically possible but much less likely if X is less electronegative than Y. In the example of Eq. 8.3a the result would be:

$$\overset{\diagup}{\underset{\diagup}{C}}\overset{\frown}{-}C=\overset{+\,\cdot}{\underset{\cdot\cdot}{O}} \rightarrow \overset{\diagup}{\underset{\diagup}{C}}{}^{-}\cdot + -\overset{+}{C}=\overset{+\,\cdot}{\underset{\cdot}{O}}. \qquad (8.3b)$$

The doubly-charged carbon monoxide species would be highly unstable. In mass spectrometry only heterolytic cleavages corresponding to Eq. 8.3a are observed as these avoid creating adjacent positive charges.

Eq. 8.3a shows the decomposition of a molecular ion, an odd electron species. Daughter ions of even electron composition may also fragment in this way. For example, the benzoyl ion in the spectrum of 1-phenylethanone (Fig. 8.2d) loses CO as already described. The mechanism is:

$$C_6H_5\overset{+}{\underset{\smile}{-C}}=O \rightarrow C_6H_5^+ + CO$$

This produces two even electron fragments. Such heterolytic cleavages producing fragments in which *all the electrons are paired* are energetically favoured and are correspondingly common in mass spectra.

The heterolytic cleavage (Eq. 8.3a) can be generalised to the situation $-C-Z$, where Z is a $-I$, a $-I$, $-M$, or even a $-I$, $+M$ atom or group (Section 8.2) and ionisation has occurred on Z, ie:

$$\overset{\diagup}{\underset{\diagup}{C}}\overset{\frown}{-}\overset{+}{Z}{\underset{\cdot}{}} \rightarrow \overset{\diagup}{\underset{\diagup}{C}}{}^{+} + Z\cdot$$

Fig. 8.2d will give you some idea of the vast range of compounds whose $M^{\ddagger}$ might fragment in this way. Cleavage of the bond next to the atom or group carrying the positive charge is often called α-cleavage.

∏   Is there any other heterolytic cleavage which may occur?

Yes there is, a β-cleavage. In this the next bond in the chain (usually a C—C bond) breaks, as follows:

$$R \overset{|}{\underset{|}{-C}} \overset{+.}{-Z} \rightarrow \overset{+}{R} + >C=Z^{.}$$

β-cleavage is usually less prevalent than α-cleavage, but will occur to a minor extent. It is promoted if $\overset{+}{R}$ or C=Z$^{.}$ are specially stable, eg if $\overset{+}{R}$ is $C_6H_5CH_2^+$ which is stabilised by resonance, as shown by the following example.

$$C_6H_5-CH_2-CH_2-\overset{+.}{B}r \xrightarrow{\beta} C_6H_5-\overset{+}{C}H_2 + H_2C=Br^{.}$$
$$m/z \; 91, \; 100\%$$

$$C_6H_5CH_2-CH_2-\overset{+.}{B}r \xrightarrow{\alpha} C_6H_5CH_2\overset{+}{C}H_2 + Br^{.}$$
$$m/z \; 105, \; 50\%$$

In this case, the m/z 105 ion formed by α-cleavage is *not* resonance stabilised and is only half as intense as the m/z 91 ion.

∏   Do you know of any other general way in which we can break a covalent bond?

There is one other way to break a covalent bond, often called *homolytic* cleavage:

$$X \div Y \longrightarrow X^{.} + Y^{.}$$

In this mechanism, when the bond stretches and breaks one of the two electrons forming the bond is returned to X forming an X˙ radical, and the other to Y giving a Y˙ radical. Applying this idea to a molecular ion, we have

$$X \mathbin{\mkern-2mu\div\mkern-2mu} Y^{+\cdot} \longrightarrow X^\cdot + :\overset{+}{Y}$$

In this case, X is released as a radical, but Y carries off the positive charge with the two electrons paired up, ie Y is an *even-electron species*.

Of course, in some cases $:\overset{+}{X}$ may be stable, so we may observe the opposite result:

$$Y \mathbin{\mkern-2mu\div\mkern-2mu} X^{+\cdot} \longrightarrow Y^\cdot + :\overset{+}{X}$$

When this happens the mass spectrum will contain both $:\overset{+}{X}$ and $:\overset{+}{Y}$, which together will add up to the relative molecular mass of the compound. Such pairs of ions are said to be *complementary*. They are very useful diagnostically. You should always search a spectrum for such ion pairs. Bear in mind their RA values may differ enormously because one is likely to be more stable than the other (Section 8.2).

Returning to our example of phenylethanone again (Fig. 8.2d) m/z 43 and 77 are complementary ions formed by homolytic α-cleavage:

$$C_6H_5 \mathbin{-}\overset{\displaystyle O}{\overset{\|}{C}}\mathbin{-}CH_3 \longrightarrow C_6H_5^+ + CH_3\dot{C}{=}O$$

$$m/z\ 77$$

$$C_6H_5\mathbin{-}\overset{\overset{\cdot+}{O}}{\overset{\|}{C}}\mathbin{-}CH_3 \longrightarrow CH_3C{\equiv}\overset{+}{O} + C_6\dot{H_5}$$

$$m/z\ 120 \qquad m/z\ 43$$

(we have already noted that $C_6H_5^+$ can also be formed from m/z 105, $C_6H_5C\overset{+}{O}$ by heterolytic α-cleavage,

$$C_6H_5-\overset{+}{C}=O \rightarrow C_6H_5^+ + C=O).$$

In the case of 1-phenylethanone, $C_6H_5C\overset{+}{O}$ is formed from $M^{\dagger}$ by *homolytic* α-cleavage:

$$
\begin{array}{ccc}
\overset{\displaystyle +O\cdot}{\underset{\displaystyle}{\overset{\displaystyle \|}{C_6H_5-C-CH_3}}} & \rightarrow & C_6H_5-C\overset{+}{\equiv}O + CH_3^{\cdot} \\
& & \updownarrow \\
m/z\ 120 & & C_6H_5\overset{+}{C}=O \leftrightarrow \text{etc} \\
& & m/z\ 105
\end{array}
$$

The formation of acyl ($RC\overset{+}{O}$) and aroyl ($ArC\overset{+}{O}$) ions from molecular ions by homolytic α-cleavage is a very important mechanism of fragmentation of carbonyl compounds.

β-Homolytic cleavage is also possible though much less common, for example:

$$
\underset{\displaystyle R'}{R-CH_2-\overset{\displaystyle}{C}=\overset{+}{\ddot{O}}} \rightarrow R\cdot + \underset{\displaystyle R'}{CH_2=C-\ddot{O}^+}
$$

The $(M - R)^+$ species produced is not very stable; on the other hand the same process occurring in an acylium ion will produce ketene ions, $CH_2=C=\overset{+}{\ddot{O}}$:

$$
R-CH_2-C\equiv O^+ \rightarrow R\cdot + CH_2=C=\overset{+}{\ddot{O}}
$$
$$
m/z\ 42
$$

These are more stable and are sometimes observed.

The mass spectra of C, H and O-containing compounds tend to be dominated by fragment ions of *odd* mass. This is because of the *even-electron rule*, which follows from the bond cleavages we have just described. This rule states that *odd-electron* ions ($M^{+\cdot}$ is such an ion) decompose by loss of radicals or *even-electron* molecules, but *even-electron* ions may fragment only by loss of neutral molecules, but not of radicals ie

$$\overset{+\cdot}{A} \longrightarrow \overset{+}{C} + N_1^{\cdot}$$

$$\overset{+\cdot}{A} \longrightarrow \overset{+\cdot}{D} + N_2$$

$$\overset{+}{B} \longrightarrow \overset{+}{E} + N_3$$

$$\overset{+}{B} \xrightarrow{\quad\times\quad} \overset{+\cdot}{F} + N_4^{\cdot}$$

In this scheme, $\overset{+\cdot}{A}$ may be a molecular ion or a daughter ion formed from it by loss of a neutral molecule, $N_1^{\cdot}$ and $N_4^{\cdot}$ are neutral radicals, and N_2 and N_3 are neutral molecules.

A simpler statement of the rule is that successive losses of radicals are forbidden. In practice they do occur, but they are not energetically favoured processes because an even-electron ion is more stable than an odd-electron ion so there is no overall reduction in energy when

$$\overset{+}{B} \rightarrow \overset{+\cdot}{F} + N_4^{\cdot}$$

Radicals lost from C, H and O-containing compounds are always of odd masses, so since the relative molecular mass of such compounds is even, the positive ion produced must have an odd mass too. This applies for heterolytic or homolytic cleavages and explains why the mass spectra of C, H and O-containing compounds have so many odd-mass ions.

For example in the spectrum of phenylethanone,

$$C_6H_5COCH_3^{+\cdot} \xrightarrow[-15\ \text{amu}]{-CH_3^{\cdot}} C_6H_5C\overset{+}{O} \xrightarrow[-28\ \text{amu}]{-CO} C_6H_5^{+} \xrightarrow[-26\ \text{amu}]{-C_2H_2} C_4H_3^{+}$$

m/z 120   *m/z* 105   *m/z* 77   *m/z* 51

Once the even-electron *m/z* 105 has been formed it fragments by successive losses of even-mass molecules of CO and HC≡CH so the positive ions are all odd mass apart from M^{+} itself.

∏   Would you expect the typical positive ions in the mass spectrum of a C, H, O and N-containing compound to be of odd or even mass when the number of N atoms is a (*i*) odd number (*ii*) even number?

(*i*)   When there are an odd number of N atoms in the molecule, the relative molecular mass will be odd (Nitrogen Rule). Hence the loss of an odd-mass radical from M^{+} would lead to an *even*-mass fragment by the Even-Electron Rule, and subsequent losses of even-mass molecules from this would give further *even*-mass fragments ie M^{+} odd-mass, fragments even-mass.

(*ii*) When there are an even number of N atoms in the molecule, the relative molecular mass will be even (Nitrogen Rule), so the Even-Electron Rule predicts fragments of *odd* masses as in a C, H and O-containing compounds.

Now try these revision SAQs.

SAQ 8.3a   How would you expect the molecular ion of 1-bromo-1-phenylethane, $C_6H_5CHBrCH_3$, to fragment? Which of the ions formed by heterolytic and homolytic cleavages, α and β, would be stabilised by +M and/or +I effects?

SAQ 8.3a

SAQ 8.3b How would you expect propan-2-ol molecular ions, $CH_3—CHOH—CH_3$, to fragment? Which of the ions formed by heterolytic and homolytic cleavages, α and β, would be stabilised by $+M$ and/or $+I$ effects?

Summary

Molecules fragment when the vibrational energy, concentrated in a particular bond, exceeds the bond energy. The QET assumes that the transfer of vibrational energy through the ion is rapid relative to the rates of subsequent fragmentations, hence any bond may break. In practice the weaker bonds break first. The process of fragmentation continues (cascade effect) until the last ion no larger possesses sufficient energy to break a bond. The reason why some ions are formed in preference to others, although bonds of similar energy are involved is explained by the influence of inductive (I) and mesomeric (M) effects on the stability of positive ions. The relative abundance of aromatic and aliphatic ions is discussed and the reason why the $C_6H_5^+$ ion is not as stable as might at first be expected.

The four main ways of cleaving covalent bonds are described, heterocyclic α- and β-cleavage, and homolytic α- and β-cleavage. The α-cleavage is the commonest process observed in mass spectrometry. A useful generalisation which follows from the modes of cleavage is the Even-Electron Rule.

Objectives

You should now be able to:

- appreciate that molecular ions are unstable, higher energy species which fragment by breaking bonds until the last daughter ion no longer has sufficient excess energy to cleave again;

- explain qualitatively what happens to the energy given to the ionised molecules before fragmentation occurs;

- predict which bonds in a molecular ion are most likely to break based on their relative bond strengths ie weakest break most readily;

- recognise and give examples of molecules which show that bond strength is not the only factor governing fragmentation, and that ions are often formed due to their enhanced stability and/or because of the expulsion of small, stable multiply bonded molecules;

- define what is meant by a $\pm I$ and a $\pm M$ group, and give examples of each;

- appreciate that $+I$, $+M$, and $+M$, $-I$ groups stabilise positive ions by donating electron density to them and thereby enhance the stability of such ions in mass spectrometry;

- recognise and give examples of positive ions stabilised and destabilised by common functional groups found in organic molecules;

- recognise correct uses of ⤵ and ⤷ symbols to show the movement of electrons through the bonds of an ion, both to (de)stabilise or to cleave a bond;

- (optional) demonstrate how $\pm M$ effects work by use of ⤵ and ⤳ symbols to show the movement of electrons through the double bonds of an ion;

- predict which ions from a simple compound would be expected to be give rise to intense peaks in the mass spectrum on the basis of their stabilisation by $\pm M$ and $\pm I$ effects, as well as ease of bond cleavage on energetic grounds;

- recognise situations where there are no clear-cut predictions possible because all the expected fragmentations lead to ions of similar stabilities;

- describe the four main modes of bond cleavage which occur in mass spectrometry:

  (i) heterolytic α-cleavage,
  (ii) heterolytic β-cleavage,
  (iii) homolytic α-cleavage,
  (iv) homolytic β-cleavage,

  and appreciate that (i) and (iii) are the most common;

- recognise the results of the four types of bond cleavage in the mass spectra of simple compounds;

- state the Even-Electron Rule, and use it to decide the most likely sequence of ions in a cascade process from both even and odd-mass molecular ions.

9. Fragmentations of Common Functional Groups

In this section we will be looking at the mass spectral fragmentations of some selected functional groups commonly met with in organic chemistry. These are alcohols, ethers, phenols, carbonyl compounds, amines, hydrocarbons, halocompounds, nitrocompounds, heterocycles and sulphur derivatives. This is necessarily a restricted list and it may not contain the compounds you are particularly familiar with, or perhaps deals in much more detail with some than you need.

We advise that you study Section 9.1 carefully, as this shows you a general approach to identifying unknown compounds from their mass spectra using correlation tables. Next go on to Section 9.2 which covers fragmentation of alcohols in some detail. You may not be particularly interested in alcohols, but in this section we have introduced a number of basic concepts which are common to the fragmentations of very many organic compounds. These you will see over and over again in the spectra of other classes of compounds. We have written the later sections on the assumption that you have understood these concepts. So, if you are tempted to go straight to say nitrocompounds because you are involved with them in your present work, you will not find it easy to understand the rationale of their fragmentation without studying Section 9.2 first.

Sections 9.1 and 9.2 are quite long but your reward in completing them will be a good basic understanding of mass spectra and the origin of common fragment ions. It would be quite in order for you to proceed to whichever section interests you most after Section 9.2, or just flip through Sections 9.3 to 9.10 to get a general idea of the main features of these classes of compound if you are not likely to need to interpret their mass spectra in the near future. These sections are there to be used if and when required – it is up to you to make the best possible use of the material.

Section 9.5.1 explains a common rearrangement process occurring particularly in carbonyl compounds, it is well worth some of your time because ions resulting from it are both diagnostically very useful and confusing if you are not aware of this possibility.

9.1. INTERPRETATION PROCEDURES

Before we deal with the features associated with common functional groups we will give a procedure for interpreting a mass spectrum, and two types of correlation tables you can use. We should stress here that these are only suggestions which you need not slavishly follow in all cases. Especially as you build up experience with interpreting mass spectra you will find it possible to ignore some of the steps and proceed quickly to identify the class of compound you are dealing with.

You should also bear in mind that the correlation tables cannot possibly include all the types of functional groups and structures which have been investigated. Those important in your particular job might not be generally so common. We suggest in that case you look for reviews of mass spectral papers in journals which cover your field, and add entries to the tables in the appropriate places. In other words, do-it-yourself!

9.1.1. Procedure for the Interpretation of a Mass Spectrum

(*a*) Identify the molecular ion M^{+}.

(*b*) Identify typical background peaks present in all spectra eg H_2O (18), N_2 (28), O_2 (32), CO_2 (44) and count the spectrum from low to high mass.

(*c*) Identify the most intense fragment (base peak).

(*d*) Using an expanding scale eg Gerber rule if possible, measure the intensities of the ions. Set the most abundant ion as 100 arbitrary units and measure all other ions > 2% relative to it. Tabulate this 'normalised' spectrum.

(*e*) Identify and tabulate any metastable peaks, and 'half mass' peaks.

(*f*) Plot a bar graph of relative abundance against m/z on square graph paper. It is also convenient to make a small table of m.s. transitions in one corner. This forms the permanent record of the mass spectrum – photographic traces fade.

(*g*) Mark abundant odd-electron ions ie those at even mass numbers if the molecule contains even numbers of N atoms, and check that the molecular ion is one of these.

(*h*) Examine general appearance of the spectrum for molecular stability, labile bonds.

(*i*) Identify neutral fragments accompanying high mass ion formation from the (M − X) Table – correlate these with metastable transitions if observed.

(*j*) Use the Mass Composition Table to assign likely formulae to the major fragment ions.

Coupled with information about the origin of the sample compound this should lead to structures for the most abundant ions and the molecular ion. Remember to allow for hydrogen transfers and rearrangements which lead to ions one mass unit greater than expected (more about these under individual functional groups later).

If this is not conclusive check the following.

(k) The counting of the original mass spectrum.

(l) Is the highest mass ion observed the M^{+}? Have the accurate mass determined and assign the molecular formula and/or have the spectrum run with chemical ionisation to reveal the MH^{+} ion.

(m) Peaks 14 amu greater or smaller than the supposed M^{+} may be due to the presence of minor amounts of homologues of the compound.

This procedure assumes that you have been given the spectrum in the form of a trace on photographic paper. Steps (d), (e) and (f) relate to the handling of this trace. The Gerber rule mentioned is a precision expanding scale with a spring of 100 turns. This is expanded until the 100th turn of the spring coincides with the end of the base peak (step (c)). The other ions can then be read off the trace directly as a % relative to this.

These days you are more likely to get the spectrum in the form of a normalised bar chart plotted by the mass spectrometer's data system, with the more intense ions 'flagged' with their m/z values. Steps (c), (d) and (e) will have been done for you. However, metastable ions and half mass peaks (M^{2+}) are not usually processed by data systems. If you need these, you will have to ask the mass spectrometer operator for an old fashioned trace, and do the plotting yourself.

After Steps (g) and (h) we can start to work identifying the neutral fragments lost from the M^{+}, and the more intense daughter ions, using the table in Fig. 9.1a. We will call this the (M − X) table (M minus X).

X	Compounds
1	Aldehydes, acetals, compounds with aryl—CH_3 groups, compounds with N—CH_3, compounds with —CH_2CN, alkynes
2	Fused ring aromatic compounds
15	Acetals, methyl derivatives, *t*-butyl and *i*-propyl compounds, compounds with aryl—C_2H_5 groups, $(CH_3)_3SiO$ derivatives
16	Aromatic nitro compounds, N-oxides, S-oxides, aromatic amides
17	Carboxylic acids, aromatic compounds with a functional group containing oxygen *ortho* to one containing hydrogen eg *o*-nitrotoluene
18	Straight-chain aldehydes (C_6 upwards), primary straight-chain and aromatic alcohols, alcohol derivatives of cyclic alkanes, steroid alcohols and ketones, aliphatic ethers with one group having 8 or more carbons, carboxylic acids
20	Aliphatic alcohols (thermal product), *n*-fluoroalkanes
26	Aromatic hydrocarbons
27	Aromatic amines, nitrogen heterocycles, aromatic nitriles
28	Diarylethers, phenols and naphthols, aldehydes, keto derivatives of cycloalkanes, quinones and polycyclic ketones, aliphatic nitriles, ethyl esters
29	Aromatic aldehydes, keto derivatives of cycloalkanes, phenols, naphthols, polyhydroxybenzenes, propionals, diarylethers, aliphatic nitriles, ethyl derivatives

30	Aromatic nitro compounds, aromatic methyl ethers
31	Methoxy derivatives, methyl esters
32	*o*-Methylbenzoates
33	Short chain unbranched primary alcohols, alcohol derivatives of cycloalkanes, steroid alcohols, secondary and tertiary thiols, thio derivatives of cycloalkanes, RNCS
34	Thiols
35	Secondary and tertiary chloroalkanes, chloroaryls
36	*n*-Chloroalkanes (loss of HCl)
40	Aliphatic nitriles and dinitriles
41	Propyl esters
42	Acetates, N-acetyl compounds (loss of $CH_2=C=O$), butyl ketones
43	Propyl derivatives, aliphatic nitriles, *t*-amides, methyl ketones, acetyl derivatives
44	Aliphatic aldehydes containing γ-hydrogens, esters, anhydrides, acids
45	Carboxylic acids, ethoxy derivatives, ethyl esters
46	Aromatic esters with ethyl *o*- to COOH, nitro compounds, long-chain unbranched alcohols, ethyl esters
47	Alkyl nitro compounds, phosphoryl compounds
48	Sulphoxides
55	Butyl esters

56	Pentyl ketones, ArOBu
57	Ethyl and butyl ketones, butyl-X
58	Straight-chain primary mercaptans, RCNO and RNCO, methyl ketones containing γ-hydrogens
59	*n*-Propyl esters, methyl esters of 2-hydroxycarboxylic acids, acetyl derivatives
60	Methyl esters of short-chain dibasic carboxylic acids, *o*-methyltoluates, acetates
63	Methyl esters of short-chain dibasic carboxylic acids
64	Methyl esters of all aliphatic dibasic carboxylic acids (not dimethyl adipate), sulphones, sulphonyl derivatives
73	*n*-Butyl esters, methyl esters of all aliphatic dibasic acids
77	Phenyl derivatives
79 (& 81)	Bromo compounds
87	Pentyl esters
91	Benzyl and tolyl compounds
93	Phenoxy derivates
107	Benzyloxy derivates
127	Iodo compounds

Fig. 9.1a. *Table of correlations between compounds and neutral fragments produced in the process $M^{+\cdot} \rightarrow (M - X)^+$ where X is the neutral fragment*

This is essentially an arithmetic approach to solving a mass spectrum. It assumes no knowledge of the mechanisms of the fragmentation processes, and that any process may occur if that structural unit is present in the molecule. The extent to which a given X is lost will of course depend on energetic and resonance factors as discussed in Section 8, but there must also be a reasonable mechanism to account for the bonding changes which have occurred. The arithmetic and the mechanistic approaches to solving a mass spectrum go hand in hand, but it is natural to start with the arithmetic. Indeed, you can often identify a compound of simple structure using the (M − X) table alone without any consideration of the mechanisms of how the fragment ions have been formed.

You are probably used to the application of correlation tables in spectroscopy, especially infrared and nuclear magnetic spectroscopy, so will not need much help in using them. You should also be aware that they are not infallible, for a number of reasons.

∏   Can you think of any reasons why correlation tables may fail to identify an unknown compound from its spectrum?

You may have thought of some of the following reasons.

(*a*) The table has no entry for that class of compound. There are so many classes of organic compound that it is not possible for a compact table to have an entry for all of them.

(*b*) Two or more classes of compound may show the same X loss. This is particularly so in mass spectrometry where different classes of compound may lose the *same* X.

(*c*) The same numerical value of X may arise from more than one molecular formula eg CO and C_2H_4, both 28 amu, CH_3CO and $(CH_3)_2CH$, both 43 amu. The X loss might indicate two (or more!) quite different structures.

(*d*) The structure concerned might exist in a number of isomeric forms, each showing the same X loss eg 2-, 3- and 4-disubstituted benzene derivatives.

(*e*) Where a compound has two (or more) structural units the characteristic X loss of one (or more) may be pre-empted by a lower activation energy fragmentation of the other substituent. That is, the X losses associated with the intense ions in the spectrum are characteristic of only *one* of the substituents.

So, we can conclude that the (M − X) table gives useful clues to the type of compound we have, but not necessarily to its precise structure. In attempting to obtain that, we should seek evidence from another sort of correlation table to be discussed next, and link this to evidence from the (M − X) table, metastable ions and molecular formulae.

The second correlation table we are going to use is shown in Fig. 9.1b. This is a mass-composition table. It lists the m/z values of ions commonly observed in mass spectra with possible groups associated with that particular mass and some possible inferences.

This table also suffers from the disadvantages mentioned in connection with the (M − X) table, namely it is selective, and many quite different compounds can give isomeric or isobaric ions of the same nominal mass. Nevertheless, it is a useful follow-up in the arithmetical process of solving a mass spectrum.

m/z	Possible Formulae	Possible Compounds/Group Type
26	C_2H_2	hydrocarbon, especially unsaturated
27	C_2H_3	hydrocarbon, perhaps unsaturated
28	CO, C_2H_4, N_2	carbonyl, ethyl, azo compounds
29	CHO, C_2H_5	aldehyde, ethyl compound
30	$CH_2{=}NH_2$, NO	primary amines, nitro compounds
31	$CH_2{=}OH$	primary alcohols, methoxy compounds
35/37 (3:1)	^{35}Cl, ^{37}Cl	chloro compounds

36/38 (3:1)	^{35}ClH, ^{37}ClH	chloro compounds
39	C_3H_3	hydrocarbons, especially aromatic
40	Argon; C_3H_4	background from air, hydrocarbons
41	C_3H_5	hydrocarbon, especially unsaturated
42	$CH_2=C=O$, C_3H_4	acetates & acetyl compounds, hydrocarbons
43	CH_3CO	CH_3COX
43	C_3H_7	C_3H_7X, especially i-C_3H_7X
44	CO_2	background from air, carbonates, anhydrides
44	$CH_3CH=NH_2$	aliphatic amines
44	$O=C=NH_2$	primary amides
44	$CH_2=CH(OH)$	aldehydes having γ-H
45	$CH_3CH=OH$	secondary alcohols
45	$CH_2=OCH_3$	some ethers
45	CO_2H, OCH_2CH_3	acids, ethoxy compounds
46	NO_2	nitro compounds
47	$P=O$, $CH_2=SH$	phosphoryl compounds, primary thiols
49/51 (3:1)	CH_2Cl	chloromethyl compounds
50	C_4H_2	aromatic compounds
51	C_4H_3	C_6H_5X
55	C_4H_7, C_3H_3O	unsaturated hydrocarbons, C_5 & C_6 cyclic ketones
56	C_4H_8	hydrocarbon
57	C_4H_9, CH_3CH_2CO	C_4H_9X, CH_3CH_2COX

58	$CH_2=C(OH)CH_3$	methyl ketones having γ-H, some dialkyl ketones
58	$(CH_3)_2N=CH_2$	aliphatic amines
59	$(CH_3)_2COH, COOCH_3$	$(CH_3)C(OH)X$, methyl esters
59	$CH_2=C(OH)NH_2,$ $C_2H_5CH=OH$	primary amides, C_2H_5CHOHX
60	$CH_2=C(OH)OH$	aliphatic acids having γ-H, acetate esters
61	$CH_3CO(OH_2)$	acetate esters $CH_3COOC_nH_{2n+1}(n > 1)$
65	C_5H_5	benzyl and tolyl compounds, phenols, anilines
66	C_5H_6	aromatic compounds
68	C_4H_4N	pyrroles (monosubstituted)
69	C_5H_9	hydrocarbon, especially unsaturated
70	C_5H_{10}	hydrocarbon, perhaps unsaturated
71	C_5H_{11}, C_3H_7CO	$C_5H_{11}X$, propyl ketone, butanoate ester
72	$CH_2=C(OH)C_2H_5$	ethyl ketone having γ-H
72	$C_3H_7CH=NH_2$	amines
73	$C_3H_7CH=OH,$ $C_3H_7OCH_2$	alcohols, ethers
73	$CO_2C_2H_5$	ethyl esters
73	$CH_2=CHC(OH)=OH$	aliphatic acids
73	$(CH_3)_3Si$	trimethylsilyl derivatives
74	$CH_2=C(OH)OCH_3$	methyl esters having γ-H
75	$(CH_3)_2Si=OH$	$(CH_3)_3SiOX$
75	$C_2H_5CO(OH_2)$	$C_2H_5COOC_nH_{2n+1}$ $(n > 1)$
76	C_6H_4	benzene derivatives, mono or disubstituted

77	C_6H_5	C_6H_5X
78	C_6H_6, C_5H_4N	C_6H_5X, X having β- or γ-H, (monosubstituted) pyridines
79	C_6H_7	C_6H_5X
79/81 (1:1)	^{79}Br, ^{81}Br	bromo compounds
80/82 (1:1)	^{79}BrH, ^{81}BrH	bromo compounds
82	C_5H_6N	methylpyrroles, monoalkylpyrroles
83	C_4H_3S	thiophenes (monosubstituted)
85	C_6H_{13}, C_4H_9CO	$C_6H_{13}X$, C_4H_9COX
86	$CH_2{=}C(OH)C_3H_7$	propyl ketones having γ-H
86	$C_4H_9CH{=}NH_2$	amines
87	$CH_2{=}CHC(OH)OCH_3$	$XCH_2CH_2COOCH_3$
88	$CH_3CH_2CH_2COOH$	$C_3H_7COOC_nH_{2n+1}$, $(n > 1)$
89	C_7H_5	N and O containing heterocyclics
90	C_7H_6	N and O containing heterocyclics
91	C_7H_7	$C_6H_5CH_2X$, $CH_3C_6H_4$-X
92	C_7H_8, C_6H_6N	$C_6H_5CH_2R$, monoalkylpyridines
93/95 (1:1)	$^{79}BrCH_2$, $^{80}Br\ CH_2$	$BrCH_2X$
93	C_6H_7N	C_6H_5NHX, X containing H
93	C_6H_5O	phenols, nitrobenzenes
93	C_7H_9	mono and sesquiterpenes
94	C_6H_6O	C_6H_5OR (not $C_6H_5OCH_3$)
95	C_7H_{11}	mono and sesquiterpenes

96	C_5H_4NO	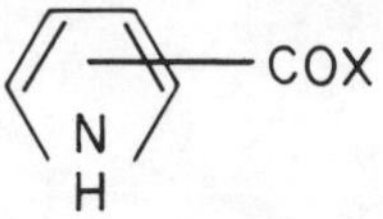 *A*
97	C_5H_5S	methylthiophenes, monalkylthiophenes
99	C_7H_{15}	$C_7H_{15}X$
103	$C_6H_5CH{=}CH$	$C_6H_5CH{=}CHX$
105	C_6H_5CO, C_8H_9	C_6H_5COX, $CH_3C_6H_4CH_2X$
107	C_7H_7O	$C_6H_5CH_2OX$, $HOC_6H_4CH_2X$
107/109 (1:1)	C_2H_4Br	$BrCH_2CH_2\text{-}X$
111	C_5H_3OS	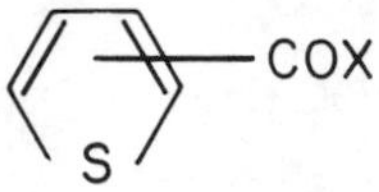 *C*
121	C_8H_9O, $C_6H_5CO_2$	$CH_3OC_6H_4CH_2X$, $C_6H_5CO_2X$
122	C_6H_5COOH	alkyl benzoates
123	$C_6H_5COOH_2$	alkyl benzoates
127	$C_{10}H_7$	naphthyl derivatives
127	I	iodocompounds
128	HI	iodocompounds
131	$C_6H_5CH{=}CHCO$	$C_6H_5CH{=}CHCOX$
141	ICH_2	ICH_2X
147	$(CH_3)_2Si{=}OSi(CH_3)_3$	$[(CH_3)_3SiO]_n$ derivatives, $n > 1$
149	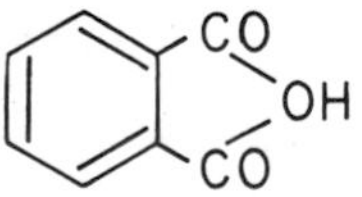 *B*	dialkyl phthalates (plasticisers)

Fig. 9.1b.   *Mass-composition table of common fragment ions*

Use the space below to add other ions as you come across them.

In using Fig. 9.1a and 9.1b, there are a number of useful points to remember. First of all, compounds containing C, H, O only or C, H, O plus an even number of N atoms, must have an *even* M_r. If an odd number of N atoms are present the M_r must be *odd*, the Nitrogen Rule).

Secondly, mass losses of 3 to 14 from *any* ion are not observed. It is easy to see that there are no common atoms or groups with relative atomic masses of 3 to 11, but you might expect to get losses of C, CH, CH_2 and perhaps N (12–14). In practice these are hardly ever observed. You should beware of trying to analyse a mass spectrum by subtracting each m/z from that of the immediately higher m/z to get X values. Remember the cascade process (Fig. 8.1a) – ions are formed by a variety of diverging pathways from M^+. The correct procedure is to subtract all the observed higher m/z values from that of M^+, to see if significant X values result, before moving down the spectrum to ions of lower masses. This is the order in which they were formed.

Thirdly, small mass losses are more likely to be fairly specific. For example, the loss of 15 amu is almost certainly a CH_3 group, but the loss of 57 amu in a C, H, O containing compound could in theory be C_4H_9, C_3H_5O, or C_2HO_2. If N is included, we can add C_3H_7N, $C_2H_5N_2$, CH_3N_3, C_2H_3NO or CHN_2O to the list. Only C_4H_9 and C_3H_5O possibilities are included in Fig. 9.1a and 9.1b because they are the commonest and we have to draw the line somewhere. On the other hand, the loss of 77 amu, or an ion m/z 77 is very indicative of the presence of a phenyl (C_6H_5) group.

This brings us to the fourth point. Inferences drawn from either table are most likely to be significant if the peak in question is of high RA. Aromatic ions in general are more stable (Section 8.2) and mass spectra of aromatic compounds tend to exhibit M^+ of fairly high intensity and a limited number of intense fragment ions of relatively high mass. Aliphatic compounds on the other hand tend to break down to a larger number of ions of relatively low mass, and their M^+ can be very weak or absent. In either case, you should work out the structures of the intense ions first, then confirm likely breakdowns from any metastable ions observed. Lastly, seek confirmatory evidence from the minor fragments.

Let us now apply the (M-X) and Mass-Composition tables to solving an unknown spectrum, shown in Fig. 9.1c. The highest mass observed in the spectrum of *Unknown 1* is 107. This is almost certainly the ^{13}C isotope of m/z 106, so M^+ is 106. The base peak is m/z 105, a loss of 1 amu, clearly H from the (M-X) table. This lists aldehydes, acetals, aryl-CH_3 compounds, $-CH_2CN$, alkynes and $N-CH_3$ compounds as possibilities. Let us discard $-CH_2CN$ and $N-CH_3$ as under the Nitrogen Rule M_r would be odd. This still leaves four possibilities, so let us turn to the 77 ion – how has this been formed and what is it?

If m/z 77 is formed direct from M^+, then X is 29, indicating an aldehyde or an ethyl compound. If it is formed from 105, X is 28, again indicating a CO or an CH_2-CH_2 compound. Now loss of H *and* 28 or 29 amu is only consistent with an aldehyde. The base peak m/z 105 could be $C_6H_5CO^+$ or $CH_3C_6H_4CH_2^+$ from the Mass-Composition table. Clearly only $C_6H_5CO^+$ fits in with the other deductions, hence *Unknown 1* is benzaldehyde, C_6H_5CHO.

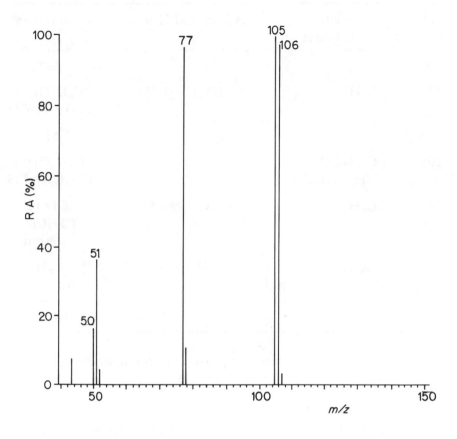

Fig. 9.1c. *Mass spectrum of* Unknown *1*

These deductions can be conveniently presented in a table, such as Fig. 9.1d. Note that in this table, the molecular ion appears at the head of each postulated fragmentation pathway, twice in this case. The loss in the X column is then derived from the entry immediately above it.

m/z	Possible Structures	Associated X loss	Inferences
106		–	$M^{+\cdot}$
77	C_6H_5	CHO, CH_3CH_2	C_6H_5CHO, $C_6H_5CH_2CH_3$
106		–	$M^{+\cdot}$
105	C_6H_5CO, $CH_3C_6H_4CH_2$	H	C_6H_5CHO, $CH_3C_6H_4CH_2$
77	C_6H_5	CO, $CH_2{=}CH_2$	CO or CH_3CH_2 present
51	C_4H_3	CH≡CH	C_6H_5 present

Deduction : $C_6H_5{-}CHO$

Fig. 9.1d. *A method for presenting the analysis of a mass spectrum*

∏   Metastable ions are observed in the *Unknown 1* spectrum at m/z 104, 56.5 and 33.8. What do these tell us about the fragmentation routes of benzaldehyde?

m^* 104 corresponds to $106 \rightarrow 105$ ($105^2/106 = 104.01$)

m^* 56.5 could correspond to $106 \rightarrow 77$, or

$105 \rightarrow 77$, so both must be tried.

$$\frac{77^2}{106} = 55.93 \qquad \frac{77^2}{105} = 56.47 \qquad \text{so the latter is correct}$$

Hence the pathway which has metastable ion support is:

$$C_6H_5CHO^{+\cdot} \xrightarrow{-H} C_6H_5\overset{+}{C}O \xrightarrow{-CO} C_6H_5^{+}$$

rather than

$$C_6H_5-CHO^{+\cdot} \xrightarrow{-CHO} C_6H_5^{+\cdot}$$

m^* 33.8 could correspond to:

$$106 \rightarrow 51, m^* = \frac{51^2}{106} = 24.53$$

which is clearly much too low, and shows that $105 \rightarrow 51$, $106 \rightarrow 50$, or $105 \rightarrow 51$ would also give m^* values below 33.8.

Next try $77 \rightarrow 51$, $m^* = \dfrac{51^2}{77} = 33.78$.

This is the answer, indicating that

$$C_6H_5^{+} \xrightarrow{-CH\equiv CH} C_4H_3^{+}$$

Hence the likely fragmentation route of the benzaldehyde $M^{+\cdot}$ is:

$$C_6H_5CHO^{+\cdot} \xrightarrow{-H\cdot} C_6H_5\overset{+}{C}O \xrightarrow{-CO} C_6H_5^{+} \xrightarrow{-CH\equiv CH} C_4H_3^{+}$$

All this has been deduced without any use of mechanisms. For simple compounds you will often be able to do this, but for the sake of completness and practice the bond breaking mechanisms are:

(*i*)

$$C_6H_5-\overset{\overset{\displaystyle H}{|}}{C}\overset{\cdot\cdot}{=}\overset{+}{O} \rightarrow C_6H_5-C\equiv\overset{+}{O} + H\cdot$$

(*ii*)

$$C_6H_5-C\equiv\overset{+}{O} \rightarrow C_6H_5^{+} + :C=O$$

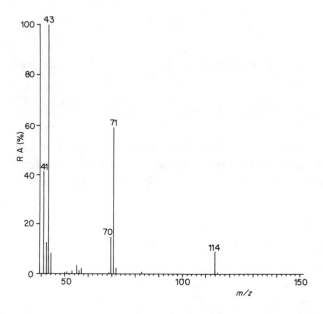

(*i*)   is a homolytic cleavage,
(*ii*)   is a heterolytic cleavage,
(*iii*) is a homolytic cleavage

Since the double bonds in the $C_6H_5^+$ ion are delocalised any consecutive pair of carbon atoms can be eliminated as $HC\equiv CH$. Where you are able to devise a mechanism for a step in the path it can conveniently be given in the Inferences column of your analysis table (Fig. 9.1d), but this is not essential.

∏   Let's try a second example. Have a look at the mass spectrum of *Unknown 2*, Fig. 9.1e and try to analyse it in the same way as *Unknown 1*.

Fig. 9.1e.   *Mass spectrum of Unknown 2*

You should have obtained an analysis table similar to Fig. 9.1f below.

m/z	Possible Structures	Associated X loss	Inferences
114	–	–	$M^{+\cdot}$
71	C_5H_{11}, C_3H_7CO	43	propyl, methyl ketone, acetyl
70	C_5H_{10}	1	stable alkenic structure?
114	–		$M^{+\cdot}$
70	C_5H_{10}	44	aldehyde with γ-H, ester, acid anhydride
114	–		$M^{+\cdot}$
43	CH_3CO, C_3H_7	71	Complementary ion to m/z 71
71	C_5H_{11}, C_3H_7CO	–	Parent for m/z 43 or 41?
43	CH_3CO, C_3H_7	28	CO or $CH_2{=}CH_2$ present?
41	C_3H_5	30	not sensible from m/z 71
41	C_3H_5	2	loss of H_2 from m/z 43, implies m/z 43 is $C_3H_7^+$

Deduction: $(CH_3CH_2CH_2)_2CO$ or $[(CH_3)_2CH]_2CO$

Fig. 9.1f. *Analysis table for* Unknown 2

The analysis suggests the presence of propyl (Pr) or isopropyl (*i*Pr) groups, CO, or CH_3CO. The relatively few ions in the spectrum show that there are only two or three easily fragmented bonds, which suggests cleavage α- to the C=O group. One possible process, 114 $\rightarrow$ 70, suggests the presence of $-CH_2CHO$ or $-CO_2-$, but since there is no other support for an aldehyde, ester, acid or anhydride and m/z 70 is only 17% RA this can be dismissed. The loss of one or two hydrogens from an aliphatic ion to give an unsaturated species is quite common, so m/z 70 is probably formed from m/z 71.

If *Unknown 2* contained both Pr and CH_3CO it would have to be $PrCOCH_3$ but this has a M^+ of only 86. Any other structure involving $COCH_3$ and the correct number of carbon atoms (now seen to be seven) to give $M_r = 114$ would have a hydrocarbon chain of five carbons ie $CH_3-CH_2-CH_2-CH_2-CH_2^+$, m/z 71. Although this could lose $CH_2=CH_2$ and give $CH_3CH_2CH_2^+$, m/z 43, by the process shown, it would also be expected to fragment in other ways eg by losses of $CH_3^{\cdot}$ and $CH_3CH_2^{\cdot}$ from M^+ and m/z 71, and these are not observed. In particular, m/z 99 is *not* found. This *should* form in a methyl ketone because it is an acylium ion and fairly stable:

$$CH_3-\overset{\overset{\cdot\,+}{O}}{\overset{\|}{C}}-(CH_2)_4CH_3 \longrightarrow CH_3(CH_2)_4\overset{+}{C}=O + CH_3^{\cdot}$$
$$m/z\ 99$$

Hence the structures $(iPr)_2CO$ or $(Pr)_2CO$ are most likely. In fact, *unknown 2* is $(iPr)_2CO$. Its main fragmentations are:

$$iPr-\overset{\overset{\cdot\,+}{O}}{\overset{\|}{C}}-iPr \longrightarrow iPr^{\cdot} + iPr-C\equiv\overset{+}{O} \longrightarrow i\text{-}Pr^+ + :C=O$$
$$\qquad\qquad\qquad m/z\ 71 \qquad\quad m/z\ 43$$

You could not be certain from the mass spectrum alone which structural isomer you had. Indeed it could have been $PrCOiPr$! To decide which of the three you had you could (*a*) consult reference lists of mass spectra; (*b*) use a computer library database to get a match or (*c*) obtain the nmr spectrum.

∏ Our third exercise is *Unknown 3*, which contains a nitrogen atom, the mass spectrum is shown in Fig. 9.1g. See if you can suggest a structure for *Unknown 3* using the correlations in Fig. 9.1a and 9.1b.

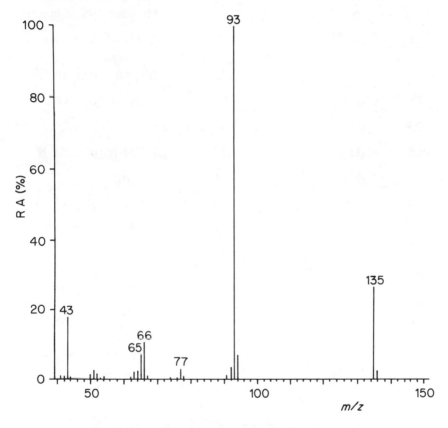

Fig. 9.1g. *Mass spectrum of* Unknown 3

The sort of analysis that you should have obtained is shown in Fig. 9.1h.

m/z	Possible Structures	Associated X loss	Inferences
135	–	–	M^{+}-odd, $\therefore$ 1,3,5 ... N (In fact 1 N is given)
93	C_6H_7N	42	C_6H_5NHX, N-acetyl compound
	C_7H_9	42	terpene derivative?
77	C_6H_5	58	RNCO, RCNO?
93	C_6H_7N	–	parent for m/z 66, 65?
66	C_5H_6	27	HCN from C_6H_7N
65	C_5H_5	1	H from C_5H_6
43	CH_3CO, C_3H_7	50	Not likely! m/z 43 not from m/z 93
135	–	–	M^{+}
43	CH_3CO, C_3H_7	92	Complementary ion to m/z 93, with H transfer?

Deduction: $C_6H_5NHCOCH_3$

Fig. 9.1h.   *Analysis table for* Unknown *3*

From the general appearance of this spectrum, the compound is aromatic. The appearance of a m/z 77 ion, although small, indicates a C_6H_5 unit. This leaves 68 amu to account for. The loss of 42 amu from M^{+} (CH_2=C=O) is very characteristic of N-acetyl compounds. Hence *Unknown 3* is $C_6H_5NHCOCH_3$. This is confirmed by the rest of the spectrum looking very like that of aniline, since m/z 93 *is* $C_6H_5NH_2^{+}$, with the confirming presence of CH_3CO^{+}, m/z 43. If this were a propyl compound, it could not fragment by loss of ketene, so this possibility can be dismissed straight away. The mechanisms of the processes involved are complex, apart from formation of CH_3CO^{+}

$$C_6H_5-NH-\overset{\overset{\displaystyle \cdot \overset{+}{O}}{\parallel}}{C}-CH_3 \longrightarrow C_6H_5\overset{\cdot}{N}H + CH_3C\equiv\overset{+}{O}$$

The other processes will be dealt with in the following sections.

You may not have been entirely convinced from these examples that mass spectrometry leads to unambiguous structure determination. You were quite right to be sceptical! It is often the case that no one spectroscopic method can solve the problem. What is important is that you can use the mass spectrum to its full advantage, but also recognise the limitations so that you will know when to look to other techniques for information. Infrared and nuclear magnetic resonance spectroscopy are the two which complement mass spectrometry best, as you probably know.

We will end this introductory section by giving two SAQs of mass spectra of unknowns for you to solve as completely as possible using the (M − X) and Mass-Composition tables. You should not go on until you have tried these and studied the solutions given.

SAQ 9.1a

Analyse the spectrum of *Unknown 4* Fig. 9.1i which contains no nitrogen. Suggest a possible structure for the compound, using the correlations in Fig. 9.1a and 9.1b (there may be more than one possible structure which is in accord with the data).

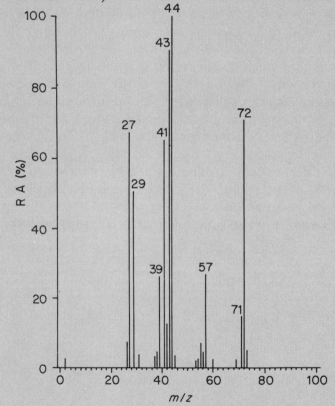

Fig. 9.1i. *Mass spectrum of* Unknown 4

SAQ 9.1a

SAQ 9.1b   Analyse the spectrum of *Unknown 5*, Fig. 9.1j, which contains a nitrogen atom. Suggest a possible structure for the compound, using the correlations in Fig. 9.1a and 9.1b.

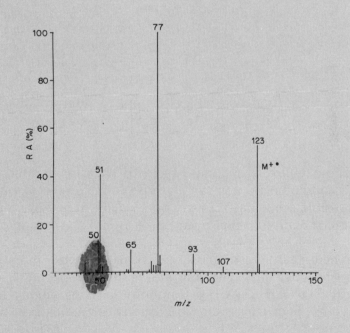

Fig. 9.1j.   *Mass spectrum of* Unknown 5

SAQ 9.1b

Summary

You have used two tables of data to interpret mass spectra, the (M − X) table (Fig. 9.1a) and the Mass-Composition Table (Fig. 9.1b).

These tables are necessarily selective bearing in mind the vast range of compounds which can give rise to mass spectra, but allowing for possible omissions they are very useful in getting to grips with the interpretation of the spectrum of an unknown compound.

When analysing a mass spectrum the most intense ion should be assigned first, and any ambiguities of structure recognised eg $CH_3CO^+/CH_3CH_2CH_2^+$. Compounds giving intense M^+ and a few intense higher mass ions are usually aromatic derivatives.

Metastable ions should be used if available to determine fragmentation routes as this often helps to distinguish alternative structures.

Objectives

After completing this section, you should be able to:

- follow a logical procedure for the plotting and interpretation of a mass spectrum;

- usc $(M - X)$ and Mass-Composition Tables to suggest what structural units might be present in a molecule from its mass spectrum;

- give reasons why correlation tables might not be successful in identifying a molecule from its mass spectrum;

- recognise that spectra of aromatic compounds are usually distinguished from aliphatic by having an abundance of higher mass ions;

- suggest structures for simple molecules of low M_r from their mass spectra;

- present your analyses of such mass spectra in a clear tabulated form;

- appreciate that recourse should be had to other types of spectra to achieve complete identification when necessary.

9.2.   FRAGMENTATIONS OF ALCOHOLS

The mass spectra of methanol and ethanol were described in Section 1. Recall that the base peak in both is m/z 31, $CH_2=\overset{+}{O}H$. In methanol this is due to the loss of H'

$$H_2C-\overset{\cdot +}{O}H \rightarrow H_2C=\overset{+}{O}H + H'$$

It is conventional to shorten this mechanism to:

$$\overset{\displaystyle H}{\underset{\displaystyle H_2C}{|}}\overset{\curvearrowright}{-}\overset{\cdot\cdot +}{O}H \rightarrow H_2C=\overset{+}{O}H + H^{\cdot}$$

showing the movement of only one of the electrons involved, in order to simplify the picture. We will use this convention where it helps to clarify what might become a rather cluttered formula. In ethanol, Fig. 9.2a, the same process can occur.

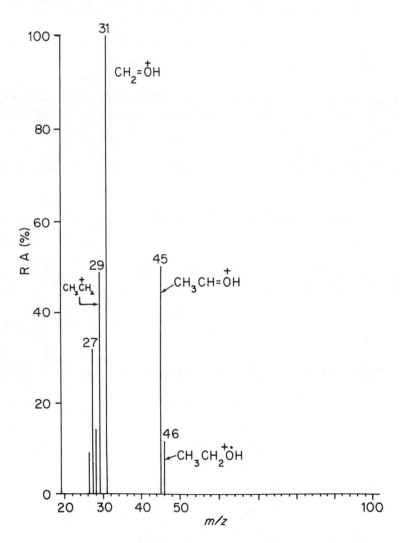

Fig. 9.2a.   *Mass spectrum of* CH_3CH_2OH

$$\text{CH}_3\overset{\overset{\displaystyle H}{\displaystyle |}}{\text{CH}}\!-\!\overset{\cdot+}{\ddot{\text{O}}}\text{H} \rightarrow \text{CH}_3\text{CH}\!=\!\overset{+}{\ddot{\text{O}}}\text{H} + \text{H}^{\cdot}$$

$$m/z\ 45,\ 20\%$$

but the base peak is formed by the alternative loss of $\text{CH}_3^{\cdot}$

$$\text{H}\!-\!\overset{\overset{\displaystyle \text{CH}_3}{\displaystyle |}}{\underset{\underset{\displaystyle H}{\displaystyle |}}{\text{C}}}\!-\!\overset{\cdot+}{\ddot{\text{O}}}\text{H} \rightarrow \text{H}_2\text{C}\!=\!\overset{+}{\ddot{\text{O}}}\text{H} + \text{CH}_3^{\cdot}$$

$$m/z\ 31,\ 100\%$$

These processes of homolytic α-cleavage are competitive. In longer chain primary alcohols eg butanol, Fig. 9.2b (i), the amount of $(\text{M} - \text{H})^+$ is much smaller, around 1% and $\text{CH}_2\!=\!\overset{+}{\ddot{\text{O}}}\text{H}$ is still the base peak.

∏   Why should the loss of the alkyl fragment be preferred over the loss of H˙ in the α-cleavage of primary alcohol molecular ions?

The radical $\text{CH}_3\text{CH}_2^{\cdot}$ would be more stable than $\text{CH}_3^{\cdot}$, just as CH_3CH_2^+ is more stable than CH_3^+.

This question was explained by Stevenson in the early 1950's. He was the first to notice this tendency for the largest alkyl fragment to be preferentially lost as a radical. The molecular ion is a high energy species trying to get rid of as much energy as quickly as possible. An alkyl radical has several bonds which can be in higher vibrational states, thus absorbing the excess energy of the molecular ion and leaving with it. The hydrogen atom can only depart with kinetic energy. The more bonds the radical species has, the better. Thus we have a very useful empirical rule named after Stevenson:

In a fragmentation, the largest radical is lost preferentially.

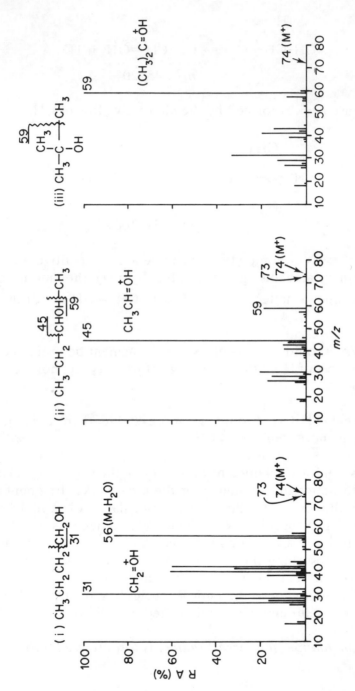

Fig. 9.2b.   *Mass spectra of the three isomeric butanols*

Examine the spectra of the isomeric butanols now and see if this works (Fig. 9.2b (i), (ii) and (iii)). Butan-2-ol (Fig. 9.2b (ii)) cleaves to give 100% (M-CH$_3$CH$_2$)$^+$ m/z 45, and 20% (M-CH$_3$) m/z 59, a ratio of 5:1 in favour of the loss of CH$_3$CH$_2$. In fact the ion CH$_3$CH=$\overset{+}{\text{O}}$H is typical of 2-ols because in the structure CH$_3$CHOHR if R=CH$_3$, CH$_3$ will be lost because CH$_3$ > H$^.$. For any other R, it will be larger than CH$_3$ so R$^.$ is lost according to Stevenson's Rule. 2-Methylpropan-2-ol (t-butyl alcohol), Fig. 9.2b (iii), is really analogous to methanol because it has three identical groups round the α-carbon and can only lose CH$_3$ to form (CH$_3$)$_2$C=$\overset{+}{\text{O}}$H as base peak. These cleavages are summarised below:

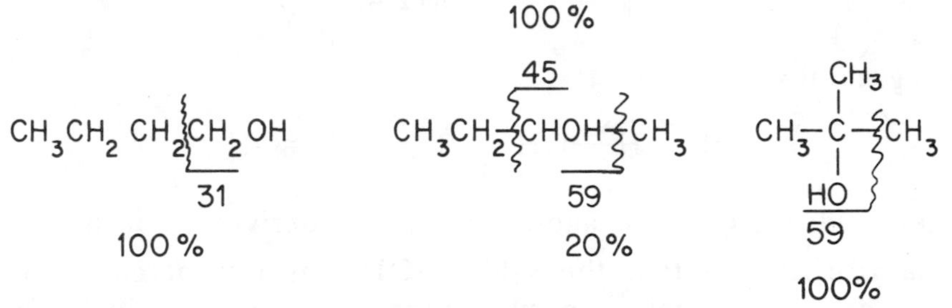

$$
\begin{array}{ccc}
 & 100\% & \\
 & 45 & CH_3 \\
CH_3CH_2\ CH_2\!\!\not|\!CH_2\,OH & CH_3CH_2\!\!\not\!CHOH\!\not\!CH_3 & CH_3\!\!-\!\!C\!\not\!CH_3 \\
31 & 59 & HO \\
100\% & 20\% & 59 \\
 & & 100\%
\end{array}
$$

In this shorthand presentation, the bond cleaved is shown broken by a wavy line, and the ion formed and its intensity are shown above or below the horizontal straight line. This is a very useful way of summarising the main features of a spectrum.

We can summarise the application of Stevenson's Rule to aliphatic alcohols by saying that the base peak is usually formed by the loss of the largest radical attached to the α-carbon:

$$
\begin{array}{ccc}
R_3 & & R_1 \\
| & & \diagdown \\
R_1\!-\!\overset{.+}{\underset{|}{C}}\!-\!\ddot{O}H & \rightarrow & C=\overset{+}{O}H + R_3^., R_3 > R_2 \geq R_1 \\
R_2 & & \diagup R_2
\end{array}
$$

The α- and β- bonds in alcohols can also break heterolytically to give hydrocarbon ions, which can be seen in the spectra in Fig. 9.2a and 9.2b:

$$CH_3-CH_2-\overset{+\cdot}{O}H \rightarrow CH_3CH_2^+ + \,{}^{\cdot}OH$$

$$m/z\ 29$$

$$CH_3-CH_2-\overset{+\cdot}{O}H \rightarrow CH_3^+ + CH_2{=}\overset{\cdot}{O}H$$

$$CH_3CH_2CH_2-CH_2-\overset{\cdot +}{O}H \rightarrow CH_3(CH_2)_3^+ + \overset{\cdot}{O}H$$

$$m/z\ 57$$

$$CH_3CH_2CH_2-CH_2-\overset{+\cdot}{O}H \rightarrow CH_3CH_2CH_2^+ + CH_2{=}\overset{\cdot}{O}H$$

$$m/z\ 43$$

and generally

$$R-CH_2-\overset{+\cdot}{O}H \rightarrow \overset{+}{R} + CH_2{=}\overset{\cdot}{O}H$$

These hydrocarbon ions and daughter ions derived from them are usually of less RA than the $R_1R_2C{=}\overset{+}{O}H$ ions, but increase in importance in long chain, especially branched alcohols, which come to resemble the corresponding hydrocarbons (see Section 9.7).

∏   In a secondary or tertiary alcohol there might be two or three different R^+ which could be formed. Try to predict what would happen if the R groups were CH_3, CH_3CH_2, and $(CH_3)_2CH$ in the structure $R_1R_2R_3COH$.

The ion $(CH_3)_2\overset{+}{C}H$ would have the greatest intensity, followed by $CH_3CH_2^+$, then CH_3^+ very much the least. I hope you said this was because the order of stability of carbocations is tertiary > secondary ≫ primary – if so, well done! If you did not remember the stability order of carbocations, revise Section 8.2. In some alcohol spectra, the R^+ ions are as intense or more intense than the $R_1R_2C{=}\overset{+}{O}H$ ions if R is secondary or tertiary.

There is one other feature of the alcohol spectra which we have yet to explain – the loss of H_2O This gives rise to the prominent m/z 56 in butan-1-ol (Fig. 9.b(i)), but only minor peaks in ethanol

(m/z 28 in Fig. 9.2a) and the other two isomeric butanols. Since alcohols thermally dehydrate to alkenes, and the inlet systems of mass spectrometers are run at about 200 °C, we might expect some dehydration prior to ionisation. The m/z 56 ion might then arise by direct ionisation of butene formed in the inlet system.

∏ From the evidence of the (M-H_2O) peaks seen in Fig. 9.2a and 9.2b, do you think these ions have originated from alkenes formed in the heated inlet of the mass spectrometer, ie they are derived from artefacts?

The answer is no. If this was the case, we would expect *all* the spectra to show (M-H_2O) peaks about as intense as that in butan-1-ol, but they do not. Indeed, the order of ease of dehydration of alcohols is tertiary > secondary > primary, so 2-methylpropan-2-ol would be expected to show the most intense m/z 56, followed by butan-2-ol. This does not mean that thermal dehydration does not occur, but it is minimised in modern instruments by the use of inactive glass-lined inlet systems.

There is in fact a mechanism for the loss of H_2O from M^+ of alcohols which involves hydrogen transfer from the third or fourth carbon atom down the chain to the $-\overset{+\cdot}{O}H$ group. The transfer from the fourth carbon is favoured because it involves a six-membered ring intermediate, but it can also occur to a limited extent from carbon-3 in a chain. Butan-1-ol has the favoured H at carbon-4, as well as carbon-3. Butan-2-ol has a carbon-3, from which a hydrogen could be transferred, but shows little loss of H_2O in practice. The mechanisms are:

1,4-elimination

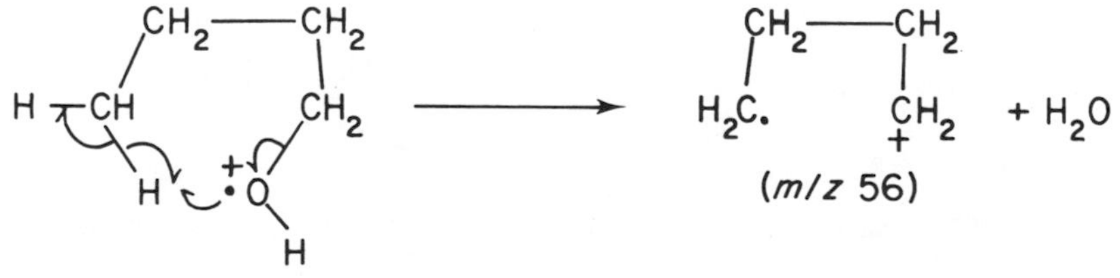

1,3-elimination

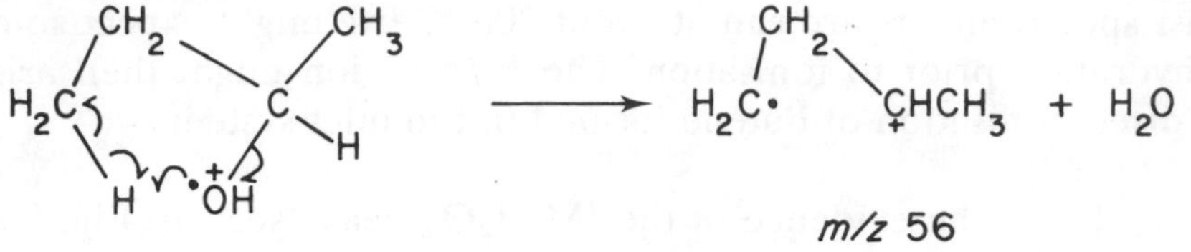

$$H_2C \xrightarrow{\quad} H_2C \cdot \quad \overset{+}{C}HCH_3 + H_2O$$

m/z 56

Since ethanol and 2-methylpropan-2-ol have neither 3 nor 4 carbons let alone hydrogens, they show little loss of H_2O.

Alcohols with *5 carbons* or more can lose $CH_2{=}CH_2 + H_2O$ in a single concerted step (as shown by a metastable ion for the loss of 46 amu all at once). The mechanism for this is:

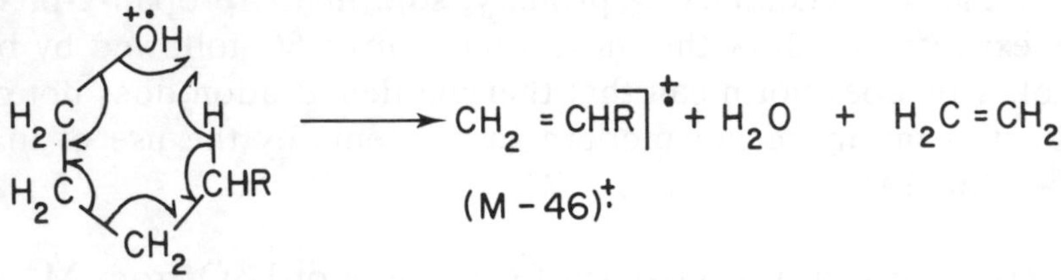

$$\xrightarrow{\quad} CH_2 = CH\overline{R}\Big]^{+\bullet} + H_2O \ + \ H_2C = CH_2$$

$$(M-46)^{+}$$

Both this and the 1,4-H_2O elimination are typical examples of an electron-impact induced hydrogen transfer process involving a six-centred ring transition state.

A similar six-centred ring process involving the loss of $CH_2{=}CH_2 + CH_3^{-}$ occurs in the fragmentation of cyclopentanols and hexanols, which behave distinctively. The fragmentation starts by α-cleavage next to the $-OH$ group but further bond reorganisation and cleavages must follow in a cyclic compound before daughter ions can be produced. For cyclohexanol the mass spectrum is shown in Fig. 9.2c and the proposed fragmentation is shown in Fig. 9.2d.

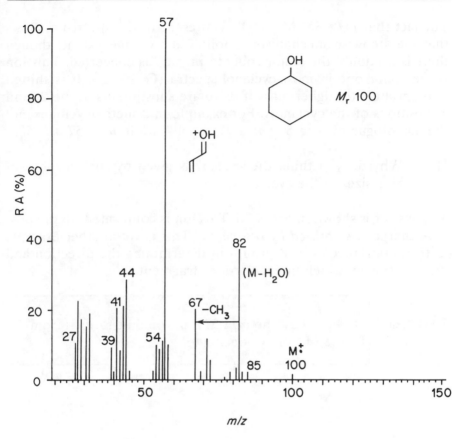

Fig. 9.2c. *Mass spectrum of cyclohexanol*

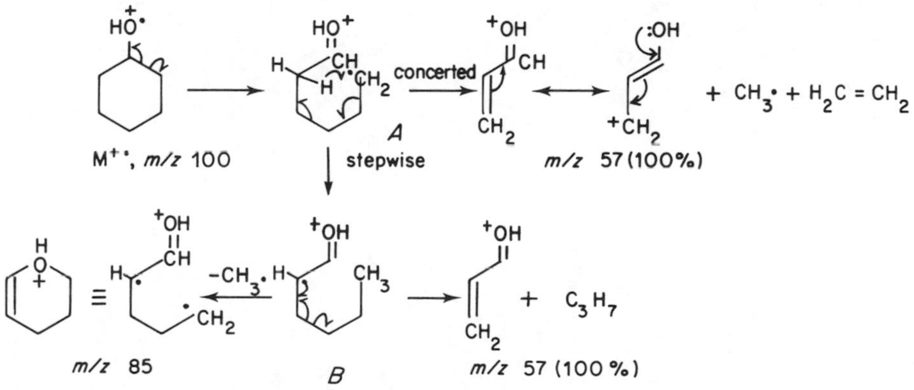

Fig. 9.2d. *Fragmentations of cyclohexanol*

The fact that m/z 85 $(M-CH_3)^+$ is present in the spectrum shows that the stepwise mechanism is followed to some extent, though there is no doubt the major route to m/z 57 is concerted. This ion is the major one in most cyclanol spectra ($C_5 \rightarrow C_8$). It is shifted appropriately to higher mass if there are substituents at the 2- and 3- positions of the cyclanols. For example in 3-methylcyclohexanol the homologue of m/z 57 (m/z 71) is 64% while m/z 57 is 35%.

∏   Why do you think the m/z 57 is given by such a range of ring sizes in the cyclanols?

The answer is shown in Fig. 9.2d. This ion is conjugated, so the positive charge is *stabilised by resonance*. This is yet another example of the importance of $+M$ groups in determining the direction and mechanisms by which molecular ions fragment.

SAQ 9.2a   Show how the m/z 57 ion would be formed from the M^+ of cyclopentanol.

SAQ 9.2b   Show how the M^{+} of 3-methylcyclohexanol can give rise to both m/z 57 and 71. Why is m/z 71 more intense than m/z 57 in this spectrum?

Another feature of cyclanol spectra is the loss of H_2O, which forms m/z 82 in Fig. 9.2c. Deuterium labelling has shown that the second hydrogen in the H_2O eliminated (the first being the —OH hydrogen) comes fairly equally from the 3- and 4- positions of the ring. Clearly this is impossible in the cyclised molecular ion, but the ring-opened M^{+} (B in Fig. 9.2d) is free to rotate about, since it is acyclic, and can bring the 3- and 4-hydrogens into close proximity to the $=\overset{+}{O}H$, just as happened in the elimination of H_2O from 1-butanol (Fig. 9.2b (i)). The fact that CH_3^{+} loss then occurs from m/z 82 supports the belief that B is the M^{+} structure involved in this process.

You will have noticed that the molecular ions are generally of low abundance in the mass spectra of saturated aliphatic alcohols. In many cases they are too low to be seen at all. There is one major exception to this rule that alcohols show weak to non-existent $M^{+\cdot}$ peaks. This is found in the spectra of aromatic alcohols which show quite intense $M^{+\cdot}$ peaks. Look at Fig. 9.2e and 9.2f now.

These are the spectra of benzyl alcohol and 2-phenylethanol. You would not mistake these for aliphatic alcohols. In benzyl alcohol $M^{+\cdot}$ is also the base peak and both show intense typically aromatic ions such as $C_6H_5^+$, $C_7H_7^+$ and $C_7H_7O^+$, m/z 77, 91 and 107. Equally they do not behave like primary aliphatic alcohols, or observe Stevenson's Rule.

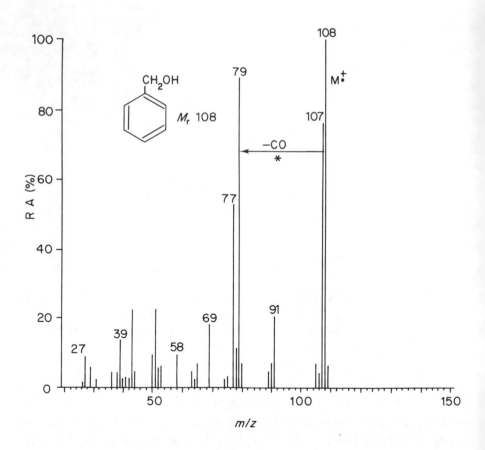

Fig. 9.2e.   *Mass spectrum of benzyl alcohol*

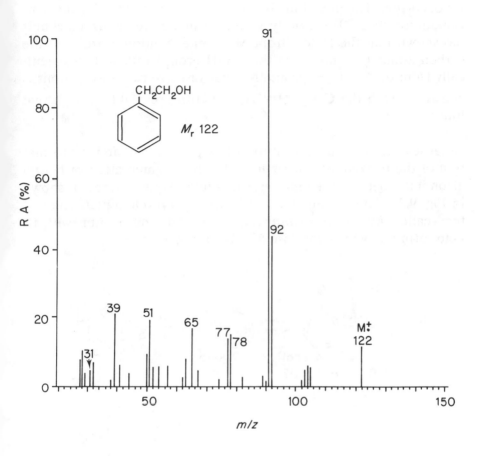

Fig. 9.2f. *Mass spectrum of 2-phenylethanol*

Π   Which fragments would you expect to be lost as radicals from the molecular ions of $C_6H_5CH_2OH$ and $C_6H_5CH_2CH_2OH$ by Stevenson's Rule?

The rule would predict loss of $C_6H_5^{\cdot}$ from benzyl alcohol and $C_6H_5CH_2^{\cdot}$ from 2-phenylethanol by α-cleavage since these are the largest radicals. Instead β-cleavage in both molecules gives $C_6H_5^+$ and $C_6H_5CH_2^+$; $CH_2=\overset{+}{O}H$ is very weak. Loss of $H^{\cdot}$ is preferred in benzyl alcohol to give m/z 107, $C_7H_7O^+$.

You might logically expect this ion to be $C_6H_5-CH=\overset{+}{O}H$, but this is not correct. The next thing it does is to lose $:C=O$ to give the benzenium ion m/z 79. Secondly a study of deuterated benzyl alcohols has shown that the H lost from $M^{\ddagger}$ comes randomly from all the carbon atoms (but not from the $-OH$ group) rather than specifically from the $-CH_2-$ group. Neither of these two observations is consistent with the $C_6H_5CH=\overset{+}{O}H$ structure, plausible though that looks!

Because of these unusual features of its breakdown, and the formation of the benzenium ion at m/z 79, the fragmentation of benzyl alcohol has been looked into very carefully. What emerges is shown in Fig. 9.2g. The ion at m/z 107 is a hydroxycycloheptatriene system, called *hydroxytropylium* (*A*). This is in equilibrium with its keto form (*B*) which can lose $:C=O$ to give m/z 79.

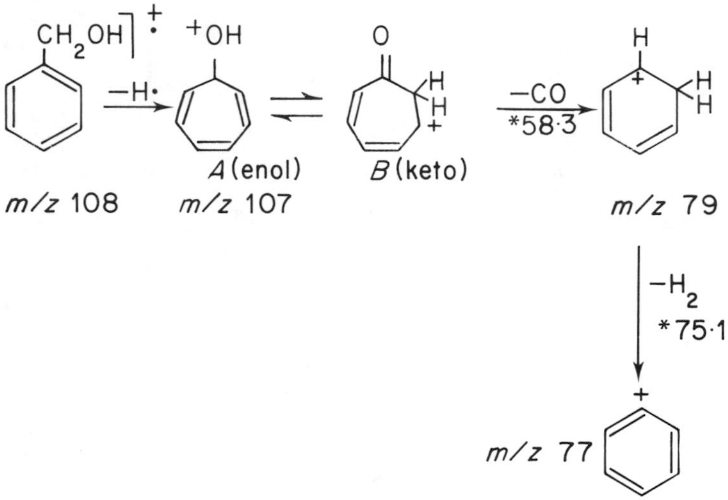

Fig. 9.2g. *Fragmentation of benzyl alcohol (* indicates a metastable ion is observed)*

Tropylium ions $RC_7H_6^+$ are now known to be very stable species, especially when R is a $+M$ group. They are formed in the mass spectra of many types of compounds in preference to more conventional structures eg when $R=H$, the 'tropylium' ion itself is formed in preference to its isomer $C_6H_5-CH_2^+$.

∏ Assuming that ion *A* in Fig. 9.2g is planar, show how it can be stabilised by resonance.

Tropylium ions have a larger number of resonance forms than other aromatic species of isomeric structure, stabilised as follows.

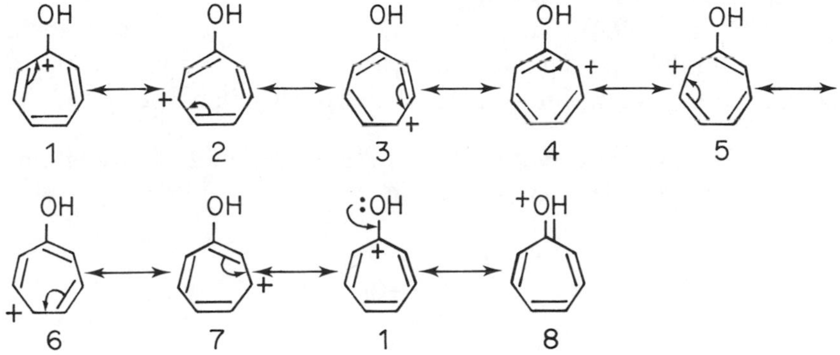

I hope you got most of these. Did you remember to use the lone pair electrons on the −OH in order to get structure 8 and explain why a +M group would further stabilise such an ion? This ion has eight resonance forms, more than you have seen in any previous example, so explaining its surprising formation and stability.

The 'tropylium' ion itself, m/z 91, has seven *identical* structures. This ion is formed very readily from benzyl and tolyl compounds and is a characteristic feature of their spectra.

∏ See if you can derive the seven structures of the tropylium ion from the formula below.

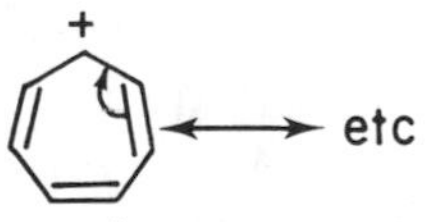

$C_7H_7^+, m/z$ 91

Your structures should be like those for hydroxytropylium, but without the −OH.

In the mass spectrum of 2-phenylethanol (Fig. 9.2f) $C_7H_7^+$ is the base peak, which metastable ions at m/z 90.0 and 67.9 show is formed by two routes, direct from M^+:

$$PhCH_2 \overset{\cdot}{\underset{}{\diagup}} CH_2 \overset{\cdot}{\underset{}{\diagdown}} \overset{\cdot +}{OH} \longrightarrow CH_2 = \overset{\cdot}{O}H \ + \ PhCH_2^+ \ \overset{-O}{\longrightarrow} \ \bigcirc\!\!\!\!\!\oplus \quad C$$

and also by loss of $H^{\cdot}$ from m/z 92, itself formed from M^+ by loss of CH_2=O (m^* 69.4). The symbol used for C ie a circle inside the seven-membered ring and the positive charge shown within that circle indicates a completely delocalised positive charge.

∏   Can you see any evidence in the spectrum of 2-phenylethanol that the M^+ in fact consists of

ions?

To answer this, you should have considered likely modes of breakdown of the tropylium ring. The (M-X) and Mass-Composition tables show that aromatic compounds break down by loss of 26 amu, HC=CH, and that benzyl and tolyl compounds show m/z 65 ions, as does this spectrum. This shows that tropylium ions typically fragment by loss of HC=CH units, formed randomly from two adjacent CH units. If the M^+ of 2-phenylethanol did this, we would expect to get a peak at m/z 96. It is entirely absent. There is no evidence that M^+ of this compound behaves like a tropylium ion. It is believed that such ions are formed *following* fragmentation although it is impossible to *prove* the structure of any M^+ ion. We cannot reach into the vacuum system of a mass spectrometer and take these out to see!

This may seem a bit nebulous and frustrating to you, but on the other hand it does mean that we can make a number of postulates about the structure of M^+ ions. We then select those which help us to explain the fragments observed. *Different* M^+ structures can be and are used as parent ions in the various fragmentation pathways of all but the simplest molecules. You have seen this on a number of occasions now and you should not let it worry you. Ionisation creates a variety of M^+ ions because the ionising electron may approach the molecule at any point and in more complicated structures can remove one of several electrons. Thus there will be several potential M^+ structures.

A good case in point is the formation of the m/z 92 ion in the spectrum of 2-phenylethanol. In order to explain this it is best to assume the benzene ring is ionised, and the $-OH$ hydrogen is transferred to it. This can be easily shown by determining the mass spectrum of $C_6H_5CH_2CH_2OD$, when the m/z 92 ion moves to m/z 93 as it should if D is transferred (Fig. 9.2h)

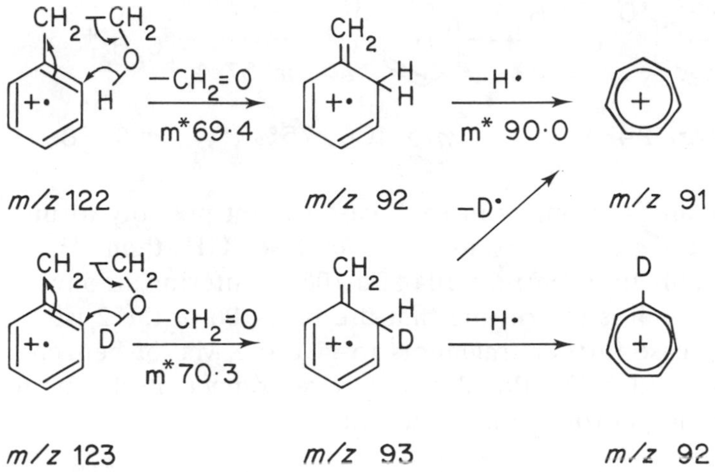

Fig. 9.2h. *Fragmentation of 2-phenylethanol and 2-phenylethanol-O-D*

Notice that m/z 92 is still found in the spectrum of the deuterated compound, but is is now $D-C_7H_6^+$, since m/z 93 can lose either H or D to form the tropylium ion. This hydrogen atom transfer is yet another example of a six-membered ring transition state.

Two other very interesting examples of this sort of rearrangement occur in 2-methylbenzyl alcohol and 2-hydroxybenzyl alcohol, causing the loss of water directly from the M^+. This direct loss is not found in the corresponding 3- and 4-methyl and hydroxy compounds.

What happens is:

m/z 122   m/z 104,100%

m/z 124   m/z 106, 75%   m/z 78,100%

The 3- and 4- isomers in each case cannot possibly form such six-membered transition states, so they lose ˙OH, then ˙H more conventionally to reach m/z 104 and 106 in differing amounts from the 2-isomers. It is interesting that the m/z 106 of the 2-isomer in the second case further fragments to give the M^{+} of benzene, $C_6H_6^{+}$ by loss of :C=O˙. Possibly it has rearranged itself to the M^{+} of tropolone in order to facilitate this:

m/z 106   tropolone   m/z 78

Such distinct features of ortho compounds (1,2-disubstituted aromatics) caused by the availability of cyclic fragmentation routes are quite common and are called *ortho effects*. They often distinguish this substitution pattern and you should look out for them.

Now try the SAQ 9.2c and 9.2d as a means of revision.

SAQ 9.2c
Fig. 9.2i shows the mass spectrum of an alcohol, *Unknown 6*. Interpret this spectrum and identify the alcohol. Note that m/z 31, not shown in Fig. 9.2i, is actually 45%.

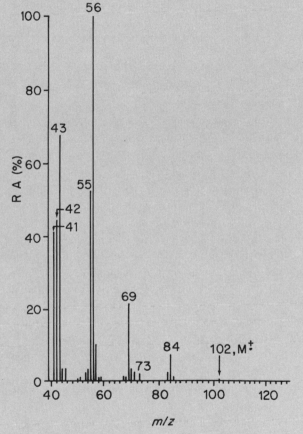

Fig. 9.2i. *Mass spectrum of* Unknown 6

SAQ 9.2c

SAQ 9.2d   Fig. 9.2j shows the mass spectrum of an alcohol, *Unknown 7*. Interpret this spectrum and identify the alcohol.

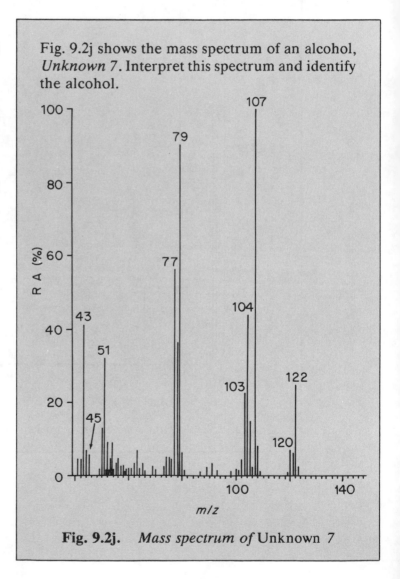

Fig. 9.2j.   *Mass spectrum of* Unknown 7

SAQ 9.2d

Summary

In this section on alcohols, we have considered in some detail the distinguishing features of acyclic, cyclic aliphatic, and aromatic alcohols. In order to understand how they fragment, we have had to introduce some new concepts such as Stevenson's Rule, tropylium ions and six-centered rearrangement processes which you might have found heavy going, especially if you are not well-versed in mechanistic organic chemistry. You are probably thinking of giving up if it is going to continue like this! I wouldn't blame you if you had found Sections 9.2 hard work, but you have now broken the back of the theory needed to understand how molecules fragment.

Objectives

After completing this Section, you should be able to:

- recognise the main features of the mass spectrum of an aliphatic acyclic alcohol;

- explain how α- and β-cleavages work to produce the main ions in aliphatic acyclic alcohol spectra;

- quote and explain Stevenson's Rule;

- use Stevenson's Rule to predict and explain the ions formed in the mass spectra of aliphatic acyclic alcohols;

- recognize and use various shorthand notations of fragmentations;

- predict and explain the loss of H_2O from acyclic aliphatic alcohols;

- predict and explain the loss of $(H_2O^{\cdot} + H_2C{=}CH_2)$ from acyclic aliphatic alcohols having five or more carbon atoms;

- recognise cyclanol mass spectra by the presence of $CH_2{=}CH{-}CH{=}\overset{+}{O}H$, m/z 57, or homologous ions;

- explain the results of deuterium labelling experiments and appreciate their importance in verifying fragmentation mechanisms;

- recognize the distinctive features of an aromatic alcohol eg much more intense M^{+}, presence of tropylium ions, hydrogen transfer processes, and suppression of fragmentations predicted by Stevenson's Rule;

- explain why tropylium ions are stabilised and recognise them in unknown spectra;

- appreciate that a variety of M^{+} may exist for a molecule, and select those which are useful for explaining the subsequent daughter ions observed;

- recognise and write mechanisms for the elimination of H_2O in the mass spectra of 2-substituted benzyl alcohols, and appreciate that such ortho effects generally are useful in distinguishing 1,2-disubstituted aromatic compounds from their 1,3- and 1,4-disubstituted isomers.

9.3. FRAGMENTATIONS OF ETHERS

Ethers and alcohols are isomeric compounds eg diethyl ether, $CH_3CH_2OCH_2CH_3$ and butanol, $CH_3CH_2CH_2CH_2OH$; benzyl alcohol $C_6H_5CH_2OH$ and methoxybenzene $C_6H_5OCH_3$, so it is interesting to ask the question, can mass spectrometry readily distinguish between them?

Firstly ethers tend to show more intense $M^{\overset{+}{\cdot}}$ than the isomeric alcohols, though they are still rather weak. They fragment by primary cleavages similar to those occurring in alcohols, ie

to give oxonium ions A or B of m/z 45, 59, 73, 87 ... , which are isomeric with the $R_1R_2C\overset{+}{=}OH$ ions found in alcohol spectra.

However, ethers show a further easy fragmentation which is *not* found in most isomeric alcohols. This is the loss of a neutral alkene derived from the remaining alkyl substituent in A or B:

In this process transfer of a β-H occurs, hence for this to be observed R^1 (or R) must have at least one carbon atom with a hydrogen on the β-carbon. You can see that in general, the final products of this cascade are still members of the series m/z 31, 45, 59, 73, 87 ....

If this process were to occur in the oxonium ions derived from alcohols the result would be quite different:

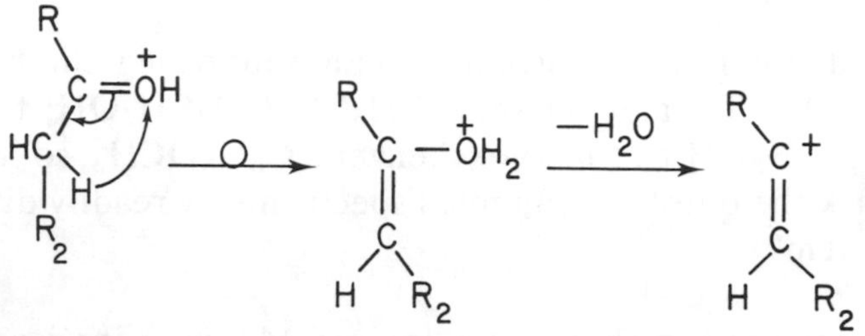

$$C \qquad\qquad D \qquad\qquad E$$

$C \rightarrow D$ is a rearrangement, presumably the likely fragmentation of D would be by loss of H_2O to give E.

To illustrate these features, let us consider the spectrum of ethyl 1-methylpropyl ether, $CH_3CH_2OCH(CH_3)CH_2CH_3$ (Fig. 9.3a). Note that this compound is isomeric with hexanol (Unknown 6, Fig. 9.2i). In accordance with Stevenson's Rule the α-cleavage of the largest radical of the ether, CH_3CH_2; occurs preferentially to give m/z 73, Fig. 9.3b (i). The alternative $CH_3^\bullet$ loss from the 1-methylpropyl group to give m/z 87 occurs to a minor extent (4%), Fig. 9.3b (ii).

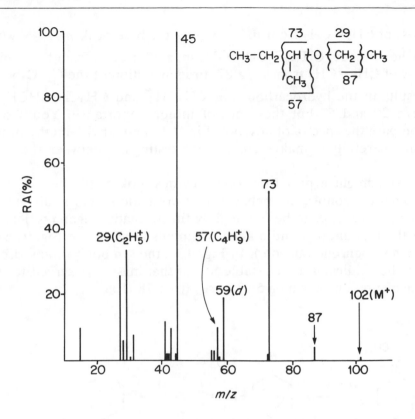

Fig. 9.3a.   *Mass spectrum of ethyl 1-methylpropyl ether*

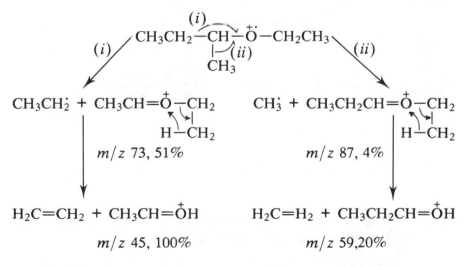

Fig. 9.3b.   *Fragmentations of ethyl 1-methylpropyl ether*

Loss of $CH_2{=}CH_2$ from m/z 73 gives the base peak m/z 45, while its homologue, $CH_3CH_2CH{=}\overset{+}{O}H$, m/z 59, results from the similar loss of $CH_2{=}CH_2$ from m/z 87. Fragmentation of the C—O bonds results in the hydrocarbon ions $CH_3CH_2^+$ and $CH_3CH_2\overset{+}{C}HCH_3$ at m/z 29 and 57 but these are of minor importance. You should compare the spectra of hexanol (Fig. 9.2i) and ethyl 1-methylpropyl ether carefully to make sure you can distinguish between them.

As you might expect from our previous look at the behaviour of aromatic alcohols, the behaviour of aromatic ethers is distinctive. The charge tends to be retained by the aromatic fragments, leading to the loss of alkyl and alkene fragments as shown in Fig. 9.3c for methoxybenzene (anisole). In Fig. 9.3c, the symbol * followed by a number indicates a metastable ion of that mass is observed for the fragmentation shown eg 56.3 arises from $78^2/108$.

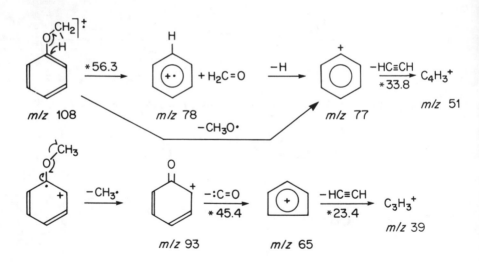

Fig. 9.3c.   *Fragmentations of methoxybenzene*

Π   In Fig. 9.3c, the mechanism of the loss of $CH_2{=}O$ has been
        left rather vague. Can you supply it?

If you started off with the positive charge on the O atom:

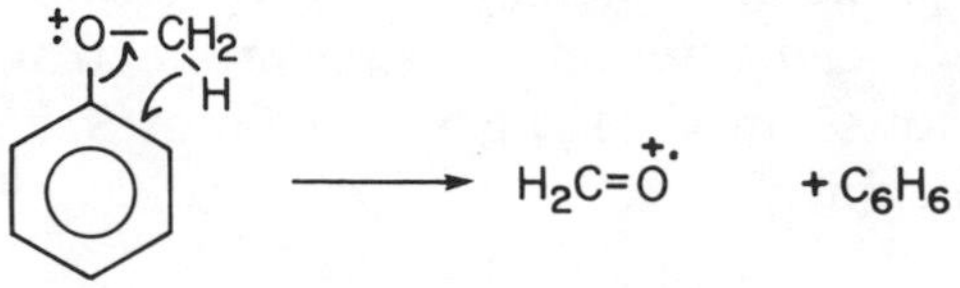

then this leads to *benzene* as the *neutral* species. I hope you realised this! You would be right if you started with the positive charge on the 2-position of the benzene ring ie

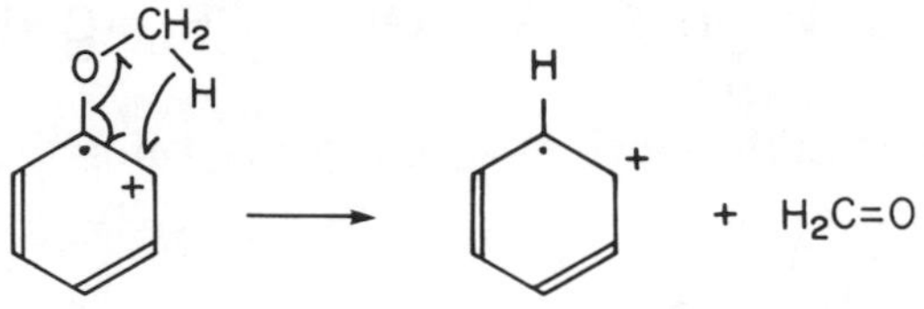

This emphasises that the fragmentations of aromatic ethers tend to start from M⁺ where the positive charge is on the aromatic ring, and explains why they differ from purely aliphatic ethers.

∏ Can you explain the stability of the $C_5H_5^+$ ion at m/z 65 in the mass spectrum of methoxybenzene?

This ion is effectively stabilised by resonance, as follows:

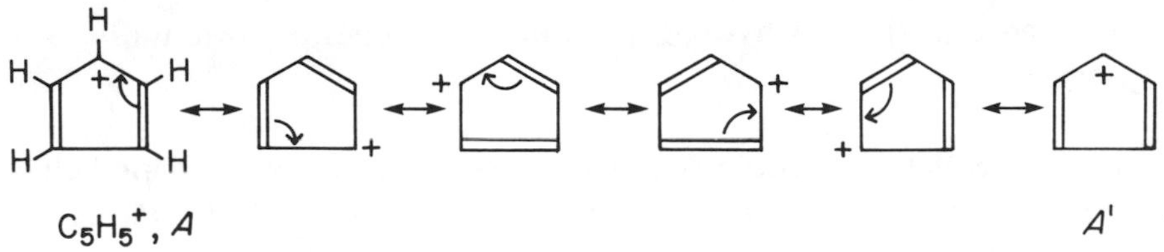

$C_5H_5^+$, *A*   *A'*

Fig. 9.3d. *Resonance forms of* $C_5H_5^+$

There are five equivalent forms here ($A = A'$), making for a very stable structure (Section 8.2). In case you are wondering how this ion can be so stabilised, when the apparently similar $C_6H_5^+$ can't, notice that in $C_5H_5^+$ there are hydrogens at each corner occupying positions in the plane of the ring. The positive charge is in an empty p-orbital where it can interact with the π electrons of the double bonds. This is not so in $C_6H_5^+$. Look again at Section 8.2.5, if you are unsure about this.

Longer chain ArOR compounds (R = CH_3CH_2 etc) show similar fragmentations plus one other, the elimination of an alkene from the aliphatic group and the formation of the molecular ion of the corresponding phenol:

$$Ar\overset{+\cdot}{O}CH_2CH_2R^1 \rightarrow Ar\overset{+\cdot}{O}H + CH_2{=}CHR^1$$

There are two possible mechanisms for this, one 4-centred and the other 6-centred:

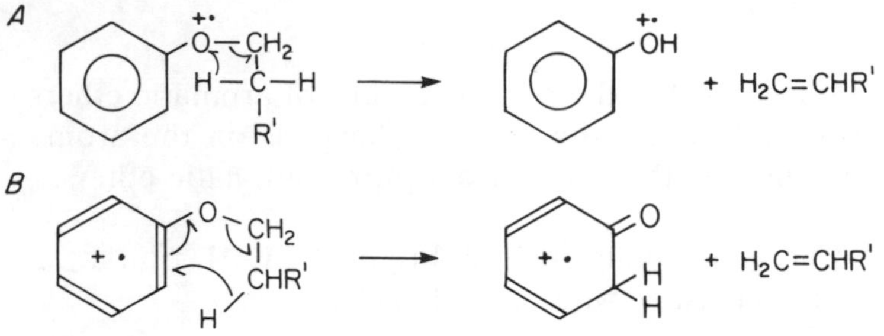

Since the subsequent fragmentations of the daughter ions closely resemble those of the corresponding phenols, (Section 9.4), it is believed that the 1,4-hydrogen transfer of scheme *A* actually takes place.

Now try SAQ 9.3a and SAQ 9.3b to see if you have grasped these fragmentations. Read the answers to SAQ 9.3a and 9.3b before you go on to Section 9.4.

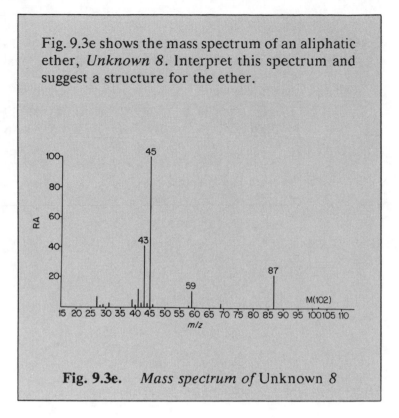

SAQ 9.3a   Fig. 9.3e shows the mass spectrum of an aliphatic ether, *Unknown 8*. Interpret this spectrum and suggest a structure for the ether.

Fig. 9.3e. *Mass spectrum of* Unknown *8*

SAQ 9.3b

Unknown 9 is an aromatic ether of M_r 136. The mass spectrum shows the following peaks and intensities:

m/z	136	95	94	77	66	51	43	41	39
RA(%)	25	6	100	8	7	6	5	5	6

Suggest a structure for *Unknown* 9 compatible with these data.

Summary

Although ethers and alcohols are isomeric compounds their mass spectra usually allow them to be distinguished. Ethers tend to have more intense molecular ions than alcohols, although they are still quite weak.

Aliphatic ethers fragment by losing alkyl groups from the α-carbon atoms according to Stevenson's Rule giving rise to a series of oxonium ions of m/z 45, 59, 73, 87 ... which are isomeric with the $R_1R_2C=\overset{+}{O}H$ found in the spectra of alcohols.

If these oxonium ions have a hydrogen on a β-carbon atom, they fragment again by loss of a neutral alkene:

$$H_2C=\overset{+}{O}-CH_2 \rightarrow H_2C=\overset{+}{O}H + H_2C=CHR$$
$$H-C-R \quad m/z\ 31$$
$$H$$

This fragmentation distinguishes them from most alcohols and explains why m/z 31 is also found in ether spectra as well as primary alcohols.

In the fragmentation of aromatic ethers, the charge is retained by the aromatic fragments, leading to the loss of alkyl and alkene neutral species. Many aryl alkyl ethers eliminate the alkyl group as an alkene and form the $M^{\ddot{+}}$ of the corresponding phenol.

Aryl methoxy ethers have a different, unique fragmentation. They eliminate the $-OCH_3$ group as $CH_2=O$ and transfer the other hydrogen to the aromatic ring forming the corresponding hydrocarbon $M^{\ddot{+}}$.

Objectives

Now that you have finished Section 9.3 you should now be able to:

● distinguish most ethers from isomeric alcohols by looking for the loss of alkenes from the $(M-R)^+$ species;

● write the fragmentation of M^+ of an ether, by loss of the alkyl group(s) from the α-carbon atoms to give fairly stable oxonium ions in the series m/z 45, 59, 73, 87 ... ;

● predict which alkyl group of an aliphatic ether would be most readily lost as a radical by applying Stevensons's Rule and therefore which series of daughter ions would be the most intense;

● recognise that the (M-R-alkene) fragmentation sequence can result in the formation of m/z 31 from an ether as well as from a primary alcohol and therefore consider the possibility that a compound having an intense m/z 31 might also be an ether;

● recognise an aromatic ether by its intense $(M-R)^+$ and (M-alkene)$^+$ ions;

● describe why aryl methoxy ethers lose $CH_2=O$ from M^+ and use this feature to identify such compounds;

● suggest structures for simple aliphatic and aromatic ethers from their mass spectra.

9.4. FRAGMENTATIONS OF HYDROXYBENZENES (PHENOLS)

Like many aromatic compounds, hydroxybenzenes usually have intense molecular ions (see Fig. 9.4a). They do show some unexpected fragmentations though.

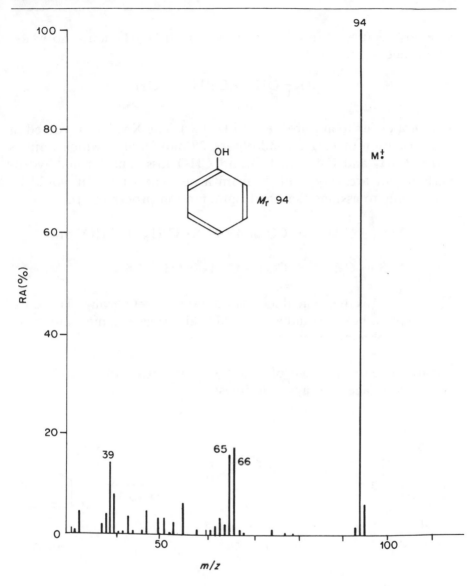

Fig. 9.4a. *Mass spectrum of hydroxybenezene (phenol)*

∏   Have a closer look at Fig. 9.4a, the mass spectrum of hydroxybenzene itself. What strikes you as peculiar about it, and what neutral fragments do you think are being released when m/z 66 and 65 are formed?

The odd feature of this spectrum is that the $C_6H_5^+$ ion is so weak. The process

$$C_6H_5\!-\!\overset{+\cdot}{\underset{\curvearrowleft}{O}}H \rightarrow C_6H_5^+ + \cdot OH$$

does not occur to any great extent (<1%). The X values involved in the formation of m/z 66 and 65 are 28 and 29 amu which implies either C_2H_4 and C_2H_5, or CO and CHO loss. The former hydrocarbon pair are very unlikely from a benzene ring so it would be reasonable to assume that the fragmentation processes are:

$$M^+_\cdot \rightarrow C_5H_6^+_\cdot + CO \text{ and } M^+_\cdot \rightarrow C_5H_5^+ + CHO^\cdot \text{ or}$$

$$M^+_\cdot \rightarrow C_5H_6^+_\cdot + CO \rightarrow C_5H_5^+ + H^\cdot$$

∏   Hydroxybenzene does not contain a C=O group. Try to explain how it could arise in $M^+_\cdot$, and suggest a mechanism for its elimination.

Clearly a rearrangement of hydrogen away from the —OH group must take place, perhaps as follows:

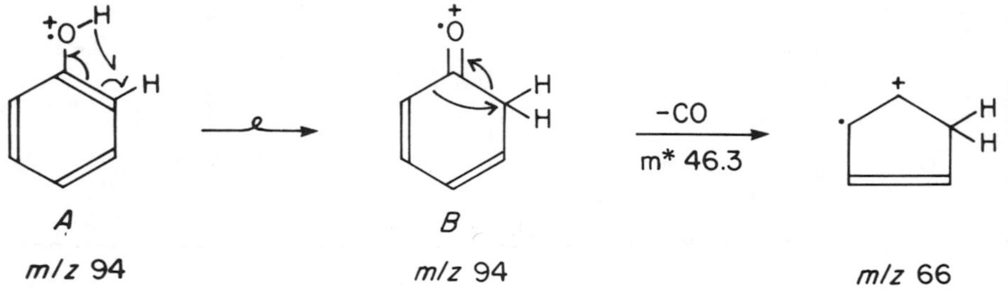

Ion *B* is the keto form of hydroxybenzene. *A* and *B* are related as enol ⇌ keto tautomers. It is interesting to note that thiophenol and aminobenzene (aniline) exhibit similar types of fragmentations, showing $(M - CS)^+$, $(M - CSH)^+_\cdot$ and $(M - HCN)^+$, $(M - CNH_2)^+_\cdot$ ions, respectively.

Most substituted hydroxybenzenes show fairly intense $M^+_\cdot$ and lose CO and CHO. Alkyl substituted compounds show an additional characteristic feature. Consider the spectra of 2-, 3- and 4-methylhydroxybenzene, Fig. 9.4b, c and d, the 'cresols'. All three show intense $(M - H)^+$ ions. You might think these are $CH_3C_6H_4O^+$ isomers, formed by loss of the hydrogen from the —OH group, but D-labelling of this reveals that it is *not* lost in this process.

∏   Suggest which hydrogen is lost from say 4-methylhydroxy-benzene and explain why this is preferred over loss of the —OH hydrogen.

It you thought of the formation of a hydroxytropylium ion such as *D*, well done!

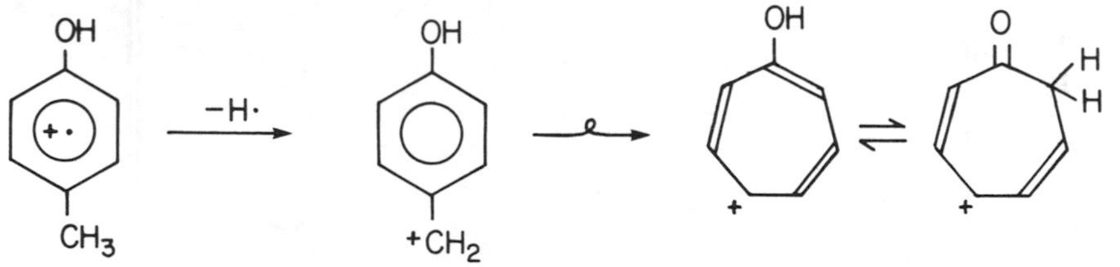

OH		OH		OH	O

m/z 108   *m/z* 107   *D*   *E*

C

Tropylium ions such as *D* would be highly stabilised by resonance involving the —OH group as well as the ring carbons, as described in Section 9.2. Refer back to Fig. 9.2g now and you will see that *D* and *E* are *identical* with the ions *A* and *B* generated from benzyl alcohol. Not surprisingly, the next fragmentation is by loss of CO to *m/z* 79. However, the spectra of the various methylhydroxyben-zenes and benzyl alcohol are not identical as can be seen from Fig. 9.2g and 9.4b c and d. This is because the *m/z* 107 formed from the four isomers will have differing amounts of excess energy and will fragment to differing degrees.

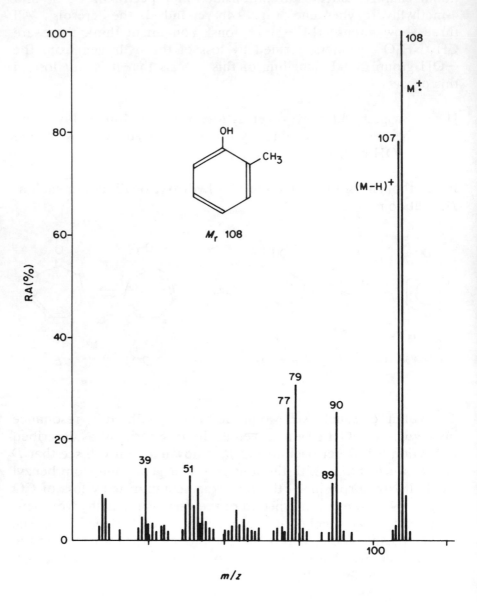

Fig. 9.4b.   *2-Methylhydroxybenzene*

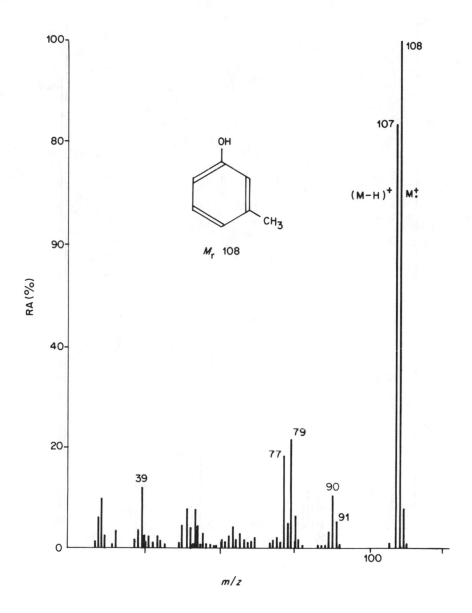

Fig. 9.4c. *3-Methylhydroxybenzene*

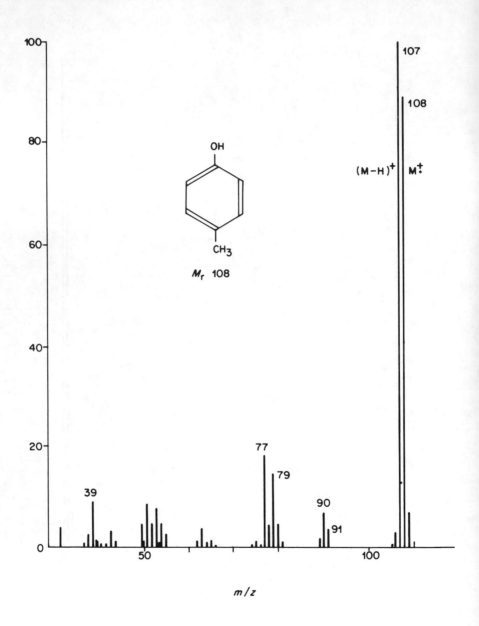

Fig. 9.4d.   *4-Methylhydroxybenzene*

2-Methylhydroxybenzenes show an ortho effect which causes the elimination of H_2O:

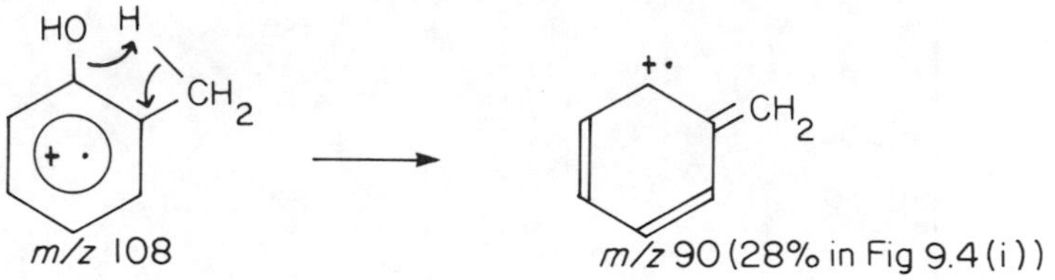

m/z 108   *m/z* 90 (28% in Fig 9.4 (i))

while hydroxybenzenes with longer saturated carbon chains undergo both *benzylic cleavage*:

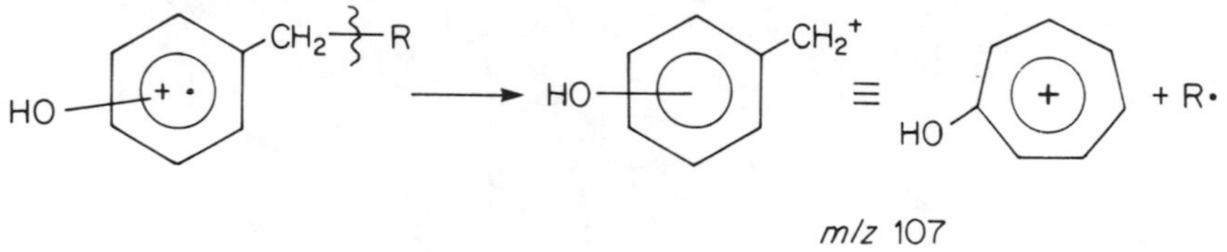

m/z 107

and a *six-centred hydrogen transfer process* involving a γ-hydrogen from the alkyl chain, if it is long enough to have one:

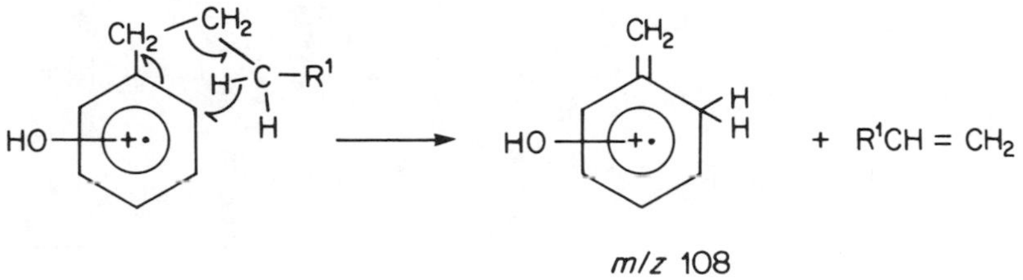

m/z 108

This means that both *m/z* 107 and 108 may be present in the spectra of such compounds.

Now try SAQ 9.4a to see if you can identify an unknown phenol from its spectrum.

SAQ 9.4a

Fig. 9.4e shows the mass spectrum of a phenol, *Unknown 10*. Interpret this spectrum and suggest a structure for *Unknown 10*.

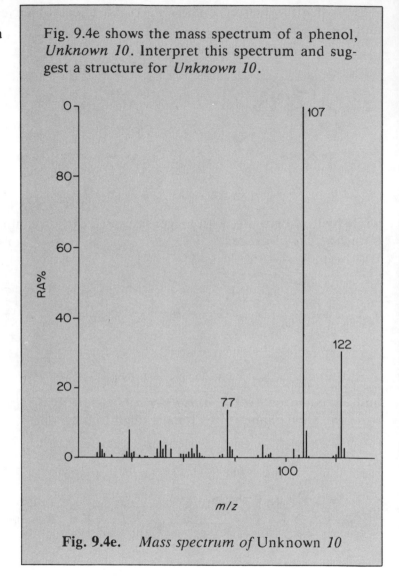

Fig. 9.4e.   *Mass spectrum of* Unknown *10*

SAQ 9.4a

Summary

Phenols like most aromatic compounds have intense molecular ions. They do not usually fragment by the loss of either the —OH hydrogen or the —OH group itself, but by loss of the neutral molecule CO and the radical CHO.

Methyl substituted phenols show a further characteristic feature. They lose α-hydrogen atoms from the methyl group to generate hydroxytropylium ions, which then fragment by loss of CO or CHO.

Alkyl substituted phenols also from tropylium ions but by cleavage of the α-β bond in the alkyl chain with loss of R· instead of H· (Stevenson's Rule). This is called benzylic cleavage.

Longer alkyl chains having γ-hydrogens eliminate an alkene, with hydrogen transfer to the aromatic ring, giving m/z 108. This means that both m/z 107 and 108 may be present in alkyl phenol spectra.

2-Alkyl substituted phenols will often show an (M-H$_2$O) ion due to an ortho effect, which helps to distinguish them from the 3- and 4-alkyl isomers.

Objectives

You should now be able to:

- recognise the characteristic features of the mass spectrum of a simple phenol;

- explain why phenols typically fragment by loss of CO and CHO rather than OH;

- explain why methyl phenols fragment by loss of H atoms followed by loss of CO and CHO;

- recognise alkyl phenols by the formation of m/z 107 and m/z 108 ions;

- distinguish 2-substituted phenols by the loss of H_2O from $M^{\ddagger}$;

- identify simple unknown phenols from their mass spectra.

9.5.   FRAGMENTATIONS OF CARBONYL COMPOUNDS

The principle cleavage of carbonyl compound is α- to the C=O group, which in most $M^{\ddagger}$ will carry the positive charge, as described already in Sections 8.2 and 9.1:

$$R - \overset{\overset{+}{\underset{\|}{O^{\bullet}}}}{C} \wedge X \longrightarrow R - C \equiv \overset{+}{O} + X\cdot$$

The formation of acylium ions by loss of $H^{\cdot}$, $R^{1\cdot}$, $R^{1}O^{\cdot}$, $HO^{\cdot}$ and $H_2N^{\cdot}$ is an important diagnostic feature of the mass spectra of aldehydes, ketones, esters, acids and amides respectively. There is one other feature which carbonyl compounds have in common which is very useful in identifying them. This is the concerted loss of an alkene from the R group, with H transfer to the carbonyl oxygen. This rearrangement is almost unique to mass spectrometry and is the one fragmentation which is a *named* reaction – the McLafferty Rearrangement. We consider it next.

9.5.1. The McLafferty Rearrangement

In the late 1950's McLafferty was studying the mass spectra of a series of aliphatic methyl esters, in fact all the methyl esters he could get his hands on from simple acids such as ethanoic to long chain fatty acids such as tetraeicosanoic acid, $C_{24}H_{49}COOH$. To his surprise he discovered that from C_6 acids onwards, the base peak of the mass spectrum of each of the methyl esters was always the same ion, m/z 74. A typical example of what he found is shown in Fig. 9.5a.

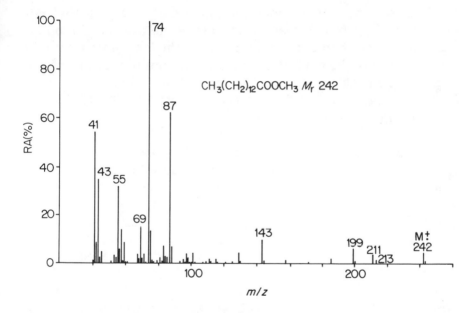

Fig. 9.5a.   *Mass spectrum of methyl tetradecanoate*

∏   What is so surprising about these observations?

Three things should surprise you. One is that such large esters as methyl tetradecanoate would give such a relatively low mass ion as base peak. Secondly that the base peak is of even mass. Most importantly, as you move up a homologous series of compounds you would expect the same fragmentation to occur and to give rise to base peak moving up by a —CH_2— unit each time. McLafferty was able to show that the formula of his m/z 74 was $C_3H_6O_2$ and he postulated the following mechanism to account for its formation:

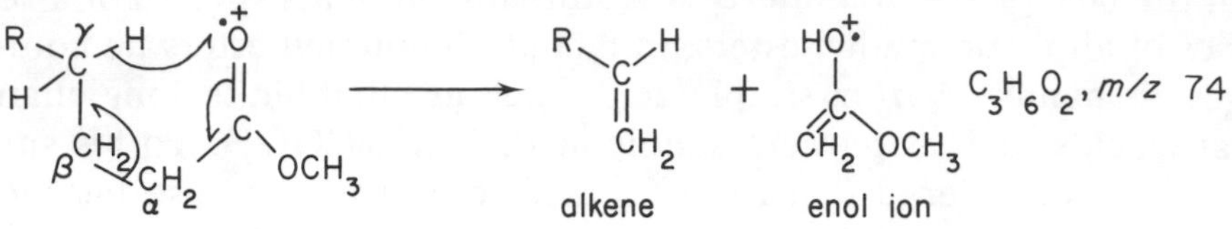

Fig. 9.5b. *McLafferty rearrangement of methyl esters*

You will see that this mechanism can occur in *any* methyl ester possessing a chain of at least three carbon atoms. The compounds McLafferty was examining only differed in the R group. This is eliminated as part of the neutral alkene so variations in the mass of R do not show up in the spectrum. The m/z 74 ion, which is the molecular ion of the enol form of methyl ethanote CH_3COOCH_3, *is the same for all of them.*

A great deal of work has been done on this rearrangement (which I will now call the 'McL' for short) and the following points have emerged;

(*a*) It is very regiospecific. Only the α-, β- and γ-carbons are involved and the hydrogens transferred *must* be from the γ-carbon. This means that the 'McL' process is strictly six-centred.

(*b*) A very wide variety of C=O compounds with a γ-H will undergo 'McL' rearrangement. It is a most useful diagnostic tool for them.

(*c*) Other $-Y=X$ groups such as $-S=O$, $-P=O$, $C=C-C=N -$ which ionise to $-Y=X^{+\cdot}$ show 'McL' rearrangements. It is not restricted to C=O compounds.

(*d*) The α-, β- and γ-atoms need not be carbons. They can be various combinations of C, O, N and S, so long as there is a γ-H present to transfer. For example, an ethyl or higher ester can rearrange using the alcohol moiety:

In such a molecule, the α-carbon is replaced by an oxygen but the 'McL' still works, forming the $M^{\ddagger}$ of the corresponding acid and releasing ethene. What this amounts to is provided a six-membered ring transition state can be formed easily using the flexibility of the side chain(s) a 'McL' rearrangement is likely to be observed.

(*e*) Since the eliminated group is a neutral molecule (usually an alkene) 'McL' rearrangements give rise to *even* mass ions from C,H,O compounds ie to molecular ions of a smaller molecule. Even-mass ions are unusual in mass spectra of C,H,O compounds, so their presence is usually distinctive even if they are weak. In the case of odd-mass compounds eg amides, the 'McL' rearrangement ions will stand out because they will have *odd* m/z values.

(*f*) H atoms on double bonds are not transferred in the 'McL' rearrangement eg compounds with side chains such as

$$\overset{\gamma}{RCH}=\overset{\beta}{CH}-\overset{\alpha}{CH_2}-CO; \quad \overset{\gamma}{R^1CH}=\overset{\beta}{CHCH_2}\overset{\alpha}{CH_2CO};$$

$$\overset{\gamma}{RCH}=\overset{\beta}{CHOCO} \text{ or } \overset{\gamma}{R^1CH}=\overset{\beta}{CHCH_2}\overset{\alpha}{OCO} \text{ do not rearrange.}$$

(*g*) The 'McL' rearrangement still works if there are a variety of substituents on the α-, β- or γ-atoms.

These points are summarised in Fig. 9.5c as far as C=O compounds are concerned.

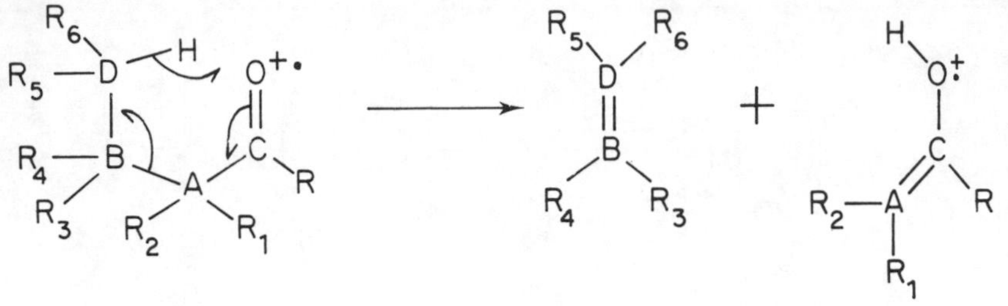

Fig. 9.5c.   *General representation of the 'McLafferty'
rearrangement*

In Fig. 9.5c R will be a typical group found in a carbonyl compound
eg R_1–R_6 can be a variety of alkyl, aryl, halo, alkoxy, aryloxy and
other groups, and A, B and D could be combinations of carbon,
oxygen, nitrogen or sulphur. It is hard to conceive of a system where
at least one of A, B or D is not carbon (most likely two of them will
be) but you would be wise not to assume too much or the order
in which these come in the A–B–D chain. This really is a very
general process! So long as double bonds can form B=D and A=C
it can work. Fig. 9.5d summarises the common 'McL' peaks found
in the spectra of carbonyl compounds.

Compound Type	Substituent R	'McL' Peak (m/z)	Structure*
Aldehyde	H	44	$CH_2{=}C\begin{smallmatrix}\dot{\overset{+}{O}}H\\ \\H\end{smallmatrix}$
Methyl ketone	CH_3	58	$CH_2{=}C\begin{smallmatrix}\dot{\overset{+}{O}}H\\ \\CH_3\end{smallmatrix}$
Amide	H_2N	59	$CH_2{=}C\begin{smallmatrix}\dot{\overset{+}{O}}H\\ \\NH_2\end{smallmatrix}$
Acid	HO	60	$CH_2{=}C\begin{smallmatrix}\dot{\overset{+}{O}}H\\ \\OH\end{smallmatrix}$
Ethyl ketone	CH_3CH_2	72	$CH_2{=}C\begin{smallmatrix}\dot{\overset{+}{O}}H\\ \\CH_2CH_3\end{smallmatrix}$
Methyl ester	CH_3O	74	$CH_2{=}C\begin{smallmatrix}\dot{\overset{+}{O}}H\\ \\OCH_3\end{smallmatrix}$
Propyl ketone	$CH_3CH_2CH_2$	86	$CH_2{=}C\begin{smallmatrix}\dot{\overset{+}{O}}H\\ \\CH_2CH_2CH_3\end{smallmatrix}$
		58**	$CH_2{=}C\begin{smallmatrix}\dot{\overset{+}{O}}H_2\\ \\CH_2\end{smallmatrix}$
Ethyl ester	CH_3CH_2O	88	$CH_2{=}C\begin{smallmatrix}\dot{\overset{+}{O}}H\\ \\OCH_2CH_3\end{smallmatrix}$

Compound Type	Substituent R	'McL' Peak (m/z)	Structure*
		$(M-CH_2{=}CH_2)$***	$R-C{\overset{\overset{\cdot+}{O}H}{<}}_{O}$
Phenyl ketone	C_6H_5	120	$CH_2{=}C{\overset{\overset{\cdot+}{O}H}{<}}_{C_6H_5}$
Phenyl ester	C_6H_5O	136	$CH_2{=}C{\overset{\overset{\cdot+}{O}H}{<}}_{OC_6H_5}$

Fig. 9.5d.   *Table of 'McL' peaks commonly observed in the spectra of carbonyl compounds*

* The $H_2C{=}$ group in any of the ions listed in this column may be substituted either singly or doubly eg $CH_3CH{=}$ ions would all be 14 amu higher.

** This ion arises because the propyl group itself has a γ-H, and the 'McL' rearrangement can occur *twice* consecutively.

*** This ion arises because the 'McL' rearrangement can utilise γ-H from either the RCO *or* the OCH_2CH_3 giving 'McL' ions of different structures and/or masses.

Now try the examples in SAQ 9.5a. You should not go on until you have attempted these and satisfied yourself that you can identify the relevant γ-H atoms in a molecule and work out the 'McL' ion which will result from the rearrangement.

SAQ 9.5a In which of the following compounds would you expect to see a 'McL' peak? Give its m/z and structure.

(*i*) $CH_3COCH_2CH_3$;

(*ii*) $CH_3COCH_2CH_2CH_3$;

(*iii*) $CH_3COCH(CH_3)_2$;

(*iv*) $(CH_3)_2CHCH_2CHO$;

(*v*) $CH_3CH_2COOCH_2CH_3$;

(*vi*) $CH_3CH_2CH_2CONHCH_3$;

(*vii*) $CH_3CH(CH_3)CH_2 CH(CH_3)COOH$;

(*viii*) $CH_3CH_2CH(CH_3)COC_6H_5$;

(*ix*) $CH_3CH=CHCH_2COCH_2CH_3$;

(*x*)
$$(C_6H_5)_2 \overset{\overset{\textstyle O}{\|}}{P} .SCH_2CH_3$$

SAQ 9.5a

If you are now happy you can predict and explain simple 'McL' rearrangements, we will go to explain footnote ** in Fig. 9.5d. This states the ion m/z 86 from a propyl ketone can rearrange again by a 'McL' process to eliminate a second molecule of alkene (ethene in this case) to give another even daughter ion, m/z 58. This is called a *Double McLafferty Rearrangement*. Dipropyl ketone (4-heptanone) is the simplest ketone in which this can occur because both alkyl groups have γ-H atoms:

$$\overset{\gamma \quad \beta \quad \alpha \qquad \alpha \quad \beta \quad \gamma}{CH_3CH_2CH_2COCH_2CH_2CH_3}$$

As shown in Fig. 9.5e, the first 'McL' ion, m/z 86, can have two resonance structures. It may fragment again in either of two ways depending on its structure, to give isomeric m/z 58 ions.

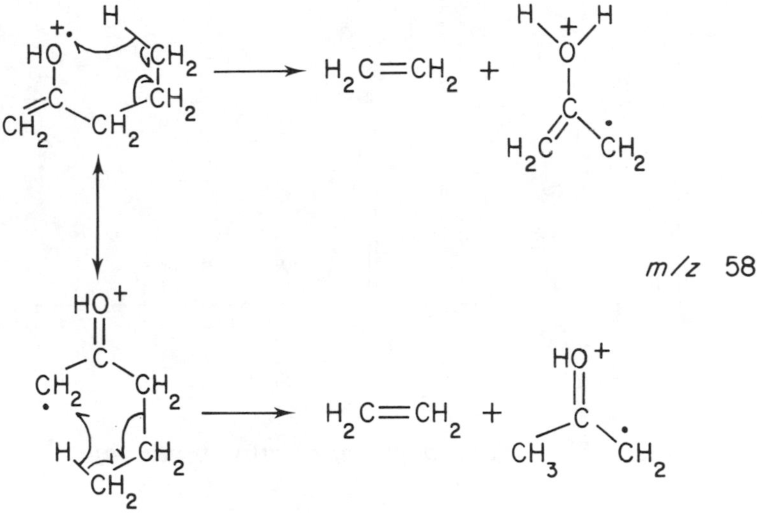

$$m/z \quad 58$$

Fig. 9.5e. *Second 'McL' rearrangement of m/z 86*

Π   Can you suggest another common type of C=O compound which would be capable of Double McLafferty Rearrangement?

The answer is an ester having γ-H in *both* the acid and the alcohol portions of the molecule, that is it must have a partial structure:

$$\overset{\gamma \qquad\quad \beta \;|\, \alpha \qquad\; \alpha \;\;|\, \beta \;\; \gamma}{-CH-C-C-CO.O-C-CH-}$$

This is mentioned in footnote *** to Fig. 9.5d. The simplest ester which could show this behaviour is ethyl butanoate, Fig. 9.5f.

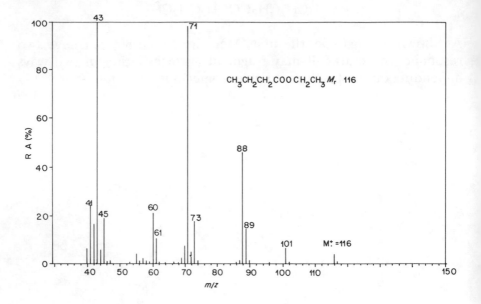

Fig. 9.5f. *Mass spectrum of ethyl butanoate*

∏   Which are the ions in Fig. 9.5f that result from Double
       McLafferty Rearrangement?

The 'McL' rearrangement ions are m/z 88 and 60. You could have
guessed this from the fact that they are the only significant even mass
ions in the spectrum, but I hoped you reasoned it out something like
this.

Either γ β α $\alpha\beta$ γ
           $CH_3CH_2CH_2CO$ or $OCOCH_2CH_3$

has γ-H atoms so the loss of $CH_2{=}CH_2$ with H transfer would occur
from either group, resulting in $(M^{+}\text{-}28) = m/z$ 88. In the next step
of the double rearrangement, the remaining sidechain's γ-H transfers with the loss of a second $H_2C{=}CH_2$ molecule, leading to m/z
60. We will leave discussion of the other ions in the spectrum to the
section on esters (Section 9.5.4.).

SAQ 9.5b	Give mechanisms for the formation of the two structures for m/z 88 in Fig. 9.5f and show that they both form the same m/z 60 in the second stage of the Double 'McL' Rearrangement.

The answer to SAQ 9.5b makes clear that the rearrangement of hydrogen in six-centred transition states does not necessarily always require a $X{=}Y^{+\cdot}$ terminus, $C{-}\overset{\cdot+}{O}{-}H$ will do for example. Nowadays, the tendency is to call any six-centred hydrogen shift a McLafferty rearrangement, even though the compound may be quite remote from the carbonyl systems originally studied 30 years ago. Incidentally, Fred, as he is known, is still active (1986) in organic mass spectrometry research.

9.5.2. Fragmentations of Aldehydes

M^{+} of aliphatic aldehydes is usually observable, although it is frequently weak. Aromatic aldehydes usually show moderately intense to very intense M^{+}. The characteristic feature of either class of aldehydes (especially aromatic) is loss of the α-H:

$$R-C{\overset{\bullet+}{\underset{\underset{H}{\overset{|}{C}H}}{\lessgtr}}}O \longrightarrow R-C\equiv\overset{+}{O} + H^{\bullet}$$
$$(M-1)$$

$$(R = \text{alkyl or aryl})$$

Benzoyl ions (R = Ar) are especially stabilised by resonance (Section 8.2). Loss of $R^{\bullet}$ or $Ar^{\bullet}$ also occurs and gives rise to $H-C\overset{+}{=}O$, m/z 29. The presence of both $(M\text{-}1)^{+}$ and m/z 29 in a spectrum is very typical of aldehydes, but note that acetals and alcohols give $(M\text{-}1)^{+}$ peaks occasionally :

$$R-C{\overset{H\overset{\bullet+}{O}R^{1}}{\underset{OR^{1}}{\diagdown}}} \longrightarrow R-\underset{\underset{OR^{1}}{|}}{C}=\overset{+}{O}R^{1} + H^{\bullet}$$

acetals   (M–1)

$$R-\underset{\underset{R^{1}}{|}}{\overset{\overset{H}{|}}{C}}\overset{\bullet+}{}OH \longrightarrow R-\underset{\underset{R^{1}}{|}}{C}=\overset{+}{O}H + H^{\bullet}$$

Alcohols   (M–1)

A second common feature of aliphatic aldehyde spectra is β-cleavage. For aldehydes with a CH_2-CHO end group this gives rise to a characteristic (M-43) peak:

$$R-\overset{\frown}{CH_2}-C\overset{+\bullet}{=}O \longrightarrow R^{+} + CH_2=CHO^{\bullet}$$
$$(M-43)$$

Homolytic β-cleavage also occurs resulting in the resonance stabilised m/z 43 and $R^{\bullet}$:

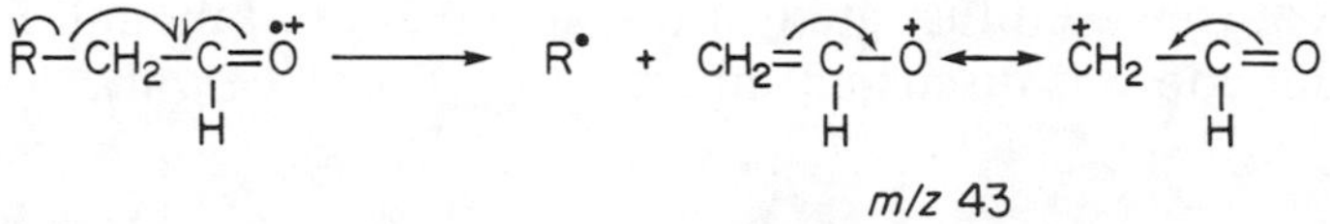

$$R\overset{\frown}{-CH_2}\overset{\frown}{-C}=\overset{\bullet\,+}{\overset{..}{O}} \longrightarrow R^{\bullet} + CH_2\overset{\frown}{=}\overset{+}{C}\overset{\frown}{-}\overset{..}{O} \longleftrightarrow \overset{+}{C}H_2\overset{\frown}{-}C\equiv O$$
$$\qquad\qquad\underset{H}{|}\qquad\qquad\qquad\qquad\underset{H}{|}\qquad\qquad\underset{H}{|}$$

m/z 43

α-Substituents will of course cause shifts in the characteristic mass loss from M‡, and the *m/z* 43 peak. The R^{+} formed may themselves fragment in the manner typical of hydrocarbons (Section 9.7). Thus the spectra of higher aliphatic aldehydes come to resemble hydrocarbons.

The third major fragmentation typical of aliphatic aldehydes is the 'McL' rearrangement (Section 9.5.1) which here forms *m/z* 44, $CH_2\overset{\bullet\,+}{=}CH\overset{..}{O}H$. The presence of this ion coupled with (M-1), (M-43) and possibly *m/z* 29 is conclusive evidence of the structure

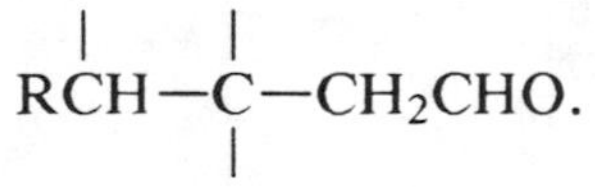

$$RCH\overset{|}{-}\overset{|}{C}\overset{|}{-}CH_2CHO.$$

It is also interesting that many higher aldehydes without branching show ions corresponding to the *alkene* eliminated in the 'McL' rearrangement, ie (M-44)‡ ions are seen. These have the formula C_nH_{2n}. Fig. 9.5g shows how this might occur starting with a γ-H transfer.

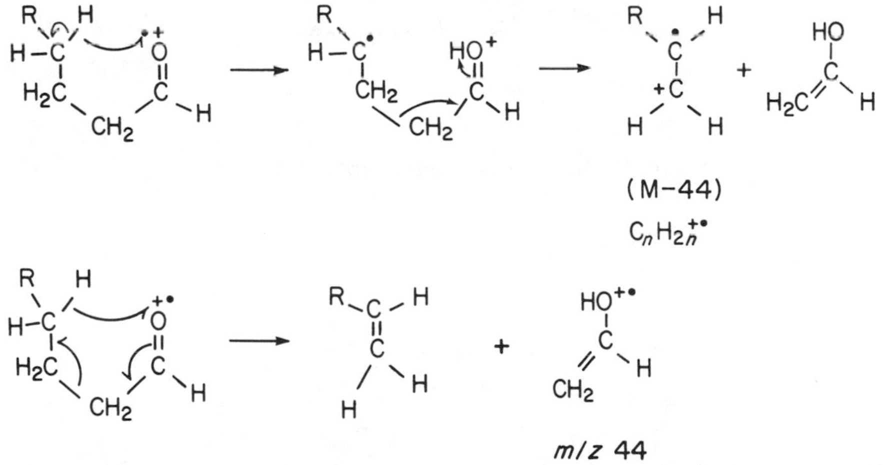

(M-44)

$C_nH_{2n}^{+\bullet}$

m/z 44

Fig. 9.5g. *Mechanism for the formation of (M-44) and 'McL' rearrangement peaks from aldehydes*

These features are illustrated in the spectrum of butanal, Fig. 9.5h.
Benzaldehyde was discussed in Section 9.1 as *Unknown 1*.

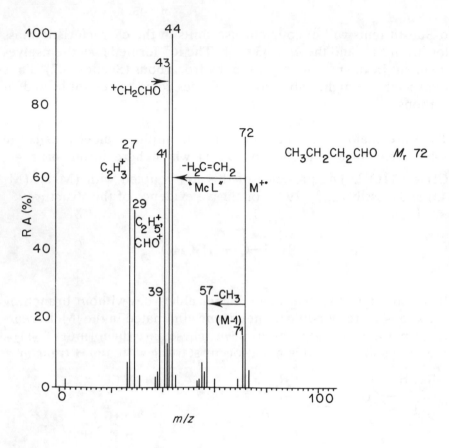

Fig. 9.5h.   *Mass spectrum of butanal*

SAQ 9.5c Fig. 9.5i shows the mass spectrum of an aldehyde, *Unknown 11*. Interpret the spectrum and
identify *Unknown 11*.

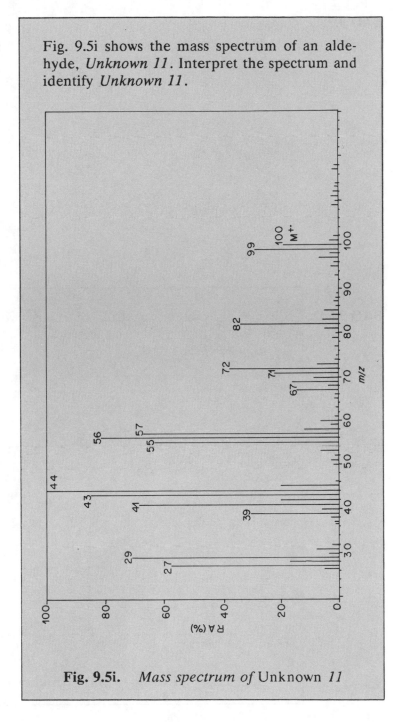

Fig. 9.5i. *Mass spectrum of* Unknown *11*

SAQ 9.5c

SAQ 9.5d   Fig. 9.5j shows the spectrum of an aldehyde *Un-known 12*. Interpret the spectrum and identify *Unknown 12*.

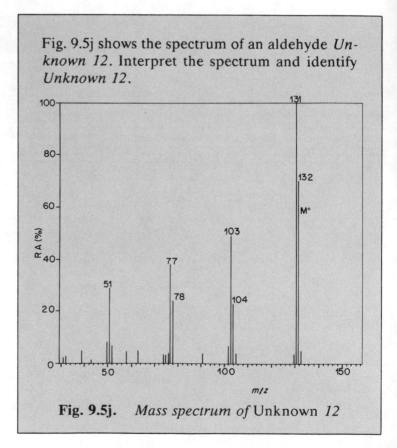

Fig. 9.5j.   *Mass spectrum of* Unknown *12*

SAQ 9.5d

9.5.3. Fragmentations of Ketones

We have already said a great deal about the ways in which ketones fragment as examples in Sections 8.2, 9.1, 9.5 and 9.5.1., so this Section will be something in the nature of a summary.

$M^{\ddagger}$ peaks of most ketones are quite intense, more so than aldehydes. Loss of alkyl groups by α-cleavage is an important mode of fragmentation. The larger of the two alkyl groups is most readily lost (Stevenson's Rule), a trend which increases if one of the R groups increases at the expense of the other. For example, in the mass spectrum of 2-butanone, the ratio of the two acylium ions m/z 43 and 57 is 10:1 (Fig. 9.5k) while in 2-hexanone the corresponding ratio of m/z 43 and m/z 85 is 20:1 (Fig. 9.5l).

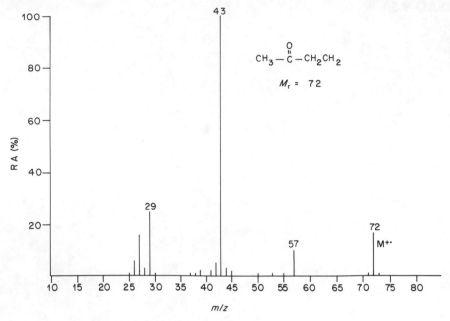

Fig. 9.5k.   *Mass spectrum of 2-butanone*

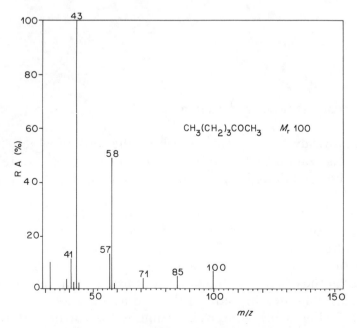

Fig. 9.5l.   *Mass spectrum of 2-hexanone*

Hydrocarbon ions, formed by heterolytic α, β and γ cleavages are usually weak unless there is a branch point in the chain eg m/z 29 in Fig. 9.5k and m/z 57 and 71 in Fig. 9.5l. Cleavage of an alkyl chain at a branch point is favoured because it will give secondary or tertiary carbocations which are more intense in a mass spectrum (see Section 9.7).

The 'McL' rearrangement of ketones was introduced in Section 9.5.1. If a γ-H is present it gives rise to ions of the type R_1R_2C $=C(OH)R^{+\cdot}$. R, R_1 and R_2 may be a great variety of substituents, and if R is long enough to possess γ-H atoms itself a double 'McL' may take place. Notice that the 'McL' peak in Fig. 9.5l is m/z 58, typical of a methyl ketone with no α-substituent, and is the second most intense ion present (50% RA).

∏ Study the spectra of 2-pentanone, Fig. 9.5m (*i*), 4-methyl-2-pentanone, Fig. 9.5m (*ii*), 4-heptanone, Fig. 9.5m (*iii*), and 2, 4-dimethyl-3-pentanone, Fig. 9.5m (*iv*). Look particularly at the 'McL' peaks (if any) and the α-cleavage peaks and make your own comments on these spectra. Do you think they agree with our predictions so far?

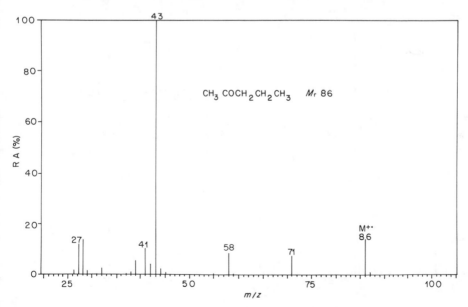

Fig. 9.5m. *(i) Mass spectrum of 2-pentanone*

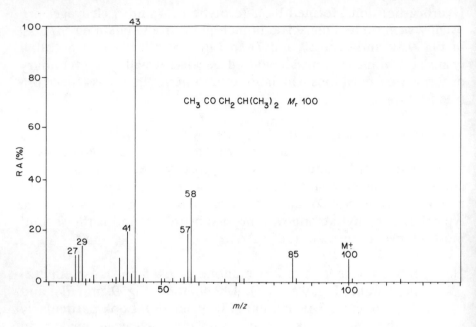

Fig. 9.5m. *(ii) Mass spectrum of 4-methyl-2-pentanone*

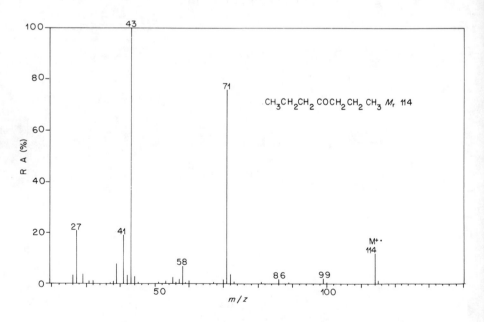

Fig. 9.5m. *(iii) Mass spectrum of 4-heptanone*

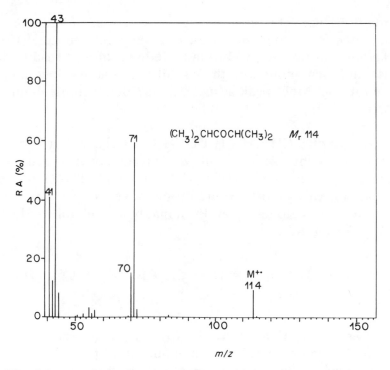

$(CH_3)_2CHCOCH(CH_3)_2$ M_r 114

Fig. 9.5m.   *(iv) Mass spectrum of 2,4-dimethyl-3-pentanone*

What you should have noted are the following:

(*i*)   *2-Pentanone* $CH_3CH_2CH_2COCH_3$. M$^+$ reasonably intense. Should show α-cleavage peaks at m/z 43 (most intense by Stevenson's Rule) and m/z 71. Also shows m/z 57, $\overset{+}{C}H_2COCH_3$ by β-cleavage. 'McL' peak should be seen at m/z 58 since it has a γ-H at C_5, and this is present though rather weak. We'll come back to this later.

(*ii*)   *4-Methyl-2-Pentanone* $(CH_3)_2CHCH_2COCH_3$. M$^+$ reasonably intense. Should show α-cleavage peaks at m/z 43 (most intense by Stevenson's Rule) and m/z 85. Does – good! Small-ish m/z 57 again, due to β-cleavage. m/z 43 could also be $(CH_3)_2\overset{+}{C}H$, a fairly stable carbocation, complementary ion to m/z 57. 'McL' peak should appear at m/z 58 since this is a methyl ketone – does, 50% RA, second most intense ion in the spectrum. This spectrum agrees very well with all the predictions for ketone cleavages. But hang on a minute! If this ketone can transfer γ-H from a methyl group so readily, why is that 2-pentanone with three similar γ-H atoms does not give an intense 'McL' peak at m/z 58 too? Better think about that later!

(*iii*) *4-Heptanone* $CH_3CH_2CH_2COCH_2CH_2CH_3$. $M^{+\cdot}$ (m/z 114) reasonably intense. Only one α-cleavage possible, since symmetrical, giving $CH_3CH_2CH_2CO^+$, m/z 71. This is the base peak, logical as most favoured α-cleavage has given base peak in the other ketones so far. Presumably m/z 43 follows by loss of CO from this:

$$CH_3CH_2CH_2 \overset{\curvearrowleft}{-\overset{+}{C}}{=}O \rightarrow CH_3CH_2\overset{+}{C}H_2 + CO$$
$$m/z \; 43$$

Very small (M-CH$_3$) at m/z 99, (M $-$ X) table predicts that for a methyl compound, but what about 'McL' peaks? Even mass ions occur at m/z 86 and m/z 58 here – could be a double 'McL' because γ-H occur in the C$_1$ *and* C$_7$ methyl groups. Let's work this out using the mechanism:

$M^{+\cdot}$, m/z 114 m/z 86 m/z 58

Yes, the mechanism predicts that two consecutive 'McL' will give m/z 86 and 58. 4-Heptanone behaves as predicted except perhaps the 'McL' peaks are rather weak.

(*iv*) *3,4-Dimethyl-3-pentanone*   CH$_3$ CH$_3$ M$\overset{+}{\cdot}$, m/z 114,

\qquad\qquad\qquad\qquad\qquad\qquad | |

\qquad\qquad\qquad\qquad\qquad CH$_3$CHCOCHCH$_3$

fairly intense, α-cleavage ion (CH$_3$)$_2$CH$\overset{+}{C}$O m/z 71 reasonably intense but noticeably less than in its isomer 4-heptanone. Base peak is m/z 43. Must be (CH$_3$)$_2\overset{+}{C}$H formed by loss of CO from m/z 71 (or perhaps from M$\overset{+}{\cdot}$ direct, can't tell without m* peaks). This is a secondary carbocation so more stable than CH$_3$CH$_2\overset{+}{C}$H$_2$ in 4-heptanone spectrum – explains why it is so intense. No 'McL' peaks seen, none expected because *no* γ-H. Shows how important the six- membered ring idea is. If a five-membered ring could work, this compound has *twelve β-*H available for transfer, yet there are no even-mass ions! Very good agreement with predicted ketone features.

I hope you noted most of these points. If not, look again at the spectra and see if you can now explain most of the ions and appreciate that they are reasonably predictable.

∏   We have seen that some CH$_3$ groups in the γ-position show little 'McL' rearrangement. We made a note to think about this. Do so now and put down as many reasons as you can as to why this could happen.

You might have put one or more of the following:

(*a*) steric factors inhibit formation of a ring;

(*b*) statistical factors – needs more than just one CH$_3$;

(*c*) energy factors – some compounds give M$\overset{+}{\cdot}$ without sufficient energy to rearrange;

(*d*) activation energy for rearrangement critical – CH$_3$ hydrogens just on borderline for process to occur in most M$\overset{+}{\cdot}$.

There is something in all of these answers but (*d*) is the most important reason. C-H bonds do differ in strength, CH$_3$ being the strongest followed by CH$_2$ then CH, so if the activation energy is high it is obvious that it will be highest for the strongest bond, that is in a CH_3 group. Rearrangements involving CH_3 even when all the other factors are favourable sometimes occur only with reluctance (as in 2-pentanone (Fig. 9.5m (i) and 4-heptanone, Fig. 9.5m (iii)). Just to confound us though, 4-methyl-2-pentanone does transfer γ-H from its methyl groups quite well since m/z 58 is about 40% RA (Fig. 9.5m (ii)). Do not be surprised if 'McL' processes involving CH_3 groups show widely varying amounts of the expected even-mass ions.

Before leaving our account of ketones we should say a few words about cyclic ketones These behave in a singular way rather like cyclanols do (Section 9.2). α-Cleavage in a cyclic ketone cannot itself give rise to fragments. It merely opens the ring and the fragments form from the ring opened $M^{\ddagger}$. These processes are illustrated in Fig. 9.5n for cyclohexanone.

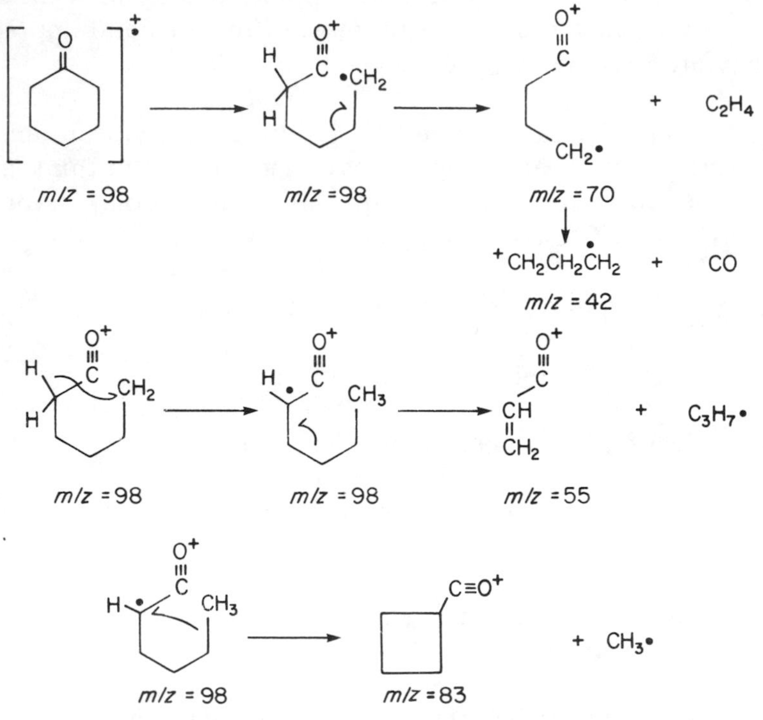

Fig. 9.5n. *Fragmentations of cyclohexanone*

Fragment ions corresponding to each of these processes may be seen in its mass spectrum (Fig. 9.5o). Cyclopentanones, cycloheptanones, and substituted cyclohexanones, behave analogously giving ions related to m/z 55.

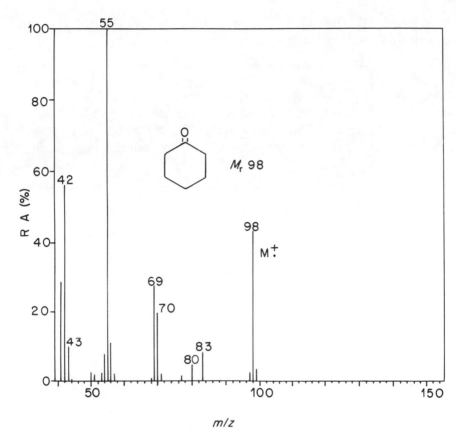

Fig. 9.5o. *Mass spectrum of cyclohexanone*

Aromatic ketones as we have already said in Sections 8.2 and 9.1 give predominantly the highly stabilised benzoyl ions:

$$Ar\overset{+\cdot}{C}OR \rightarrow Ar\overset{+}{C}O + R^{\cdot}$$
$$(100\%)$$

though the corresponding acylium ion will usually be observed

$$Ar\overset{+\cdot}{C}OR \rightarrow RC\overset{+}{O} + Ar\cdot$$
(up to 20%)

If R has γ-H's the 'McL' ion m/z 120, $C_6H_5C(OH)=CH_2^{+\cdot}$ or a substituted analogue will appear (Section 9.5.1).

SAQ 9.5e

What peak(s) would you look for in the mass spectrum of 4-methyl-3-pentanone, $CH_3CH(CH_3)COCH_2CH_3$, to distinguish it from its isomer 4-methyl-2-pentanone (Fig. 9.5m (ii))?

SAQ 9.5f

Fig. 9.5p shows the spectrum of a ketone, *Unknown 13*. Interpret this spectrum and identify *Unknown 13*.

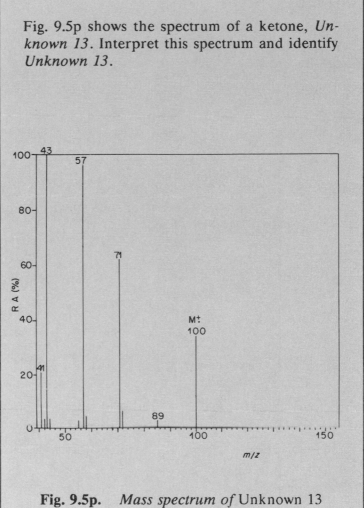

Fig. 9.5p. *Mass spectrum of* Unknown 13

SAQ 9.5f

SAQ 9.5g

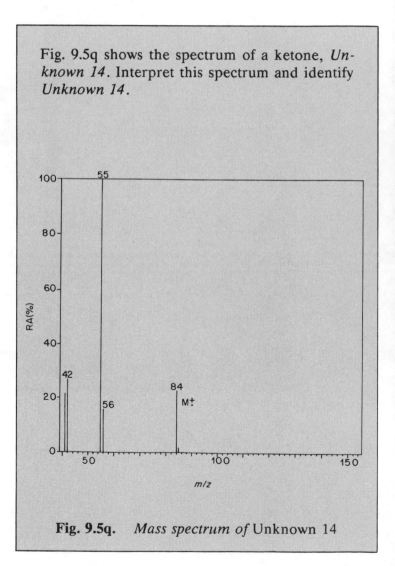

Fig. 9.5q shows the spectrum of a ketone, *Unknown 14*. Interpret this spectrum and identify *Unknown 14*.

Fig. 9.5q.   *Mass spectrum of* Unknown 14

SAQ 9.5g

9.5.4. Fragmentations of Esters

Now that you have had some experience with the fragmentation of carbonyl compounds, perhaps you can predict for yourself what types of cleavage would occur in esters.

$\prod$ What types of ions would you expect to be formed from the esters RCO_2R_1, $ArCO_2R_1$ and RCO_2Ar?

Your answer should have been something like this. Laying out the molecules so that all the α-bonds are clear, we can then fragment them and see what types of ion we can get.

$$
\begin{array}{ccccccc}
& R & & & Ar & & & R \\
R_1OCO^+ \leftarrow \!\!\!\!-\!\!\!\!\sim\!\!\!\!\!\!\!\!\overset{|}{-}\!\!\!\!\!\!\!\!\sim\!\!\!\!-\!\!\!\! \rightarrow R^+ & & R_1OCO^+ \leftarrow \!\!\!\!-\!\!\!\!\sim\!\!\!\!\!\!\!\!\overset{|}{-}\!\!\!\!\!\!\!\!\sim\!\!\!\!-\!\!\!\! \rightarrow Ar^+ & & ArOCO^+ \leftarrow \!\!\!\!-\!\!\!\!\sim\!\!\!\!\!\!\!\!\overset{|}{-}\!\!\!\!\!\!\!\!\sim\!\!\!\!-\!\!\!\! \rightarrow R^+ \\
& C=O & & & C=O & & & C=O \\
R_1O^+ \leftarrow & & \rightarrow RCO^+ & R_1O^+ \leftarrow & & \rightarrow ArCO^+ & ArO^+ \leftarrow & & \rightarrow RCO^+ \\
& O & & & O & & & O \\
RCO_2^+ \leftarrow & & \rightarrow R_1^+ & R_1^+ \leftarrow & & \rightarrow ArCO_2^+ & RCO_2^+ \leftarrow & & \rightarrow Ar^+ \\
& R_1 & & & R_1 & & & Ar
\end{array}
$$

You should also have predicted even mass 'McL' ions from those esters whose R and R_1 groups had a γ-H (see Section 9.5.1) eg $CH_2=C(OH)OR_1^{+\cdot}$ and $RCOOH^{+\cdot}$. In addition, if both R and R_1 have a γ-H, then Double 'McL' rearrangement is possible leading to a further 'McL' ion, $CH_2=C(\overset{+}{O}H_2)O^{\cdot}$ at m/z 60. If you forgot the 'McL' possibilities or put down the wrong structures for them, have another look at Section 9.5.1.

Now at first sight it seems that an ester can give rise to six α-cleavage ions plus 'McL' ions. But some of the α-cleavage ions are unstable and are either not formed, or are very weak in most mass spectra. These are R^+, R_1^+, R_1OCO^+, R_1O^+, RCO_2^+ and $ArCO_2^+$. Unless there is something to specially stabilise R or R_1, the ions containing them in this list may be present but will not be prominent in the mass spectrum. These groups are normally released as radicals.

The spectra of esters tend to be dominated by the RCO^+ and $ArCO^+$ ions, the latter especially, and the 'McL' rearrangement peaks. R_1OCO^+ ions, starting at m/z 59 for $R_1 = CH_3$ and going on to the homologous m/z 73, 87, 101 ... etc, although weak, can be useful in identifying the alcohol portion of an ester.

There are two processes found in esters which help to characterise them which I would not have expected you to predict. The first is γ-cleavage of the acyl chain to give ions starting with m/z 87 in methyl esters:

$$
\left[\underset{87}{\underbrace{\sim\!\!CH_2}} \overset{\gamma}{} \overset{\beta}{CH_2} - \overset{\alpha}{CH_2} - C \overset{\displaystyle O}{\underset{OCH_3}{\diagup}} \right]^{+\cdot} \longrightarrow {}^+CH_2-CH_2-COOCH_3 \quad m/z \ 87
$$

It is not immediately clear why this cleavage should be favoured over any other cleavage of the acyl chain of carbon atoms, but it has been shown that the structure of m/z 87 is actually $CH_2{=}CH$ $-C(OCH_3){=}\overset{+}{O}H$. As you can see, this structure is resonance stabilised. This favours γ-cleavage. This type of cleavage is found in other $C{=}O$ compounds which have the structural unit $CH_2CH_2C{=}$ OX. In esters, γ-cleavage gives ions in the series m/z 87, 101, 115 ... etc for methyl, ethyl, propyl esters etc.

The second process is related to the 'McL' rearrangement which occurs in the alcohol part of the ester, but this time *two* hydrogens are transferred. It is called *double H-transfer*, and leads to ions of the structure $R-C{=}\overset{+}{O}H(OH)$. These are 1 amu higher than the 'McL' peaks. A stepwise mechanism which accounts for them is shown in Fig. 9.5r.

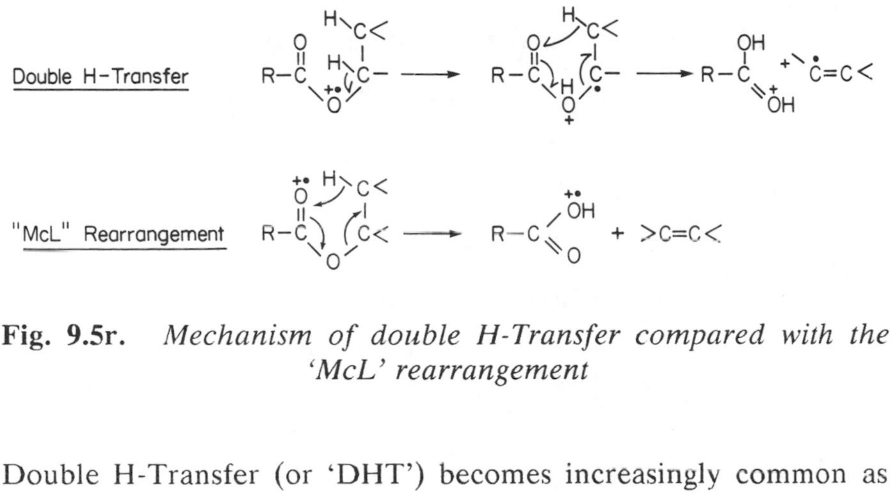

Fig. 9.5r. *Mechanism of double H-Transfer compared with the 'McL' rearrangement*

Double H-Transfer (or 'DHT') becomes increasingly common as the alkyl chain of the alcohol moiety lengthens. It can give rise to ions more intense than the 'McL' peaks themselves. 'DHT' peaks are of course *odd*-mass peaks in the spectra of C,H,O esters.

∏ Study the spectra of propyl ethanoate, Fig. 9.5s (*i*) and ethyl 2-methylpropanoate, Fig. 9.5s (*ii*). Look particularly for the 'McL' and 'DHT' peaks and α-cleavage peaks. Make your own notes on these spectra. Do you think the ions they show agree with your predictions?

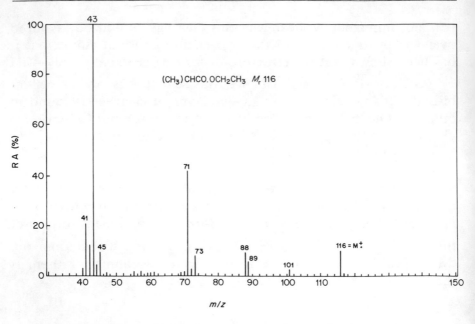

Fig. 9.5s.   *(i) Mass spectrum of propyl ethanoate*

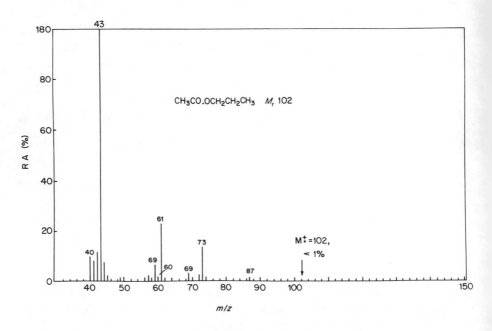

Fig. 9.5s.   *(ii) Mass spectrum of ethyl 2-methylpropanoate*

(*i*) $M^{\ddot{+}}$ is very weak, compared with most aldehydes and ketones. The base peak is CH_3CO^+, formed by α-cleavage, and dominates the spectrum. The alternative cleavage to give $CH_3CH_2CH_2O^+$, m/z 59 occurs, but as expected this ion is of low RA. Likewise $(M\text{-}CH_3)^+$ and $CH_3CO_2^+$ (isobaric with $CH_3CH_2CH_2O^+$) are weak. 'McL' would give $CH_3COOH^{\ddot{+}}$, m/z 60, while 'DHT' would give $CH_3C(OH)_2^+$, m/z 61. In this spectrum 'DHT' predominates as m/z 61 is 25% and m/z 60 only 2%. $(M\text{-}CH_3CH_2)$, m/z 73 is the third most intense ion. This could be formed by a γ-cleavage of the propyl group, assuming that $M^{\ddot{+}}$ is charged on the alkyl oxygen:

$$CH_3C-\overset{+\bullet}{\underset{\displaystyle \overset{\|}{O}}{O}}{\overset{}{}}CH_2-CH_2-CH_3 \longrightarrow CH_3C-\overset{+}{\underset{\displaystyle \overset{\|}{O}}{O}}=CH_2$$

$$m/z\ 73$$

Shows reasonable agreement with the predicted behaviour.

(*ii*) $M^{\ddot{+}}$ is more intense here, base peak m/z 43 which *cannot* be due to CH_3CO^+. Could be $(CH_3)_2CH^+$, a fairly stable carbocation formed either by α-cleavage of $M^{\ddot{+}}$, or by loss of CO from the m/z 71 acylium ion:

$$(CH_3)_2CH\;\overset{\displaystyle \overset{+\bullet}{\overset{\|}{O}}}{\underset{\underset{43}{\underbrace{\qquad}}}{COCH_2CH_3}}$$

$$(CH_3)_2CH-\overset{+}{C}O \longrightarrow (CH_3)_2CH^+$$

$$m/z\ 71,\ 45\% \qquad\qquad m/z\ 43,\ 100\%$$

Other α-cleavage ions formed in about 10% RA are $CH_3CH_2O^+$, m/z 45 and $CH_3CH_2OCO^+$, m/z 73. The acid portion has no γ-H, so cannot give a 'McL' peak, but the $-OCH_2CH_3$ loses $H_2C=CH_2$ to give the 'McL' ion m/z 88, and also $\cdot CH=CH_2$ to give the 'DHT' ion m/z 89. The ratio here is about $2:1$ in favour of the 'McL' ion. Shows good agreement with expected behaviour.

I hope you got most of these points, and convinced yourself that aliphatic ester spectra are easy to analyse. Isomers can be distinguished too. Note that 'DHT' can give ions with a higher RA than the 'McL' ions from the alkyl group of an ester.

SAQ 9.5h Which acylium ion base peak, 'McL' and 'DHT' ions would you predict for each of the following esters?

(i) $CH_3CH_2CO.OCH_2CH_2CH_3$;

(ii) $C_6H_5CO.OCH_2CH_2CH_3$;

(iii) $CH_3CH_2CO.OCH{=}CHCH_3$;

(iv) $CH_3CO.OCH_2CH_2CH{=}CH_2$;

(v) $CH_3CH_2CH_2CH_2CO.OCH_2CH_3$.

Esters of aromatic acids may also show both 'McL' and 'DHT' peaks if the alkyl group has the appropriate β-H and γ-H

$$\text{eg ArCO.O}\overset{\beta}{\underset{|}{C}}\text{H}-\overset{\gamma}{\underset{|}{C}}\text{H}-\text{R}.$$

The resulting ions are $ArCOOH^+$ and $ArC(OH)_2^+$.

Ortho-substituted esters show some interesting ortho effects leading to the loss of ROH rather than RO$^{\cdot}$. Two examples will suffice:

A

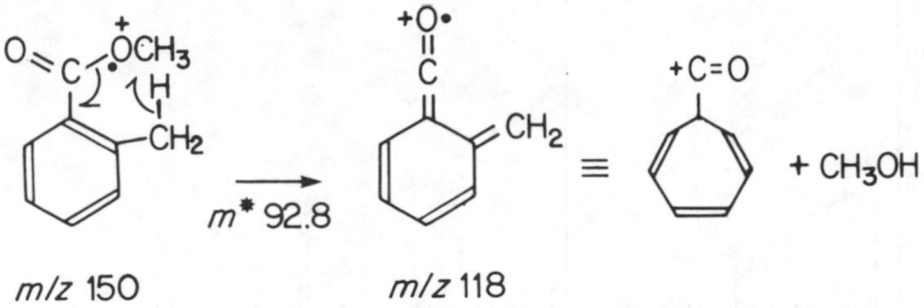

m/z 150   *m/z* 118

B

m/z 138   *m/z* 120

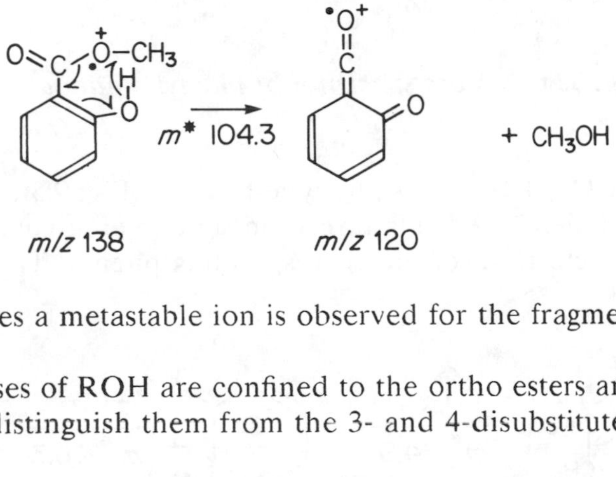

* indicates a metastable ion is observed for the fragmentation

Such losses of ROH are confined to the ortho esters and serve usefully to distinguish them from the 3- and 4-disubstituted isomers.

Another sort of aromatic ester which behaves distinctively is the benzyl ester. This rearranges to eliminate a neutral ketene molecule and form $C_6H_5CH_2OH^+$, m/z 108:

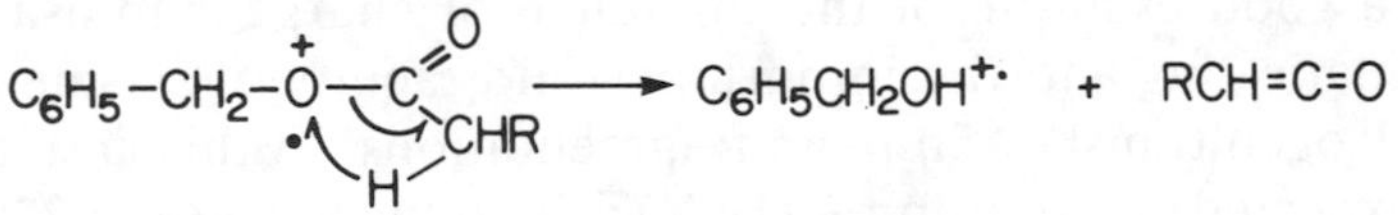

Intense m/z 108 peaks are typical of benzyl esters. Subsequent fragmentation of m/z 108 is by loss of H˙ and CO as in benzyl alcohol (Section 9.2).

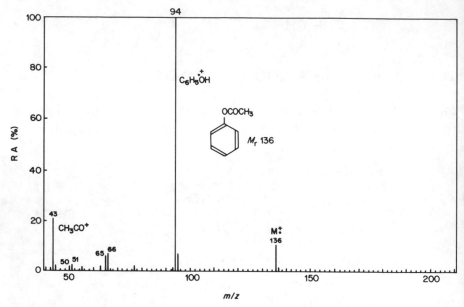

Fig. 9.5t.   *Mass spectrum of phenyl ethanoate*

Aryl esters ArOCOR such as phenyl ethanoate, Fig. 9.5t, also rearrange by a six-membered ring transition state to eliminate a ketene and give the keto form of the corresponding phenol. This is often the base peak:

This is a good example of the influence of charge stabilisation by the aromatic ring and the importance and facility of six-membered ring hydrogen transfers in mass fragmentations. In this case the RA of the expected acylium ion CH_3CO^+ is reduced to only 22%.

The fragmentation of di and polyesters is too complex to go into here, with one exception, dialkylphthalates, the esters of benzene-1,2-dicarboxylic acid. These are widely used as plasticisers and frequently turn up as background contaminants in other samples which have been stored in plastic bags or plastic containers generally. Many organic liquids, not just recognised solvents, will leach out the plasticiser from the walls or cap of their container, as well as any other plastic material they come into contact with eg tubing. You should not submit samples in plastic containers for mass spectrometric analysis. Always use glass sample tubes and avoid caps with plastic linings, which are easily overlooked sources of phthalates.

The sure sign of phthalates is an intense m/z 149 peak. This ion is the base peak of *all* phthalates other than dimethyl phthalate. It is formed by the cyclisation of the first formed acylium ion from one of the —COOR groups (Fig. 9.5u)

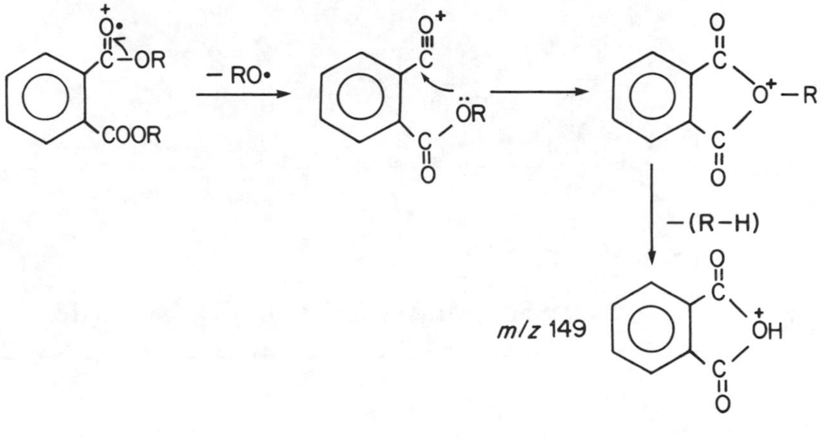

Fig. 9.5u. *Origin of* m/z *149 in the spectrum of dialkyl phthalates*

The second R group is eliminated as an alkene and a hydrogen is transferred to the central oxygen. Thus m/z 149 is protonated phthalic anhydride.

M^+ of phthalates are usually very weak but the $(M-OR)^+$ ions can usually be seen and help to indicate the M_r, as nearly all commercially used phthalates are symmetrical. Other characteristic though often weak ions are m/z 121 (149-CO), 105 (149-CO$_2$), 104 (149-CO$_2$H) and 76 (104-CO).

SAQ 9.5i

Fig. 9.5v shows the mass spectrum of an ester, *Unknown 15*. Interpret this spectrum and suggest a structure for *Unknown 15*.

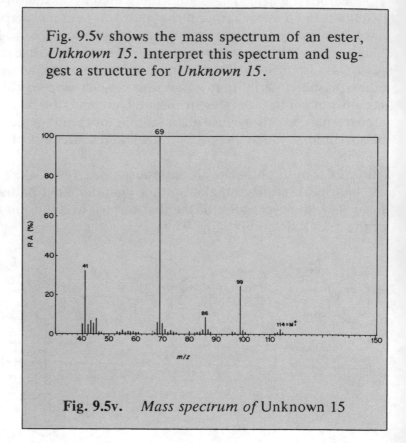

Fig. 9.5v. *Mass spectrum of* Unknown 15

SAQ 9.5j   Fig. 9.5w shows the mass spectrum of an ester, *Unknown 16*. Interpret this spectrum and suggest a structure for *Unknown 16*.

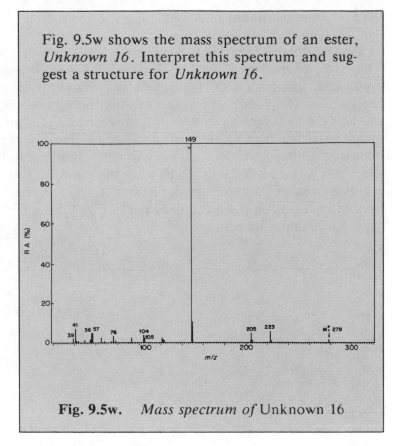

Fig. 9.5w.   *Mass spectrum of* Unknown 16

9.5.5.   Fragmentations of Acids

In general the mass spectra of aliphatic acids show weak $M^{\ddagger}$ peaks. Their fragmentations resemble those of the methyl esters. Short chain acids lose both $\cdot CO_2H$ and $\cdot OH$ but $\cdot CO_2H$ loss if favoured, ie the smaller $\cdot OH$ radical is lost less readily (Stevenson's Rule). Loss of the OH hydrogen does not normally occur. Loss of the alkyl group as $R\cdot$ leaving $COOH^+$, m/z 45 occurs. This ion is very characteristic of the spectra of carboxylic acids (but note ethyl esters often give $CH_3CH_2O^+$ ions which have the same m/z value). Once a γ-H is present the 'McL' peak $CH_2{=}C(OH)_2^{\ddagger}$ m/z 60 is usually very intense and may be the base peak. Fig. 9.5x is typical of the mass spectrum of an acid.

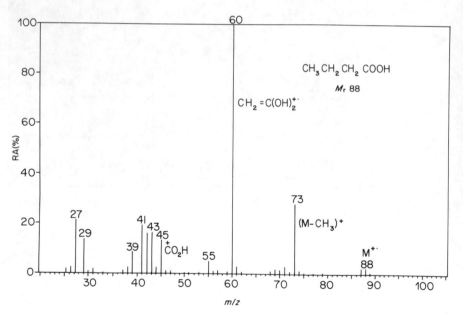

Fig 9.5x.   *Mass spectrum of butanoic acid*

Aromatic acids on the other hand produce intense $M^{\ddagger}$ peaks and usually fragment by the expected loss of $\cdot OH$, then CO from the aroyl ion to give Ar^+. Thus, it is usually easy to work out what substituents are present in the aromatic ring from the mass of this ion and its further fragmentation.

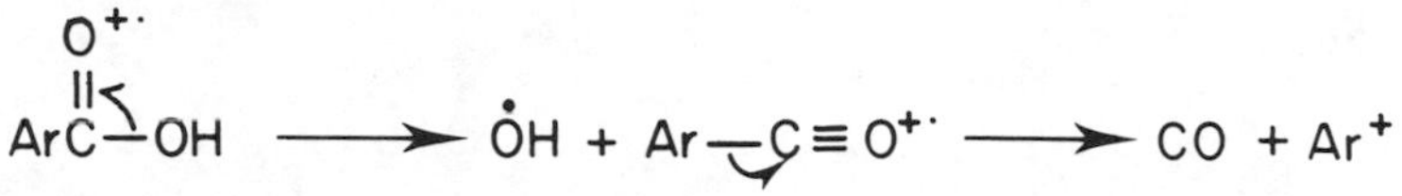

$$\underset{\substack{\| \\ ArC-OH}}{O^{+\cdot}} \longrightarrow \overset{\cdot}{O}H + Ar-C\equiv O^{+\cdot} \longrightarrow CO + Ar^+$$

Ortho-substituted aromatic acids can eliminate H_2O from $M^+_\cdot$ if there is a suitably located hydrogen in the ortho-group, eg.

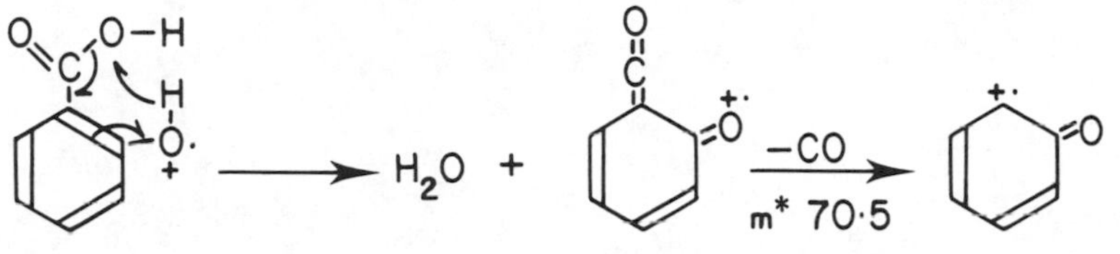

salicylic acid
m/z 138

m/z 120

m/z 92

Once again this ortho effect is useful in distinguishing between the 2- and the 3- and 4-substituted isomers.

SAQ 9.5k

Which ions would you expect to feature prominently in the mass spectra of:

(*i*) $(CH_3)_2CHCO_2H$;

(*ii*) 2-methylbenzoic acid;

(*iii*) 3-methoxybenzoic acid;

(*iv*) $CH_3(CH_2)_4CO_2H$.

SAQ 9.5k

In most analytical work acids are base-extracted, neutralised, and converted to esters, such as methyl $-OCH_3$ and trimethylsilyl $-OSi(CH_3)_3$ (TMS), because their volatility and gas chromatographic behaviour is so much better. Hence, from the methyl ester, the ions to be expected are $(M\text{-}OCH_3)^+$, the 'McL' peak m/z 74 $CH_2{=}C(OH)OCH_3^{\ddagger}$ and the β-cleavage peak m/z 87 $\overset{+}{C}H_2CH_2COOCH_3$, and from the TMS ester, ions 72 and 57 amu higher than expected for the acid. These are due to the new parent ion and its $(M\text{-}CH_3)^+$ daughter ion:

$$\underset{P}{\overset{\displaystyle \overset{O}{\|}}{-C-OH}} \rightarrow \underset{(P\,+\,72)}{\overset{\displaystyle \overset{O}{\|}}{-C-OSi(CH_3)_3}} \overset{e}{\rightarrow} \underset{(P\,+\,72)^{\ddagger}}{\overset{\displaystyle \overset{O}{\|}}{-C-O{\overset{+\cdot}{\underset{CH3}{\overset{|}{\curvearrowleft}Si(CH_3)_2}}}}} \rightarrow$$

$$\overset{\displaystyle \overset{O}{\|}}{-C-\overset{+}{O}{=}Si(CH_3)_2} + CH_3^{\cdot}$$
$$(P\,+\,57)^+$$

9.5.6. Fragmentations of Amides

Nitrogen containing compounds such as amides must follow the nitrogen rule: if they contain an *odd* number of nitrogen atoms, they will have an *odd* M_r. Aliphatic amides usually show an observable M^{+} peak (lower members quite intense) and aromatic amides an intense M^{+} peak. The fragmentation patterns are quite similar to those of the corresponding acids and esters. α-Cleavage occurs by loss of R˙ which in primary aliphatic amides gives $[O{=}C{=}NH_2]^{+}$.

$$R-C\overset{O^{+\cdot}}{\underset{NH_2}{\diagup}} \quad \rightarrow \quad R^{\cdot} + \overset{+}{O}{\equiv}C-\ddot{N}H_2$$

or $\qquad\qquad\qquad\qquad\qquad\qquad\qquad$ m/z 44

$$R-C\overset{O}{\underset{NH_2}{\diagup}} \quad \rightarrow \quad R^{\cdot} + \ddot{O}{=}C{=}\overset{+}{N}H_2$$

As you can see, there are two mechanisms leading to the m/z 44 fragment depending on whether O or N carries the charge. The two forms of m/z 44 are resonance hybrids, thus explaining the intensity usually found for the ion and N-substituted versions which form a series m/z 44, 58, 72, 86 ... .

Acylium ion RCO^{+} peaks are usually weak in aliphatic amide spectra. It is only in aromatic acid amides where this type of α-cleavage gives significant amounts of aroyl ions, $ArCO^{+}$, by loss of ˙NH_2 (Fig. 9.5y (*i*) and (*iv*)).

∏ Why do you think there is this change in the relative proportion of the α-cleavage ions, $ArCO^{+}$ and $O{=}C{=}\overset{+}{N}H_2$ in aromatic primary amides?

This is our old friend resonance stabilisation at work again. Aromatic groups can donate electrons from their π-orbital systems to help stabilise the positive charge on the carbonyl groups. Refer to Section 8.2 if you have forgotten this vital point.

'McL' rearrangement readily occurs in amides, even when the γ-H is from a CH_3 group. For example, in butanamide, Fig. 9.5y (*ii*),

$$\overset{\gamma}{C}H_3\overset{\beta}{C}H_2\overset{\alpha}{C}H_2CONH_2,$$

the 'McL' peak at m/z 59 is the base peak of the spectrum, with m/z 44 next at 70% RA.

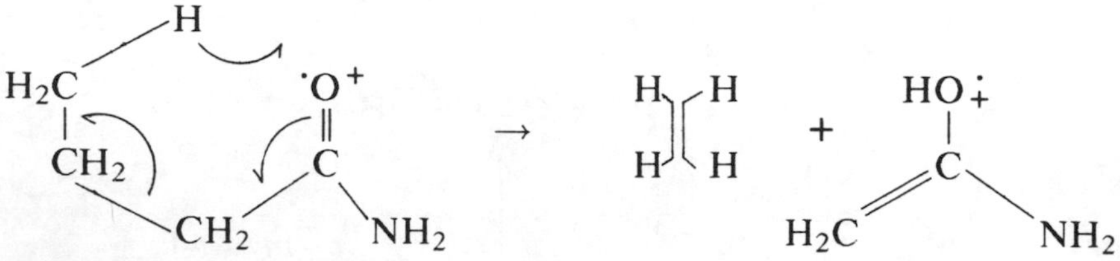

m/z 87 m/z 59, 100%

Secondary and tertiary amides give significant amounts of substituted 'McL' ions, and there is some evidence that they can also give 'DHT' rearrangement peaks using β- and γ-H atoms from the N-alkyl group, as esters do, but these ions are relatively weak. They also show ions typical of the $-NR_1R_2$ group similar to those formed in amine spectra (Section 9.6), such as m/z 58, $CH_3CH_2\overset{+}{N}H=CH_2$ from $-N(CH_3CH_2)_2$ amides.

Cleavage between the β and γ atoms in the acid portion of an aliphatic amide is usually quite important and is found in primary ($R_1=R_2=H$), secondary ($R_2=H$), and tertiary ($R_1=R_2=$ alkyl) compounds.

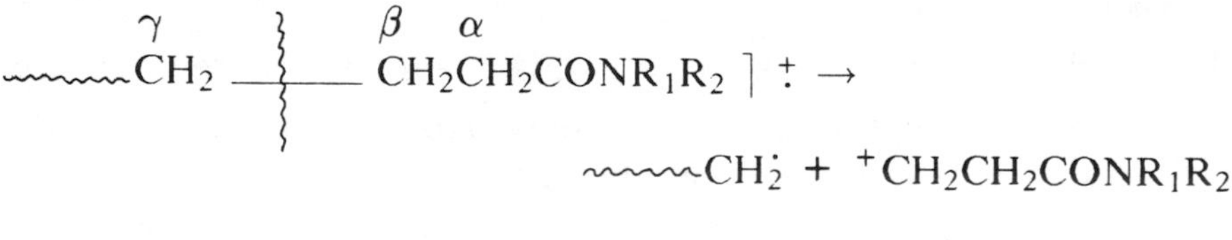

In primary amides this ion occurs at m/z 72, $\overset{+}{C}H_2CH_2CONH_2$. β-Cleavage also occurs in the N-alkyl chains.

$$CH_3CO-\overset{+\cdot}{\underset{H}{N}}-CH_2-R \rightarrow R^{\cdot} + \underset{\underset{m/z\ 72}{CH_2-H}}{\overset{\overset{O}{\parallel}}{C}-\overset{+}{N}H=CH_2} \rightarrow$$

$$CH_2=C=O + H_2\overset{+}{N}=CH_2$$
$$m/z\ 30$$

This is followed by expulsion of ketene and formation of m/z 30, which is usually the base peak of N-alkyl ethanamides (Fig. 9.5y (*iii*)).

Aromatic amides such as acetanilide $C_6H_5NHCOCH_3$ (Fig. 9.5y (*v*)) fragment almost entirely by loss of ketene or substituted ketenes giving base peaks of the corresponding anilines.

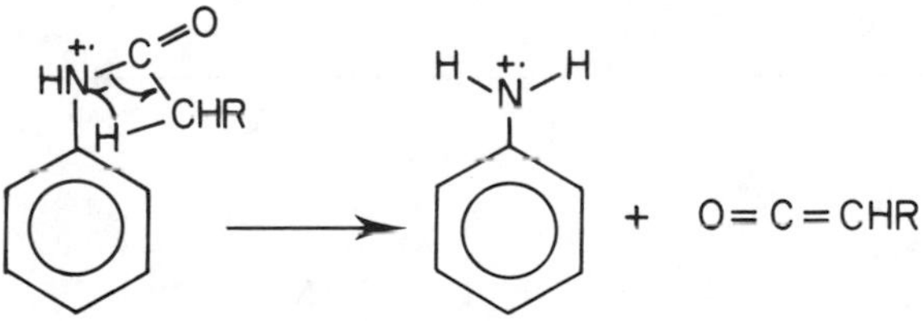

$$m/z\ 93, 100\%$$

The rest of the mass spectrum looks like that of the aniline (Section 9.6) except for a relatively small RCH_2CO^+ ion derived from α-cleavage of M^{+} (m/z 43 in acetanilide).

These common factors of amide spectra are illustrated in Fig. 9.5y.

(*i*)

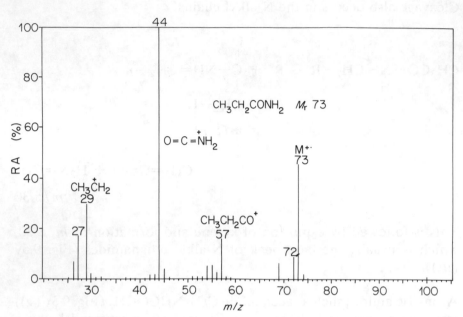

(*ii*)

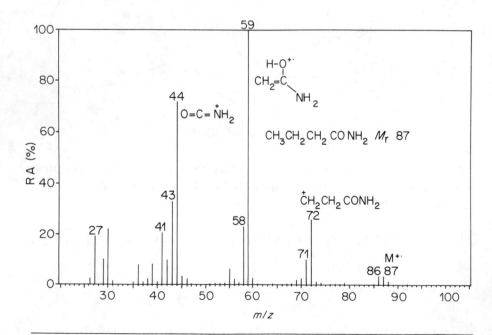

(*iii*)

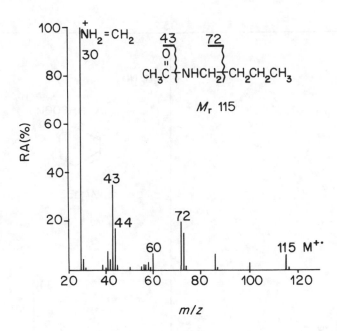

(*iv*)

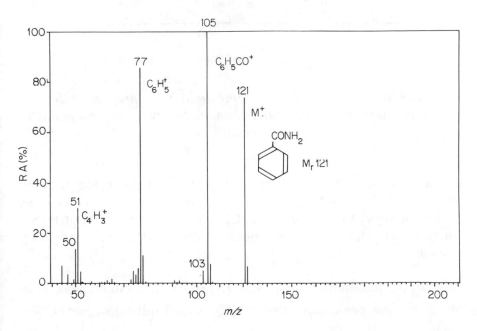

(*v*)

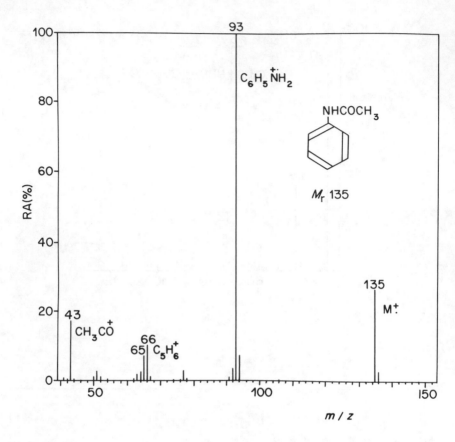

Fig. 9.5y.   *Mass spectra of some amides: (i) propanamide;*
(ii) butanamide; (iii) N-butylethanamide; (iv) benzamide and
(v) acetanilide

In the spectrum of propanamide, Fig. 9.5y (*i*) the principle ions
are m/z 44, $O=C=NH_2$; m/z 73, M^+; m/z 29, $CH_3CH_2^+$ derived
by loss of $H_2NCO\cdot$ and m/z 57, $CH_3CH_2CO^+$ from loss of $H_2N\cdot$.
There is no 'McL' peak here because the alkyl chain is not long
enough.

In contrast, butanamide, Fig. 9.5y (*ii*) shows the 'McL' peak m/z
59 as its base peak, but $O=C=\overset{+}{N}H_2$ is also quite intense (73%).

β-γ cleavage occurs here to give m/z 72 by loss of the CH_3 group, and m/z 71 is the acylium ion $CH_3CH_2CH_2CO^+$ formed by loss of H_2N. Notice how weak this ion is compared with m/z 72. Loss of $H_2NCO^{\cdot}$ gives the $CH_3CH_2CH_2^+$ ion at m/z 43.

Fig. 9.5y (*iii*) shows the typical features of an N-alkyl amide, namely β-cleavage in the N-alkyl group which here gives m/z 72, $CH_3CO\overset{+}{N}H{=}CH_2$, and this ion then breaks down by $CH_2{=}C{=}O$ elimination to give the base peak m/z 30, $CH_2{=}\overset{+}{N}H_2$. The acylium ion m/z 43, CH_3CO^+ is the second most intense ion present but is relatively weak (36%).

Fig. 9.5y (*iv*) shows the contrasting features of an aromatic primary amide. In this case the aromatic ions predominate and $O{=}C{=}\overset{+}{N}H_2$, for example, is negligible. Loss of the $H_2N^{\cdot}$ gives the benzoyl ion m/z 105 and the ions formed at m/z 77, 51 and 50 are its further daughter ions by loss of CO and C_2H_2 and C_2H_3 respectively. The features of acetanilide, Fig. 9.5y (*v*) have already been explained.

SAQ 9.51	The mass spectrum of an amide C_4H_9NO (isomeric with butanamide, Fig. 9.5y (*ii*)) shows a base peak of m/z 30, with intense ions at m/z 43 and 72. What is the structure of this isomer?

SAQ 9.5m   Which typical ions would you expect to find in
the mass spectra of:

(i) $CH_3(CH_2)_4CONHCH_3$;

(ii) $CH_3CONH(CH_2)_4CH_3$;

(iii)   CONH$_2$;

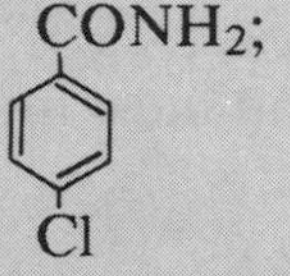

(iv)   NHCOCH$_3$

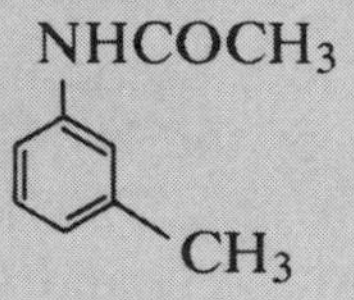

SAQ 9.5n   Would you expect to observe any ortho effects in
either ArNHCOR or ArCONHR compounds?

SAQ 9.5n

SAQ 9.5o   Fig. 9.5z shows the mass spectrum of *Unknown 17*, another amide isomeric with butanamide. Suggest its structure.

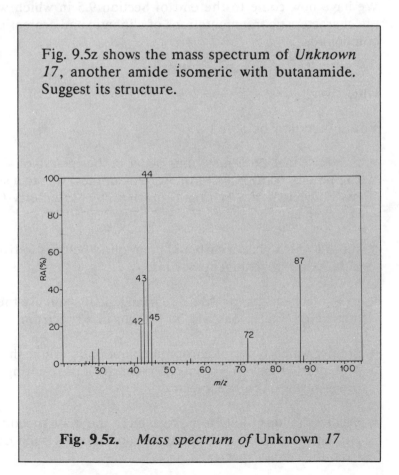

Fig. 9.5z.   *Mass spectrum of* Unknown *17*

SAQ 9.5o

Summary

We have now come to the end of Section 9.5 in which we studied the characteristic fragmentations of aliphatic and aromatic carbonyl compounds.

Objectives

You should now be able to:

- recognise that even-mass ions occur in the spectra of all carbonyl compounds (odd-mass in the case of amides) due to a six-centred rearrangement of γ-H atoms known as the McLafferty-(McL) Rearrangement;

- suggest which class a carbonyl compound belongs to from a study of its 'McL' rearrangement ions;

- write mechanisms for 'McL' rearrangements and use them to predict which 'McL' ions will be present in a spectrum;

- predict which types of carbonyl compounds would show Double McLafferty Rearrangements and identify ions resulting from such processes in unknown spectra;

- predict the ions likely to be formed in the mass spectra of simple aliphatic and aromatic aldehydes, and identify such compounds from their mass spectra;

- predict the ions likely to be formed in the mass spectra of simple aliphatic, cyclic and aromatic ketones, and identify such compounds from their mass spectra;

- predict the ions likely to be formed in the mass spectra of simple aliphatic and aromatic esters, and identify such compounds from their mass spectra;

- recognise the typical fragmentation pattern produced by a dialkyl phthalate ester;

- predict the ions likely to be formed in the mass spectra of simple aliphatic and aromatic acids, and identify such compounds from their mass spectra;

- predict the ions likely to be formed in the mass spectra of simple aliphatic and aromatic amides, and identify such compounds from their mass spectra;

- recognise 2-substituted aromatic carbonyl compounds from the distinctive fragments formed by ortho-effects, and that, conversely, mass spectra rarely enable one to distinguish between 1,3 and 1,4-disubstituted aromatic isomers.

9.6. FRAGMENTATIONS OF AMINES

You will have noticed in Section 9.5 how the features of the fragmentations of carbonyl compounds followed a common pattern of α- and β-cleavages, and 'McL' rearrangements. This made the study of each successive type of $C=O$ compound relatively easy. If we could discover a common pattern for amines based on a structural type we had studied already then we could proceed by analogy, noting any important differences as they arise. This is essentially what we did for $C=O$ compounds, wasn't it?

∏   In fact there are three structures we have already studied which are analogous to amines. Can you think what they are?

The answer is that primary amines are analogous to primary alcohols, $CH_2-\overset{+\cdot}{O}H$ cf $CH_2\overset{+\cdot}{N}H_2$; secondary and tertiary amines to ethers, $CH_2-\overset{+\cdot}{O}-R(Ar)$ cf $CH_2\overset{+\cdot}{N}HR(Ar)$; and aromatic amines to phenols, $C_6H_5\overset{+\cdot}{O}H$ cf $C_6H_5\overset{+\cdot}{N}H_2$.

∏   What were the preferred fragmentations of alcohols, ethers and phenols?

For alcohols and ethers the charge is located on the oxygen atom, and they fragment by α-cleavages, with the loss of the largest alkyl fragment where there is a choice (Stevenson's Rule).

$$CH_3-\underset{H}{\overset{\overset{\displaystyle \cdot^+OH}{|}}{C}}-CH_2CH_3 \;\rightarrow\; CH_3CH_2^+ \;+\; CH_3CH=\overset{+}{O}H$$

$$m/z\ 45$$

$$CH_3-\overset{+\cdot}{O}-CH_2-CH_2CH_3 \rightarrow CH_3CH_2^+ + CH_3\overset{+}{O}=CH_2$$

$$m/z\ 45$$

These fragments form largely due to the stability of the oxonium ions, $>C=\overset{+}{O}H$ and $R\overset{+}{O}=CH_2$.

Phenols give intense $M^{+\cdot}$ peaks, and fragment from their keto tautomers by loss of CO.

$$m/z\ 94 \qquad\qquad -CO \qquad\qquad m/z\ 66$$

If you did not remember these fragmentations have another look at Sections 9.2, 9.3 and 9.4

Amines too behave like this because the analogous immonium ions $>C=\overset{+}{N}H_2$ and $R-\overset{+}{N}H=CH_2$ are even more stable than oxonium ions. Nitrogen is better able to carry a positive charge. Consequently, since the M_r is odd, aliphatic amine spectra are dominated by ions occurring in the *even* series m/z 30, 44, 58, 72, 86, 100 etc compared with the *odd* series m/z 31, 45, 59, 73, 87, 111 etc found in the spectra of alcohols and ethers. For example, in the spectrum of $CH_3CH_2NH_2$ (M-1) is 18% ($M^+ = 17\%$) and m/z 30, $CH_2=\overset{+}{N}H_2$ is the base peak.

$$
\begin{array}{c}
H \\
| \\
H-\underset{|}{\overset{+}{C}}-\overset{+}{N}H_2 \\
CH_3
\end{array}
\quad
\begin{array}{c}
\xrightarrow{\;-H^{\cdot}\;} \quad CH_3CH=\overset{+}{N}H_2 \qquad m/z\ 44 \\
\\
\xrightarrow{\;-CH_3^{\cdot}\;} \quad CH_2=\overset{+}{N}H_2 \qquad m/z\ 30
\end{array}
$$

For most RCH_2NH_2 compounds the base peak is m/z 30 in fact. The presence of even mass ions in a spectrum should make you suspect an amine especially when the odd mass M^+ is not to be found. Unfortunately, the M^+ peak of many aliphatic amines is very weak or even absent.

Hydrocarbon ions in the series $CH_3(CH_2)_n^+$, m/z 29, 43, 57, 71, etc, ie of m/z $(14n + 1)$ where n is the number of carbon atoms in the ion, are quite weak in amine spectra, even when n is large. This is in contrast to the mass spectra of higher alcohols and ethers which come to resemble hydrocarbons (see Section 9.7). Another difference is that amines do not fragment by loss of NH_3 in analogy with the loss of H_2O from alcohols. Instead they undergo a fragmentation to form cyclic ammonium ions of variable ring sizes in the series m/z 58, 72, 86, 100 etc:

$$
\underset{\underset{(CH_2)_n}{}}{CH_2}\diagdown\overset{R}{\diagup}\overset{+\cdot}{N}H_2 \rightarrow R^{\cdot} + \underset{(CH_2)_n}{CH_2-NH_2^+}
$$

$$n = 2, \quad 3, \quad 4, \quad 5, \quad 6 \text{ etc}$$

$$m/z = 58, 72, 86, 100, 114 \text{ etc.}$$

The m/z 86 ion is usually the most prominent of these because it has a stable six-membered ring structure.

Secondary and tertiary aliphatic amines behave very much like ethers, as the example of N-ethyl-(1,3-dimethylbutyl)amine, $CH_3CH_2NHCH(CH_3)CH_2CH(CH_3)_2$ shows (Fig. 9.6a).

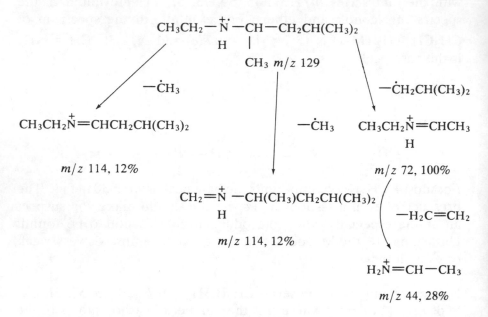

Fig. 9.6a   *Fragmentation of N-ethyl-(1,3-dimethylbutyl)amine*

The mass spectrum is dominated by the α-cleavage which gives rise to the loss of the largest radical, $\cdot CH_2CH(CH_3)_2$, an almost perfect example of Stevenson's Rule. The resulting ion further breaks down by loss of ethene and transfer of a β-hydrogen to the positive nitrogen, just as the oxonium ions in ethers do (Section 9.3).

$$CH_2-\overset{\displaystyle H}{N}=CHCH_3 \rightarrow H_2N=\overset{+}{C}HCH_3 + H_2C=CH_2$$
$$H_2C-H$$

$$m/z\ 72 \qquad\qquad\qquad\qquad\qquad m/z\ 44$$

If the first α-cleavage gives rise to an ion of the type $RCH_2CH_2\overset{+}{N}H=CH_2$, the elimination of $RCH=CH_2$ coming next will lead to $H_2\overset{+}{N}=CH_2$, m/z 30. This explains why some secondary and tertiary amines will give m/z 30, the expected base peak of a primary amine.

SAQ 9.6a   Predict which ions would be prominent in the mass spectrum of each of the following isomeric amines:

(*i*) $CH_3(CH_2)_5NH_2$;

(*ii*) $CH_3NH(CH_2)_4CH_3$;

(*iii*) $CH_3CH_2NHCH(CH_3)CH_2CH_3$;

(*iv*) $(CH_3)_2CHN(CH_3)CH_2CH_3$;

(*v*) $(CH_3CH_2)_3N$.

Could they be distinguished from their mass spectra alone?

SAQ 9.6b   Fig. 9.6b shows the mass spectrum of a tertiary amine, *Unknown 18*. Suggest a structure for this amine. $\longrightarrow$

SAQ 9.6b
(cont.)

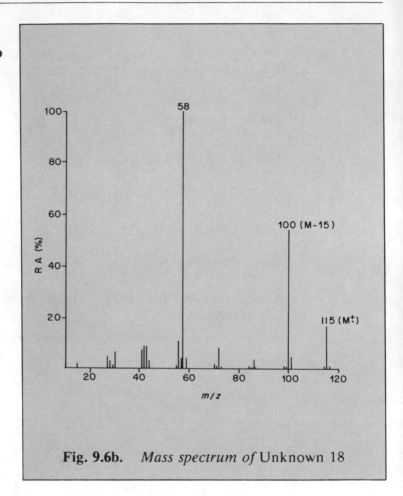

Fig. 9.6b.   *Mass spectrum of* Unknown 18

Cycloalkylamines such as substituted cyclopentyl and hexylamines show a singular mode of fragmentation similar to that seen in cyclanols (Section 9.2) which leads to ions in the series m/z 70, 84, 98 etc by the process illustrated for N-methylcyclopentylamine in Fig. 9.6c:

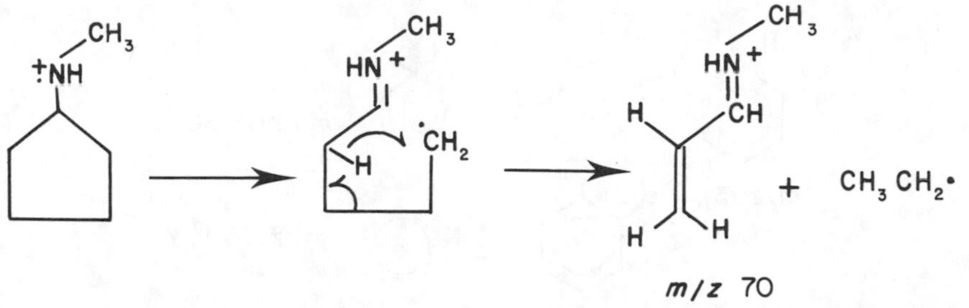

$$m/z \ 70$$

Fig. 9.6c. *Fragmentation of N-methylcyclopentylamine*

α-cleavage has occurred here as expected in the largest substituent, the ring, which then breaks down in exactly the same way as in cyclanols (SAQ 9.2a). Clearly all cycloalkylamines will give rise to the same type of ion, with the substitution of the appropriate alkyl group for the CH_3 in Fig. 9.6c.

When the nitrogen is actually in the ring, eg in pyrrolidines and piperidines two main processes occur as shown in Fig. 9.6d for N-methylpyrrolidine. The first is the loss of an α-H from the ring followed by formation of an $R-\overset{+}{N}\equiv CH$ ion (route 1), and the second is α-cleavage within the ring then expulsion of alkene to form $R\overset{+}{N}=CH_2(CH_2^+)$ ions (route 2).

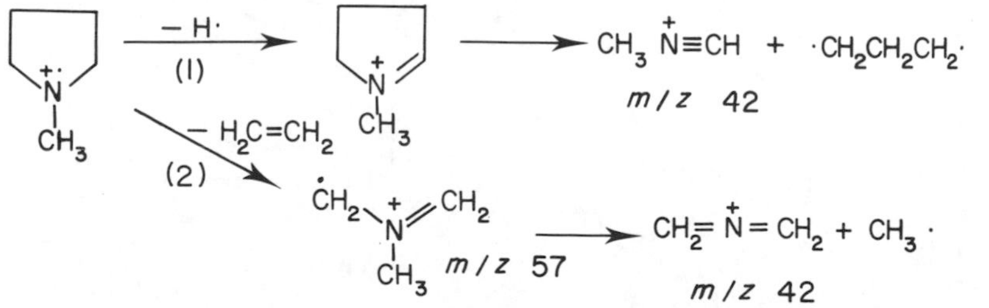

Fig. 9.6d. *Fragmentation of N-methylpyrrolidine*

We have already discussed aniline in connection with metastable ion peaks. Its behaviour is typical of aromatic amines in that they show intense M^+ peaks, and fragment relatively little by loss of HCN, either from M^+ itself or from $(M - H)^+$, which is usually a moderately intense peak.

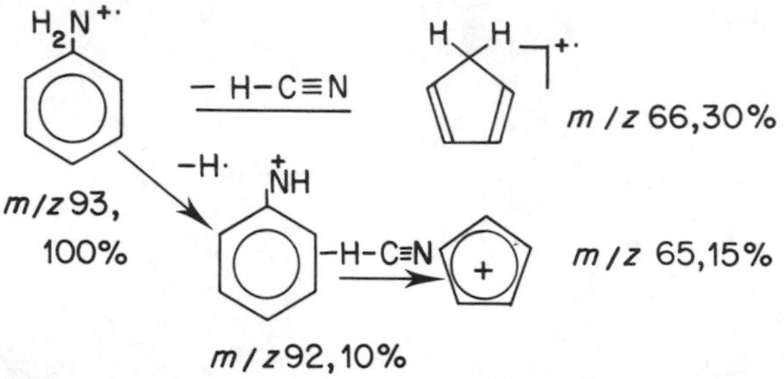

In substituted anilines it is more common for $C_6H_5\overset{+}{N}H_2$ or $C_6H_5\overset{+}{N}H$ to be formed, then release HCN to give m/z 66 and 65, than for the molecular ion itself to lose HCN. For example, in the mass spectrum of ethyl 4-aminobenzoate (benzocaine), Fig. 9.6e, the expected fragmentations of the $-COOCH_2CH_3$ group ('McL' to give m/z 137, loss of $.OCH_2CH_3$ to give m/z 120, the base peak) occur from M^+. Lower down the spectrum m/z 91, 92 and 93 are all present ($C_6H_4NH^{+\cdot}$, $C_6H_4NH_2^+$, $C_6H_5\overset{+}{O}$) and so is m/z 65. There is no ion at m/z 138 due to loss of HCN from M^+.

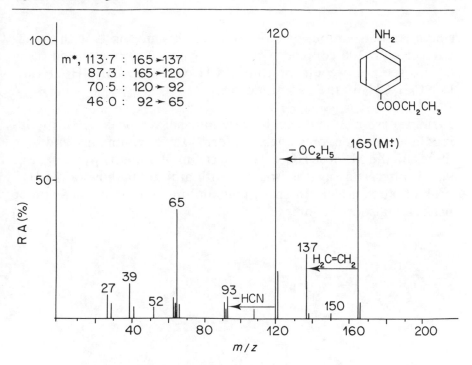

Fig. 9.6e. *Mass spectrum of benzocaine*
$(4\text{-}H_2NC_6H_4COOCH_2CH_3)$

Alkylanilines, however, behave rather like alkylphenols. The three methylanilines lose H to give m/z 106 as base peak and then lose HCN to give m/z 79. The most likely explanation of this is that an azatropylium ion is formed.

This is supported by the fact that the spectrum of N-methylaniline is remarkably similar to the spectra of methylanilines.

N-Alkylanilines all give intense m/z 106, supposedly $CH_2=\overset{+}{N}HC_6H_5$, by α-cleavage, and the mass spectra below m/z 106 look similar to the various methylanilines. This means that the presence of m/z 106 coupled with m/z 79 and possibly 92 and 65 in a mass spectrum could indicate $-CH_2NHC_6H_5$, $CH_3NHC_6H_4$, or $H_2NC_6H_3CH_3$ in the parent molecule.

Pyridines are characterised by very intense $M^{\ddagger}$ peaks. Alkylpyridines lose hydrogen, presumably to form azatropylium ions, and both these and the $M^{\ddagger}$ lose HCN. The spectrum of 3-methylpyridine, Fig. 9.6f, is interesting as it is isomeric with aniline. It shows a $(M-H)^{+}$ peak of much greater intensity than aniline and the m/z 65 and 66 peaks are also more intense.

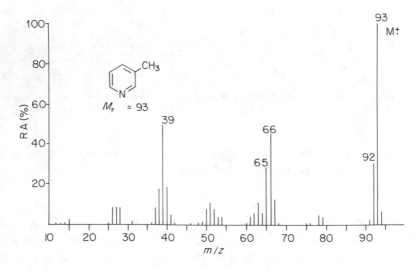

Fig. 9.6f.   *Mass spectrum of 3-methylpyridine*

Pyridyl compounds will often show $C_5H_4N^{+}$ ions, m/z 78, and both m/z 51 and 52 caused by loss of $HC{\equiv}N$ and $HC{\equiv}CH$ from this ion. If there is a side-chain possessing a hydrogen γ to the nitrogen, a 'McL'-type rearrangement can occur where the terminus of the H shift is $C{=}N^{+}$.

The rest of the spectrum below m/z 93 will then resemble Fig. 9.6f.

SAQ 9.6c

Fig. 9.6g shows the mass spectrum of an amine, *Unknown 19*. Suggest a structure for this amine.

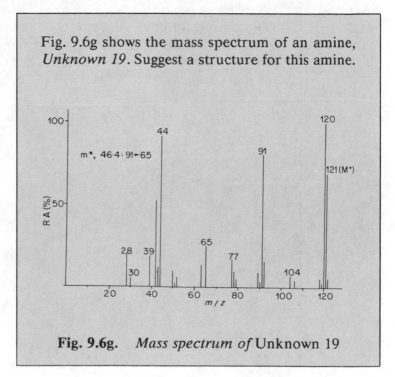

Fig. 9.6g. *Mass spectrum of* Unknown 19

SAQ 9.6d   Fig. 9.6h shows the mass spectrum of an aromatic amine of formula $C_6H_6N_2O$, *Unknown 20*.
This compound shows a strong absorption in its
IR spectrum at 1700 cm^{-1}. Suggest a structure
for this compound.

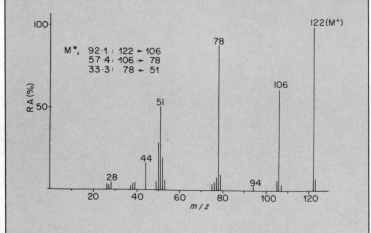

Fig. 9.6h.   *Mass spectrum of* Unknown 20

Summary

This section covers the fragmentation of aliphatic and aromatic amines.

Objectives

Now that we have completed this section, you should be able to:

- recognise that a mass spectrum which has an odd relative molecular mass and contains intense even mass ions is that of an amine;

- suggest to which class an aliphatic amine belongs from the m/z values of its even mass ions;

- predict, using Stevenson's Rule, which α-cleavages are most likely to occur in the mass spectra of secondary and tertiary amines;

- recognise and predict which further even mass ions will be formed from the α-cleavage product ions by loss of alkenes;

- recognise the typical features of the mass spectra of cycloalkyl amines, pyrrolidines and piperidines;

- recognise the typical features of the mass spectra of aromatic amines and pyridines.

9.7. FRAGMENTATIONS OF HYDROCARBONS

In most courses concerned with the study of organic compounds hydrocarbons always seem to come first. We have not done this when considering fragmentation patterns. You might have wondered why as I expect you have studied organic chemistry at one time or another. Well, hydrocarbons are not encountered as often as you might think (unless you work for an oil company)! Also, their spectra tend to be quite complex and the wood gets lost among the trees. It is easier to understand the basic types of electron impact induced fragmentations if you start with alcohols and carbonyl compounds, which show relatively few ions and fragmentation routes. Having said this, you do need to know something about hydrocarbon mass spectra.

Π   Sit back for a moment and think of a few reasons why you might find it useful to be able to recognise and identify hydrocarbons from their mass spectra.

Did your list include some of the following?

(*a*) We use hydrocarbon solvents in our laboratory. Sometimes we need to check their purity, and whether they have been removed from our final product(s).

(*b*) We never use hydrocarbons in our work, but I suppose it might come in useful one day.

(*c*) I've found the work on the other compounds interesting and would like to have some idea what hydrocarbons look like, even though I can think of no direct use we make of them at work.

(*d*) Hydrocarbon oils and greases are very commonly used in all sorts of manufacturing processes, and their residues frequently contaminate the final products. Mass spectrometry would be useful to detect them.

(*e*) Hydrocarbons from vehicles and fuel oil frequently contaminate the environment in small (or even gross!) amounts. Mass spectrometry can detect very small amounts contaminating say water supplies.

(*f*) Many widely used organic compounds have large hydrocarbon chains, or rings, eg waxes, fats, detergents, to name a few. Their mass spectra are likely to be dominated by hydrocarbon ions even though they have a functional group.

(*g*) Essential plant oils and fragrances contain hydrocarbons. Mass spectrometry could be used to characterise them.

(*h*) We make and/or analyse pure hydrocarbons in our work so knowledge of their mass spectral behaviour would be very useful.

If your answers included (*b*) or (*c*) you are probably among the majority of chemists and users of chemicals, as well as being honest with yourself. If you answered (*a*), (*g*) or (*h*) you are among the minority of chemists, probably working in a research environment, who are organic orientated. You will find this section useful. Does this mean the majority should skip it? Did your answers include any of (*d*), (*e*) or (*f*)? Hydrocarbons, or compounds with hydrocarbon chains are in very widespread use and tend to turn up in mass spectral analyses as unexpected or background peaks.

Answer (*f*) is perhaps the most important reason for an analytical chemist to know something about hydrocarbon fragmentations. Many molecules have substantial hydrocarbon rings or chains, as well as a functional group. The fragmentation of the hydrocarbon part will be observed in the spectra. Thus, the mass spectrum of such a compound consists of two types of peaks, those arising because of the presence of the functional group and those arising by fragmentation of the hydrocarbon chain or ring. You need to be able to recognise the typical m/z values they produce if only to eliminate them from further consideration. Look through this section to get an overview of the essential features of hydrocarbon spectra.

Saturated hydrocarbons as a class exhibit quite distinctive mass spectra. The trouble is that their $M^{+}_{\cdot}$ peak can be quite weak, especially as the carbon number increases and mixtures of long chain hydrocarbons such as arise from oils, greases and industrial grade solvents are difficult to sort out. The fact that saturated hydrocarbons are present is obvious, but which ones is much less easy to tell. This also applies to long chain alcohols, acids, ethers and some esters. The main features you will see are well-illustrated by the mass spectrum of dodecane, M_r 170, Fig. 9.7a.

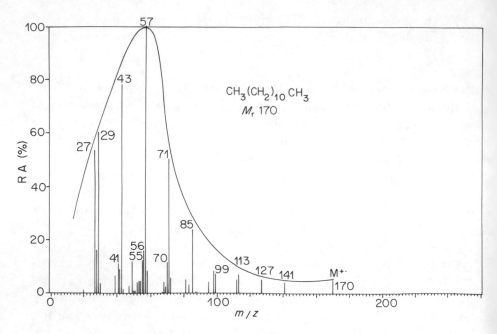

Fig. 9.7a.   *Mass spectrum of dodecane*

In the initial ionisation, an electron is removed from any carbon in the chain followed by cleavage of the adjacent C—C bond, with the loss of an R$^\cdot$ and formation of R$'$CH$_2^+$ ions. These form an odd numbered series m/z $(14n + 1)$, from CH$_3^+$ and CH$_3$CH$_2^+$ to CH$_3$(CH$_2$)$_9^+$, where n is the number of carbon atoms in the fragment. Thus the typical ions in a saturated hydrocarbon spectrum are m/z 15, 29, 43, 57, 71, 85, 99 ... (M-CH$_2$CH$_3$), differing by 14 amu. The second characteristic feature is the distribution of the relative intensity of these ions. In a saturated, unbranched chain, the most intense ion is m/z 57 (C$_4$ fragment), while the C$_3$ and C$_5$ fragments (m/z 43 and 71) are also intense. Either side of this range, the $(14n + 1)$ ion intensities fall off rapidly. The whole spectrum falls within an envelope shown by the smooth curve in Fig. 9.7a.

This arises because of the excess energies possessed by the higher mass ions. Initially the $(14n + 1)$ ions will be produced in roughly equal amounts, but the higher mass ones will fragment further:

$$CH_3(CH_2)_4 \overset{\frown}{-CH_2} \overset{+}{-CH_2} \xrightarrow[m^*50.9]{-H_2C=CH_2}$$
$$m/z \ 99$$

$$CH_3(CH_2)_2 \overset{\frown}{-CH_2} \overset{+}{-CH_2} \xrightarrow[m^*26.0]{-H_2C=CH_2} CH_3(CH_2)_2^+$$
$$m/z \ 71 \qquad\qquad\qquad\qquad m/z \ 43$$

There are successive losses of $H_2C=CH_2$ until their excess energy has been dissipated. Losses of $:CH_2$ are much less favoured and rarely occur. Repetition of this process with all the initial fragments favours the formation of the C_3, C_4 and C_5 carbocations.

A third distinguishing feature of saturated hydrocarbon spectra is the formation of two further series of ions, separated by 14 amu, due to the loss of RH from $M^{+\cdot}$ with the formation of $\dot{}(CH_2)_nCH_2^+$ ions, and H_2 from the $(14n + 1)$ series ions.

$$R(CH_2)_nCH_3 \ ]^+ \xrightarrow{-RH} \dot{}(CH_2)_nCH_2^+ \ \rightarrow \text{could cyclise or}$$
$$(A\text{-}1)^+ \qquad\qquad \text{form alkene ions}$$

$$RCH_2\overset{+}{C}H_2 \xrightarrow{-H_2} RC\overset{+}{=}CH_2$$
$$A^+ \qquad\qquad (A\text{-}2)^+$$

If the $(14n + 1)$ ions are termed A^+, these ions appear at $(A\text{-}1)^{+\cdot}$ and $(A\text{-}2)^+$ and are always of lower intensity than the A^+ ions.

When there is a branch in the hydrocarbon chain, the distribution of the $(14n + 1)$ ions changes dramatically because cleavage at a branch point would give either secondary or tertiary carbocations which are much more stable (Section 8.2). Such ions are always more intense than the isomeric primary RCH_2^+ ions and whichever is formed by the loss of the largest radical $R^{\cdot}$ will usually be the base peak (Stevenson's Rule).

$$R-\underset{\underset{H}{|}}{\overset{\overset{R_1}{|}}{C}}-R_2 \ \Big]^{\overset{+}{\cdot}} \rightarrow R_2^{\cdot} + \underset{R_1}{\overset{R}{\diagdown}}CH^+ \qquad R_2 > R_1 > R$$

base peak

∏   Now see if you can predict which ion(s) should be the base
      peak or most intense peaks in the mass spectra of:

(*i*) $(CH_3)_3C-CH_2CH_2CH_3$;

(*ii*) $(CH_3)_2CHCH_2CH-CH_2CH_3$;
        $\qquad\qquad\qquad\;\; |$
        $\qquad\qquad\qquad CH_3$

        $\qquad\qquad CH_3$
        $\qquad\qquad |$
(*iii*) $CH_3CH_2C-CH_2CH_2CH_3$;
        $\qquad\qquad |$
        $\qquad\quad CH_2CH_3$

(*iv*) $(CH_3)_2CH-\!\!\!\diagup\overline{}\Big|$

The answers are:

(*i*) m/z 57, $(CH_3)_3C^+$; (formed by loss of $CH_3(CH_2)_2^{\cdot}$).

(*ii*) m/z 43, $(CH_3)_2\overset{+}{C}H$ ($^{\cdot}CH_2CH(CH_3)CH_2CH_3$ lost) and m/z 57,
        $CH_3CH_2\overset{+}{C}HCH_3$ ($(CH_3)_2CHCH_2^{\cdot}$ lost). Since both these ions
        are secondary either could be the base peak, but m/z 43
        favoured.

(*iii*) m/z 85, $(CH_3CH_2)_2\overset{+}{C}CH_3$ by loss of $CH_3(CH_2)_2^{\cdot}$ and m/z
        99, $CH_3(CH_3CH_2)\overset{+}{C}CH_2CH_2CH_3$ ($CH_3CH_2^{\cdot}$ lost), but m/z 85
        would be favoured (Stevenson's Rule).

(*iv*) m/z 43, $(CH_3)_2CH^+$ (cyclopentyl$^{\cdot}$ lost).

These are essentially application of Stevenson's Rule (Section 9.2). The answers to (*ii*) and (*iii*) show you have to be careful not to be too dogmatic when two carbocations of the same class can be formed. Stevenson's Rule is helpful, but it is only a rule, not a law!

SAQ 9.7a   Fig. 9.7b shows the mass spectrum of a hydro-carbon *Unknown 21*. Identify the hydrocarbon.

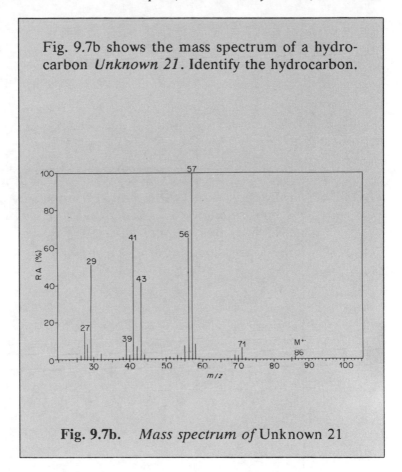

Fig. 9.7b.   *Mass spectrum of* Unknown 21

SAQ 9.7a

SAQ 9.7b   Fig. 9.7c shows the mass spectrum of a hydrocarbon *Unknown 22* isomeric with *Unknown 21*.
Identify this isomer.

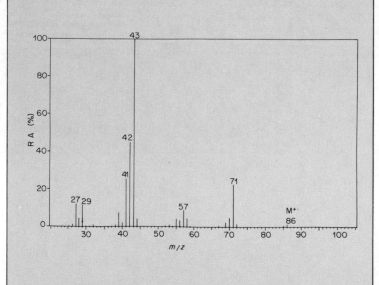

Fig. 9.7c.   *Mass spectrum of* Unknown 22

SAQ 9.7b

The answers to SAQ 9.7a and 9.7b and the material in the text should bring home to you how difficult it is to be categorical about the precise structure of a saturated hydrocarbon from its mass spectrum. In some cases of branched chains there are so many possibilities differing only slightly in the relative energies of the various carbocations that it comes down to one's gut feelings about what the peak intensities mean! It is essential, therefore, that you refer to literature spectra of authentic compounds and other spectroscopic techniques, particularly nmr spectra, to clinch an identification.

The mass spectra of cycloalkanes show prominent *even*-mass ions in addition to peaks formed from the loss of any R groups attached to the ring. This is because the rings fragment by successive losses of ethene. Note that their M_r are 2 amu less than the corresponding alkane, if monocyclic, 4 amu less if bicyclic, etc. As an example, Fig. 9.7d shows the spectrum of methylcyclopentane which has prominent $(M - CH_3)^+$, m/z 69 and $(69 - H_2C{=}CH_2)^+$, m/z 41 ions, but the base peak m/z 56 is formed by loss of $H_2C{=}CH_2$ from $M^{\ddagger}$.

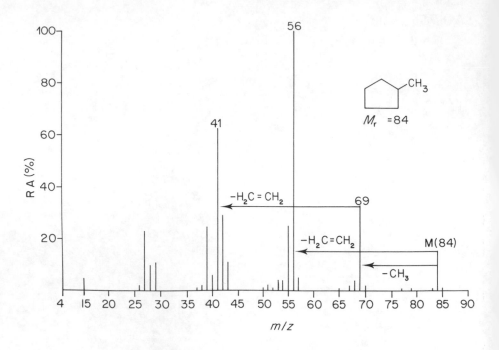

Fig. 9.7d.   *Mass spectrum of methylcyclopentane*

Alkenes also have M^{+} 2 amu less than the corresponding alkane, or 4 amu less if dienes etc, so it is important to consider what the distinctions are between these and cycloalkanes. Their M^{+} tend to be quite intense relative to their saturated analogues because ionisation occurs on the double bond by removal of one of the π-electrons, leaving the carbon skeleton relatively undisturbed.

When alkenes fragment they tend to give a series of ions corresponding to $C_nH_{2n}^{+}$ and $C_nH_{2n-1}^{+}$, ie $14n$ and $(14n - 1)$ of which the latter are the more intense. They are formed by a cleavage β to the double bond, known also as *allylic* cleavage.

$$R \overset{\frown}{-} CH_2 \overset{\curvearrowright}{-} \overset{.}{C}H - \overset{+}{C}H - R' \rightarrow R. \; + \; CH_2 = CH \overset{\curvearrowleft}{-} \overset{+}{C}H - R' \leftrightarrow$$

$$\quad\quad\quad M^{+} \quad\quad\quad\quad\quad\quad (M\text{-}R^{.}),\; C_nH_{2n-1}$$

$$\overset{+}{C}H_2 \overset{\curvearrowleft}{-} CH = CHR'$$

The allyl ions formed, of general structure $R'\overset{+}{C}H-CH=CH_2$, are stabilised by resonance, as shown above, and make the series $C_nH_{2n-1}^+$.

∏   We say 'series' here. How many $C_nH_{2n-1}^+$ ions would you predict for an unsymmetrical alkene of structure $RCH_2CH=CHCH_2R'$?

I expect you said two, $RCH_2\overset{+}{C}HCH=CH_2$ and $R'CH_2\overset{+}{C}HCH=CH_2$, since the double bond can ionise with the positive charge on either end and the allylic cleavage can take place at either CH_2 group. So how can we get a series of such ions? Two makes a pair, not a series!

Well, this points to one of the great problems with alkene mass spectra – the double bonds migrate along the chains by a series of H shifts.

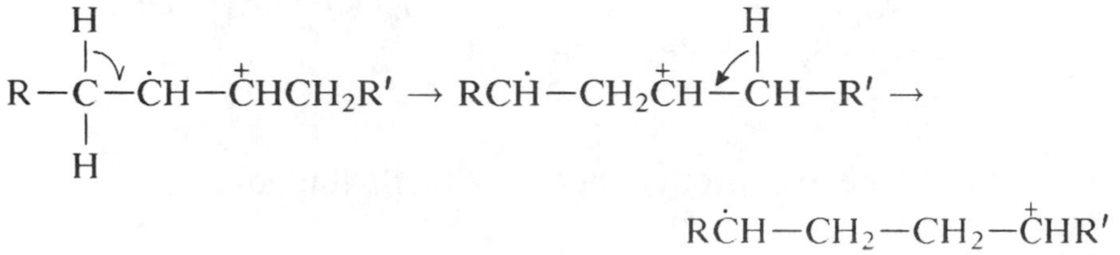

$$R-\overset{\overset{H}{|}}{\underset{\underset{H}{|}}{C}}-\overset{\cdot}{C}H-\overset{+}{C}HCH_2R' \rightarrow RC\overset{\cdot}{H}-CH_2\overset{+}{C}H-\overset{\overset{H}{|}}{C}H-R' \rightarrow$$

$$R\overset{\cdot}{C}H-CH_2-CH_2-\overset{+}{C}HR'$$

This means that more $C_nH_{2n-1}^+$ ions form than expected, and also, very important for identification, that isomeric alkenes give the same distribution of molecular ions by these shifts *and the spectra are virtually identical*. This also occurs in cycloalkenes, especially cyclopentenes and heptenes. So mass spectrometry is a poor technique to try to locate the position of a double bond in most alkenes. A further disadvantage is that once you have ionised a double bond, free rotation about it is possible, so *E*- and *Z*-isomers (*cis* and *trans*) also give nearly identical spectra.

Cyclohexenes, however, do have a characteristic fragmentation which leads to the breakdown of the ring to a dienyl ion and ethene.

$$m/z \; 54$$

The m/z 54 ion is equivalent to the $M^{\ddot{+}}$ of butadiene $CH_2{=}CH$ $-CH{=}CH_2$. This type of cleavage is a retro Diels–Alder reaction and is very useful in characterising the many different sorts of terpenes found in natural oils such as limonene which forms 2-methylbutadienyl ions, m/z 68 as the base peak in its mass spectrum.

$$m/z \; 68$$

Note that the neutral product here is 2-methylbutadiene!

Alkynes also show quite intense $M^{\ddot{+}}$ (4 amu less than the corresponding alkanes) and if terminal often show quite marked $(M - H)^+$ ions. These are probably *propargyl* ions $H-C{\equiv}C-\overset{+}{C}HR \leftrightarrows$ $H\overset{+}{C}{=}C{=}CHR$ since the propargyl ion itself, $HC{\equiv}C-\overset{+}{C}H_2$ m/z 39, formed by loss of $R^{\cdot}$ from $M^{\ddot{+}}$ is usually quite intense and characteristic of terminal alkynes.

$$H\overset{+}{C}{=}\overset{\cdot}{C}{-}\overset{\overset{\displaystyle H}{|}}{\underset{\underset{\displaystyle R}{|}}{C}}{-}H \rightarrow H\overset{+}{C}{=}C{=}CH_2 + R^{\cdot}$$

$$m/z \; 39/\text{Propargyl ion}$$

Carbocations derived from the R group (R^+ itself and its decomposition products, see alkane spectra) are present but generally weaker. If the R group possesses hydrogens γ to the $C{=}C$ a 'McL' type process results in the loss of alkenes:

$$R \quad H$$
$$H{-}\underset{\underset{CH}{|}}{\overset{|}{C}}{-}H \overset{H}{\underset{\underset{CH_2}{\diagdown\diagup}}{\overset{\diagup}{C^+}}} \rightarrow \underset{CH_2}{\overset{R\diagdown \quad \diagup H}{\overset{\|}{C}}} + \underset{\underset{CH_2}{\diagup}}{\overset{H\diagdown \quad \diagup H}{\overset{C^{\cdot}}{C^+}}} \equiv \overset{CH_2}{\underset{CH_2}{\overset{\|}{C}}} \, \Big]^{\ddagger}$$

$$m/z \ 40$$

So, to sum up, ions at m/z 39, 40 and $(M\text{-}1)^+$ tend to be typical of alkynes.

The mass spectra of most aromatic hydrocarbons show very intense $M^{+}_{\cdot}$ peaks, as the rings are so stable. When they do fragment, $(M - C_2H_2)$ and $(M - C_3H_3)$ ions are formed, and also to a limited extent there is loss of $H^{\cdot}$, then loss of C_2H_2 and C_3H_3. C_6H_5 compounds show characteristic ions at m/z 77, 52, 52 and 39 because of these processes but they are usually less than 20% RA. When an alkyl group is attached to a benzene ring, preferential cleavage occurs β to the ring (*benzylic* cleavage) to form the famous m/z 91, the tropylium ion:

$$C_6H_5{-}CH_2{-}R \, \Big]^{+}_{\cdot} \xrightarrow{-R^{\cdot}} C_6H_5\overset{+}{C}H_2 \equiv \text{⬡}^{+} \quad m/z \ 91$$

The mass spectra of the dimethylbenzene isomers are almost identical and show loss of $CH^{\cdot}_3$ to give $CH_3C_6H^{+}_4$ (again the tropylium ion m/z 91), and significantly loss of H from one of the CH_3 groups to give a medium peak at m/z 105 which is methyltropylium $CH_3C_7H^{+}_6$. This is a benzylic cleavage where R=H. The formation of a substituted tropylium ion is typical for alkyl substituted aromatics. Owing to its symmetry this usually means that the spectra are very similar and the substitution pattern cannot be found by mass spectrometry.

On the other hand, branching in the side chain may give rise to considerable differences. For example, isopropylbenzene readily loses one of the CH_3 groups by benzylic cleavage to give m/z 105 (methyltropylium) as base peak, but propylbenzene of course loses CH_3CH_2

by this process so its base peak is m/z 91, the tropylium ion itself and (M-CH$_3$) is only 5% RA.

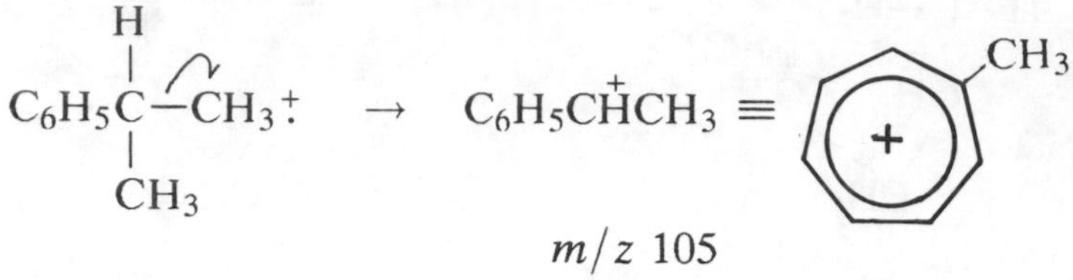

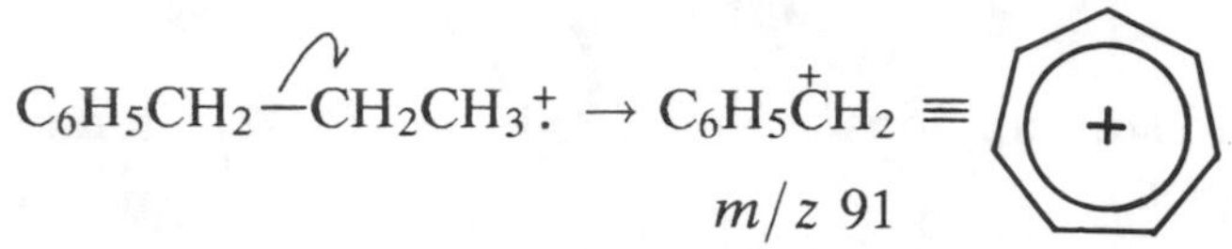

$$C_6H_5\overset{\overset{\displaystyle H}{|}}{\underset{\underset{\displaystyle CH_3}{|}}{C}}-CH_3\overset{+}{\cdot} \quad \rightarrow \quad C_6H_5\overset{+}{C}HCH_3 \equiv$$

m/z 105

$$C_6H_5CH_2-CH_2CH_3\overset{+}{\cdot} \rightarrow C_6H_5\overset{+}{C}H_2 \equiv$$

m/z 91

Another useful fragmentation is 'McL' rearrangement of γ-H from an alkyl side chain to the aromatic ring which gives rise to ions 1 amu *higher* than the tropylium ions. We have already seen something of this in the mass spectrum of 2-phenylethanol (Section 9.2). Butylbenzene is a typical example (Fig. 9.7e).

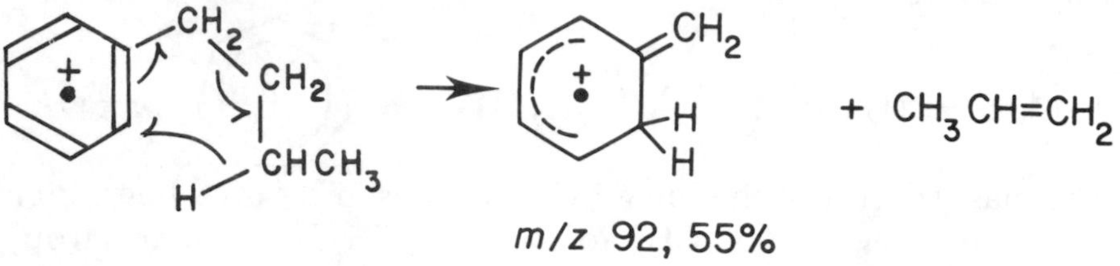

$+ \ CH_3\,CH{=}CH_2$

m/z 92, 55%

Fig. 9.7e.   *'McL' rearrangement in butylbenzene*

You should be aware though that isobutylbenzene, $C_6H_5CH_2CH(CH_3)_2$, also has a γ-H and rearranges to give 61% of m/z 92, so here again it is difficult to distinguish between such similar isomers as this by mass spectra alone. You need to compare with a mass spectrum of the authentic compound and get nmr spectra to be really sure which isomer you have.

Try the following SAQ's to see if you have understood this section.

SAQ 9.7c

Fig. 9.7f shows the mass spectrum of a hydrocarbon, *Unknown 23*. Suggest a structure for this compound and explain the main features of the spectrum.

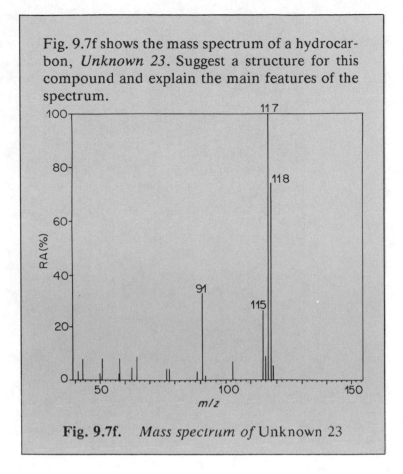

Fig. 9.7f. *Mass spectrum of* Unknown 23

SAQ 9.7d   Fig. 9.7g. shows the mass spectrum of a hydrocarbon, *Unknown 24*. Suggest a structure for this
compound and explain the main features of the
spectrum.

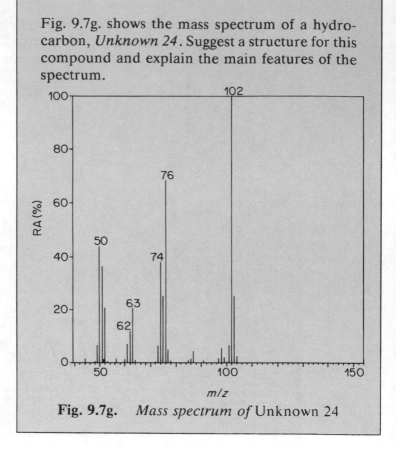

Fig. 9.7g.   *Mass spectrum of* Unknown 24

SAQ 9.7e

Fig. 9.7h shows the mass spectrum of a hydro-carbon, *Unknown 25*. Suggest a structure for this compound and explain the main features of the spectrum.

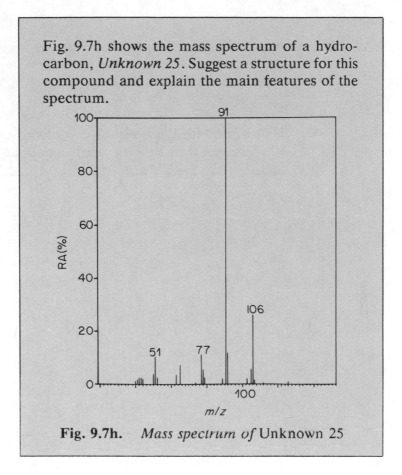

Fig. 9.7h. *Mass spectrum of* Unknown 25

SAQ 9.7f   Consider the isomers of the hydrocarbon, $C_{10}H_{14}$, a benzene derivative. Their mass spectral behaviour divides them into three main groups. What are these groups?

Would you be able to positively identify any of the isomers from its mass spectrum alone?

Summary

Pure hydrocarbon spectra are not met with that often in analytical work but the widespread occurrence of hydrocarbon groups in many products makes it useful to have an idea about the fragments they produce.

This section outlines the fragmentation patterns of saturated and unsaturated compounds. The limitations of mass spectrometry to identify all such compounds is also indicated.

Objectives

Now that you have finished this section on the mass spectra of hydrocarbons, you should be able to:

- anticipate the analytical areas where hydrocarbons might be present as back-ground contaminants;

- suggest from the principle ions observed in an hydrocarbon spectrum whether it was saturated, cycloalkyl, alkenic, alkynic or aromatic;

- predict which will be the base peaks of simple alkanes, alkenes, alkynes and aromatic hydrocarbons in pure samples;

- recognise where the ions observed in an hydrocarbon mass spectrum could result from other isomers;

- recognise that it is exceptional to be able to identify an hydrocarbon from its mass spectrum alone and be prepared to seek other confirmation, when necesssary;

- list structures of hydrocarbons consistent with the fragments seen in a mass spectrum and identify them as far as possible.

9.8. FRAGMENTATIONS OF HALOCOMPOUNDS

In Sections 7.3 and 7.4 we considered methods for the calculation of RA for the ion clusters produced by the chlorine and bromine isotopes. These were summarised in Fig. 7.3g. Turn back to Fig. 7.3g now to remind yourself of these typical patterns. We also pointed out that although fluorine and iodine were monoisotopic, they could be spotted by typical increments in the M_r of the compounds into which they were substituted.

∏   What are the increments typical of fluorine and iodine substitution?

For fluorine they are 18 amu and iodine 126 amu per hydrogen substituted. Have a look at Section 7.3.1 if you have forgotten this. From this you might think that halocompounds are fairly easy to identify from their mass spectra, especially as you will usually have a micro-analysis or be able to perform a simple qualitative test which shows that halogens are present. It is then, a matter of arithmetic to fit the number and types of halogens to the observed M_r, and cluster intensities for Cl and Br.

For haloaromatic compounds this is largely true, but they divide into two categories, fluoroaromatic and the rest. We pointed out in Section 7.3.1 that the very high C—F bond strength, about 423 kJ mol^{-1}, means that ˙F is rarely lost from a F compound. Other, weaker bonds cleave first. In general, the normal fragmentations of the other functional groups occur giving rise to fluoro-substituted daughter ions 18 amu to higher mass, per fluorine atom present, than the usual cleavage products. Thus, in the mass spectrum of 2,4-dinitrofluorobenzene, Fig. 7.3c, which typically has an intense M^{+} ion, the loss of one nitro group leads to the small ion at m/z 140 and the loss of the second to the base peak at m/z 94, $C_6H_3F^{+}$. The third prominent ion, m/z 30 is NO^{+}.

∏   Can you see any ions in the mass spectrum of Fig. 7.3c which do not contain F; other than m/z 30?

Loss of HF does occur because m/z 74, $C_6H_2^{+}$ is present, along with its daughter ion, $C_4H_2^{+}$ m/z 50:

$$C_6H_3F^+ \rightarrow HF + C_6H_2^+ \rightarrow C\equiv C + C_4H_2^+$$

HF is lost because of the even electron rule. Note that these ions are quite weak in intensity.

The other haloaromatics behave differently. They also usually have quite intense M^{+} peaks, and because the C—X bonds are weaker in the order I < Br < Cl they tend to lose the halogen atoms in successive fragmentations leading to the formation of hydrocarbon, or substituted hydrocarbon ions, which being aromatic are quite stable, eg

$$C_6H_3X_3^{+\cdot} \rightarrow X^{\cdot} + C_6H_3X_2^+ \rightarrow X^{\cdot} + C_6H_3X^+ \rightarrow X^{\cdot} + C_6H_3^+$$

∏   There is something unusual about the mechanism of this series of fragmentations. Can you see what it is?

The second fragmentation in the sequence offends the even electron rule (Section 8.2) – an even electron ion $C_6H_3X_2^+$ loses a radical $X^{\cdot}$ instead of an even electron species. This is presumably because of the weakness of the C—X bond and the stability of the halogen radicals. Nevertheless, HX is sometimes lost from such even electron ions, as the rule predicts.

When an haloaromatic contains a mixture of halogens, the fragmentation pattern can be quite useful in determining which are present. The most intense ions in the spectrum, apart from M^{+}, will be those formed by loss of the highest A_r halogen first, because this will have the weakest C—X bond, ie I before Br, Br before Cl, Cl before F. Thus, bromochloroiodobenzene will largely fragment as follows:

$$C_6H_3ClBrI^{+\cdot} \rightarrow I^{\cdot} + C_6H_3ClBr^+ \rightarrow Br^{\cdot} + C_6H_3Cl^{+\cdot}$$

Ions such as $C_6H_3ClI^+$, $C_6H_3BrI^+$, $C_6H_3I^{+\cdot}$, $C_6H_3Br^{+\cdot}$ will be weak or absent. Note that, as usual, we cannot tell much about the substitution pattern from the mass spectrum alone.

The molecular ion peak in the spectra of benzyl halides is usually observable. The halogen atom is lost extremely readily to form the tropylium ion, $C_7H_7^+$.

$$C_6H_5CH_2-\overset{\cdot+}{X} \xrightarrow{-X^\cdot} C_6H_5CH_2^+ \qquad\qquad C_7H_7^+$$

When the aromatic ring carries substituents, a substituted tropylium ion is formed, but a substituted phenyl carbocation may also be seen:

$$YC_6H_4CH_2X \,]^{\ddag} \rightarrow {}^\cdot CH_2X + YC_6H_4^+$$

SAQ 9.8a

Which fragment would you expect to be most prominent in the mass spectra of:

(*i*) $4\text{-}FC_6H_4CHO$ 4-fluorobenzaldehyde;

(*ii*) $4\text{-}FC_6H_4CH_3$ 4-fluorotoluene;

(*iii*) $4\text{-}IC_6H_4Br$ 4-bromoiodobenzene;

(*iv*) $3,4\text{-}Br_2C_6H_3Cl$
                    3,4-dibromochlorobenzene;

(*v*) $4\text{-}CH_3C_6H_4CH_2Br$
                    4-methylbenzylbromide?

For aliphatic halogen compounds, there is a marked trend in the $M^{\ddagger}$ ion intensities. They are most intense for iodocompounds, less strong for bromocompounds, weaker still for chlorocompounds and weakest for fluorocompounds. As the alkyl group becomes longer, as the amount of α-branching increases, or the number of halogen atoms increases, the $M^{\ddagger}$ ion intensity falls. This means that it may be very difficult to accurately detect the cluster intensities of the $M^{\ddagger}$ ions of aliphatic halocompounds, so a knowledge of the characteristic cleavages is useful.

Perhaps the most important of these is the simple loss of the halogen atom, leaving a carbocation. If this is a hydrocarbon, the rest of the spectrum will resemble a hydrocarbon mass spectrum (Section 9.7). This cleavage is most important when the halogen is a good leaving group, so it is most characteristic of iodo- and bromocompounds. Often the $(M-Cl)^+$ ion is weak in the spectrum of a chlorocompound, and as you might expect, $(M-F)^+$ is negligible in that of a fluorocompound. In the mass spectrum of 1-bromohexane, for example, (Fig. 9.8a) the peak at m/z 85 is due to the formation of the (hexyl)$^+$ ion. This then fragments by loss of propene to form a $C_3H_7^+$ ion, m/z 43.

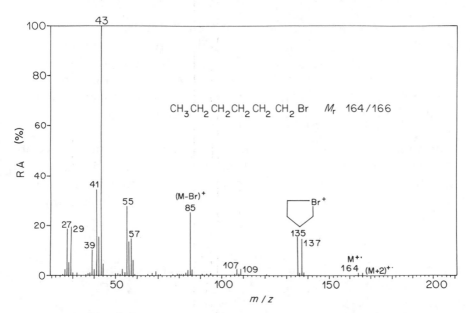

Fig. 9.8a. *Mass spectrum of 1-bromohexane*

Haloalkanes may also lose a molecule of hydrogen halide according to the process:

$$[R-CH_2CH_2-X]^{+\cdot} \rightarrow [RCH=CH_2]^{+\cdot} + HX$$

This mode of fragmentation is most important for fluoro and chloroalkanes, especially if they are secondary or tertiary, much less so for iodo and bromocompounds which more readily lose $\cdot$I or $\cdot$Br.

A less important mode of fragmentation is α-cleavage, in contrast to the behaviour of many other functional groups such as $-OH$, $C=O$ and $-NH-$:

$$R\overset{\frown}{-CH_2}-\overset{\cdot+}{X} \rightarrow R^{\cdot} + CH_2=\overset{+}{X}$$

This is because charge retention by the halogen atom is not favoured owing to its greater electron affinity than O or N atoms. Instead, the heterolytic loss of $X^{\cdot}$ is favoured:

$$RCH_2-\overset{+\cdot}{X} \rightarrow RCH_2^+ + X^{\cdot}$$

In cases where the α-position is branched, the largest alkyl group is preferentially lost (Stevenson's Rule) and α-cleavage is enhanced somewhat because the $RCH=X^+$ ion will be stabilised by the $+I$, $+M$ effect of the R group. Even so, such α-cleavage ions are of low RA, but may be useful in structure elucidation, especially if they contain Cl and Br with their distinctive isotope ratios.

The fourth characteristic fragmentation of halo-compounds is observed mainly in the spectra of chloro- and bromocompounds. It involves rearrangement to a cyclic halonium ion with loss of an alkyl radical. As in the spectra of alcohols, five-membered rings are favoured:

$$\rightarrow R^{\cdot} + \quad , C_4H_8X^+ \qquad X=Cl, Br$$

halonium ion

All the straight-chain chloro- and bromoalkanes from hexyl (C_6) to octadecyl (C_{18}) form these $C_4H_8X^{+\cdot}$ ions (m/z 91/93 and 135/137, respectively) as major or base peaks. In Fig. 9.8a, m/z 135/137, $C_4H_8Br^+$, is similar in intensity to the $(M-Br)^+$ ion, and in 1-chlorooctane, Fig. 9.8b, $C_4H_8Cl^+$ is the base peak. Branched chain chloro and bromalkanes give very much reduced amounts of substituted cyclic ions of this sort, probably because of competition with HX loss and hydrocarbon fragmentation. It is also curious that the corresponding six-membered ring ions, $C_5H_{10}Cl^+$ and $C_5H_{10}Br^+$ although frequently found in the spectra of long-chain compounds, are never very intense. Why the five-membered ring should be more stable has not been explained.

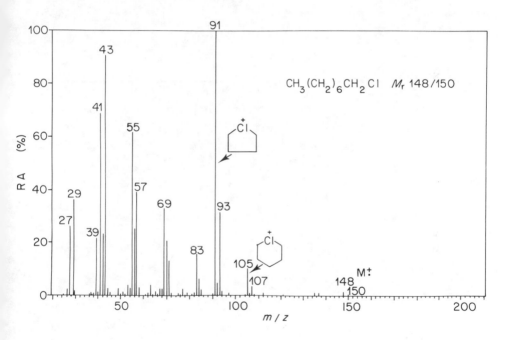

Fig. 9.8b. *Mass spectrum of 1-chlorooctane*

SAQ 9.8b	Which types of aliphatic halocompounds are likely to give significant $(M-HX)^{+\cdot}$ ions in their mass spectra?

SAQ 9.8b

SAQ 9.8c

> In the mass spectrum of fluoroethane, M_r 48, the base peak m/z 47 is CH_3CHF^+ while (M-$CH_3)^+$, m/z 33 is only 30% RA. Why are these RA values anomalous?
>
> Can you explain them?

SAQ 9.8d

> Which would you expect to give the most intense R^+ ion peak, an aliphatic primary or branched chain compound?
>
> $$(R-X)^{+\cdot} \rightarrow R^+ + X^\cdot \; (X = Br, I)$$

SAQ 9.8e

> If a chloro- (or fluoro-) alkane were branched at the carbon bearing the chlorine (or fluorine) would you expect the intensity of the (M-Cl) (or M-F) ion to increase, decrease, or stay the same compared to the primary isomer? For which type of compound and isomer would you expect to see the most intense (M-X) ion?

SAQ 9.8f   Apart from the $M^{\ddot{+}}$ ions, the mass spectra of
                   chloro- and iodoethane show one very signifi-
                   cant difference. Can you suggest what that would
                   be?

SAQ 9.8g

Fig. 9.8c shows the mass spectrum of *Unknown 26*, a haloalkane. Suggest a structure for this compound. Assign structures as far as you can to the ions m/z 77/79, 63/65, 57, 56 and 41.

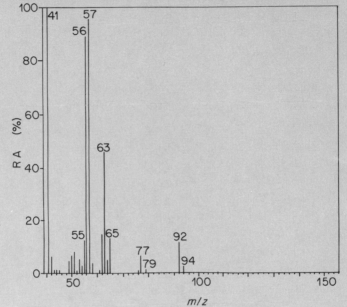

Fig. 9.8c. *Mass spectrum of* Unknown 26

SAQ 9.8h   Fig. 9.8d. shows the mass spectrum of *Unknown 27*, a polyhalo-compound. Which halogen(s) are present in *Unknown 27*? Suggest a structure for *Unknown 27*. Why cannot *Unknown 27* be conclusively identified from its mass spectrum alone?

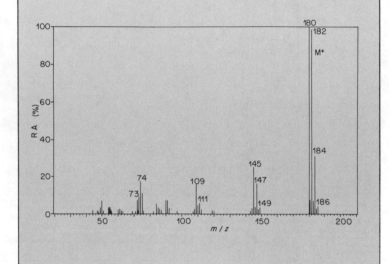

Fig. 9.8d.   *Mass spectrum of* Unknown 27

SAQ 9.8i Fig. 9.8e shows the mass spectrum of *Unknown 28*, a haloalkane. Which halogen is present in *Unknown 28*? Suggest a structure for this compound, and assign structures to the ions m/z 71, 70, 55 and 43 consistent with your structure for *Unknown 28*.

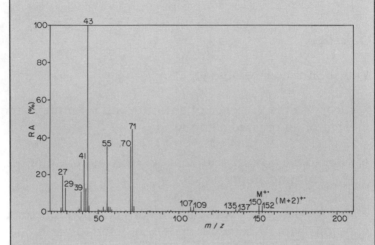

Fig. 9.8e. *Mass spectrum of* Unknown 28

Summary

Bromo- and chlorocompounds can usually be spotted from the pattern of (M + 2) isotopes present, as described in Section 7.2 and 7.3, especially when they are aromatic since their M^+ ions are usually intense. Aliphatic halocompounds, however, often give weak M^+ ions so their fragmentation patterns need to be analysed if you wish to identify them. This also applies to fluoro- and iodocompounds where there are no isotopes. Procedures for dealing with haloalkanes and haloaromatics have been explained.

Objectives

You should now be able to:

- recognise the presence of chlorine and bromine in the mass spectrum of an organic compound;

- refer to Section 7.2 and 7.3 and work out the number of chlorine and bromine atoms present;

- deduce the presence of fluorine in a compound from the existence of ions with increments of 18n, n being the number of fluorines, and losses of 20 amu due to HF;

- deduce the presence of iodine in a compound from the existence of ions with increments of 126n, n being the number of iodines, and losses of 127 amu (occasionally 128) due to $I^.$ (or HI);

- suggest structures for haloaromatics from their mass spectra;

- suggest structures for simple benzyl halides from their mass spectra;

- describe the characteristic features of the fragmentations of haloalkanes, and the distinguishing features of fluoro-, chloro-, bromo- and iodoalkanes;

- identify simple haloalkanes from their spectra, and suggest structures for more complex ones where complete identification is not possible by mass spectrometry.

9.9. FRAGMENTATIONS OF NITROCOMPOUNDS

Aliphatic nitrocompounds are sometimes met with as solvents, eg nitromethane, CH_3NO_2. Apart from CH_3NO_2, they have very weak M^{+} ions and fragment to give hydrocarbon ions.

$$R-\overset{+}{N}O_2 \rightarrow NO_2 + R^+ \rightarrow (alkyl)^+ \text{ ions}$$

The 'McL' rearrangement does occur but the 'McL' ions, eg

$$\overset{\cdot}{C}H_2-\underset{+}{N}\overset{OH}{\underset{\diagdown}{\big|}}_{O} \qquad m/z \text{ 61 are about 1\% RA or less.}$$

β-Cleavage to give $\overset{+}{C}H_2NO_2$, m/z 60 is negligible. Thus, (M-46) is typical for an aliphatic nitrocompound. NO_2^+, m/z 46 is not normally observed.

The isomeric nitrites, $R'-O-N=O$, on the other hand, show easy β-cleavage to give an ion m/z 60 and $R^{\cdot}$.

$$R-CH_2-\overset{\cdot\cdot}{O}-N=O \rightarrow R^{\cdot} + CH_2=\overset{+}{O}-N=O \quad m/z \text{ 60}$$

Nitrites also show strong NO^+, m/z 30, but NO_2^+ is not produced. There is no sign of the isomerisation

$$RNO_2^{+\cdot} \rightleftarrows RONO^{+\cdot}$$

which occurs in aromatic nitro compounds, so RNO_2 and $RONO$ compounds are readily distinguishable.

Aromatic nitrocompounds are frequently met with and show distinctive mass spectra. The M^{+} peaks are usually quite intense, and as expected loss of $^{\cdot}NO_2$ frequently gives the Ar^+ as base peak, especially if this is stabilised by $+M$ groups and/or can rearrange into a tropylium ion. Many $ArNO_2$ compounds show a small $(M-O)^+$ ion which is quite distinctive – few other functional groups lose a single oxygen atom from $M^{\ddagger}$. The most interesting fragmentation though is loss of NO, followed by CO, from $M^{\ddagger}$. It is believed the nitrocompound $M^{\ddagger}$ rearranges into the nitrite which readily loses NO to give an aryloxycation. This is shown for nitrobenzene in Fig. 9.9a. The phenoxy ion m/z 93 and its m/z 65 daughter ion are familiar from the mass spectra of phenyl ethers (Section 9.3.).

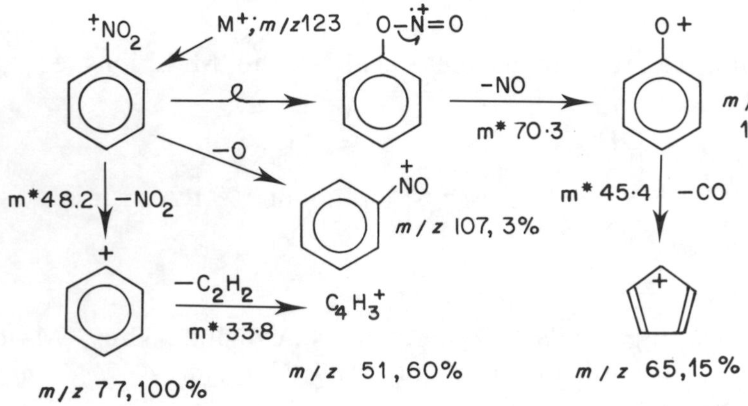

Fig. 9.9a. _Fragmentations of nitrobenzene_

Thus, the most general features of the mass spectra of aromatic nitrocompounds are M-16 (M-O), M-30 (M-NO) and M-46 (M-NO$_2$) ions, but not NO_2^+ itself. Ortho compounds frequently show a variation where they lose OH by H transfer from a suitable position in the ortho group. For example, in 2-nitroaniline, m/z 121 (M-OH) is formed as well as m/z 122.

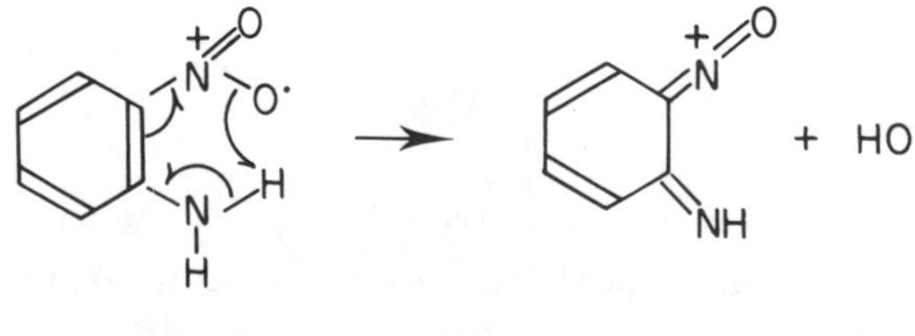

m/z 121,5%

The spectra of the three isomeric nitroanilines are shown for comparison in Fig. 9.9b (*i*), (*ii*) and (*iii*). The (M-OH) ion in Fig. 9.9b (*i*) although small distinguishes the 2-nitrocompound. In 2-nitrotoluene, $M_r = 137$, (M-OH), m/z 120 is the base peak, but such an intense ortho effect is unusual.

(*i*)

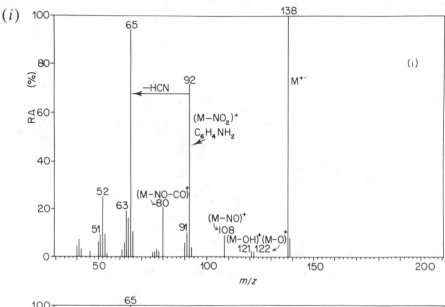

(*ii*)

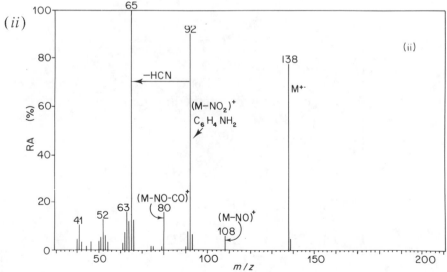

(*iii*)

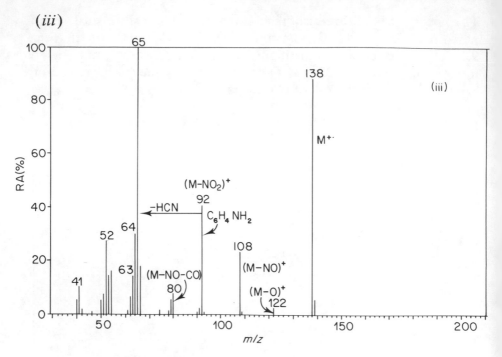

Fig. 9.9b.   *Mass spectra of nitroanilines*
(i) 2-nitroaniline, (ii) 3-nitroaniline, (iii) 4-nitroaniline.

SAQ 9.9a

A compound X shows base peak m/z 43, other intense ions at m/z 29, 41, 57, 71 and weak ions at m/z 30 and 75. The M^{+} is not discernable. X is known to be a nitro compound and could be:

A: $CH_3(CH_2)_4NO_2$

B: $CH_3CH_2CH(CH_3)NO_2$

C: $(CH_3)_2CHCH(CH_3)NO_2$

D: $(CH_3)_2CHCH_2CH_2NO_2$

Which do you think it is?

SAQ 9.9a

SAQ 9.9b   A compound Y has $M^{\ddagger}$ 151. It shows base peak m/z 91, and other intense ions at m/z 30, 51, 60, 65. There are weak ions at m/z 77, 105 and 121. Deduce which of the structures E to H is the most likely structure for Y:

E: $C_6H_5CH_2CH_2NO_2$

F: $4\text{-}CH_3CH_2C_6H_4NO_2$

G: $4\text{-}CH_3C_6H_4CH_2NO_2$

H: $C_6H_5CH_2CH_2ONO$

SAQ 9.9c

A compound Z has M^{+} 151, but shows (M-1) RA 2%. Its base peak is shown, by a metastable, to lose 28 amu to give 93, which further loses 28 amu to give 65. The m/z 104 ion is 30% RA and loses 28 amu to give m/z 76. A close inspection of the spectrum shows m/z 134, 3% and 135, 1%. Suggest a structure for Z.

Summary

Aliphatic nitrocompounds fragment by loss of $^{\cdot}NO_2$ to give strong hydrocarbon ions. 'McL' rearrangement occurs, but such peaks are weak though distinctive, as are the M^{+}. (M-46) is typical. They are easily distinguished from the isomeric nitrites because the latter show β-cleavage, and have m/z 30, NO^{+} quite intense. Aromatic nitrocompounds have much more intense M^{+} and show (M-NO), (M-NO—CO) and (M-O) ions as well as (M-NO_2).

Objectives

Now that you have completed Section 9.9 you should be able to:

- state the main fragmentations expected in the mass spectra of aliphatic nitrocompounds and nitrites, and aromatic nitrocompounds;

- recognise nitrocompounds from their mass spectra;

- identify simple nitrocompounds from their mass spectra;

- suggest structures for more complex nitrocompounds from their mass spectra;

- distinguish between aliphatic nitrocompounds and isomeric nitrites;

- distinguish 2-substituted aromatic nitrocompounds from their 3- and 4-substituted isomers.

9.10. FRAGMENTATIONS OF HETEROCYCLIC COMPOUNDS AND SULPHUR COMPOUNDS

This section completes our study of fragmentation patterns with a brief consideration of some common heterocyclic compounds ending with sulphur heterocycles and some brief notes on the sulphur compounds such as thiols, thioethers and thiophenols.

9.10.1. Heterocyclic Compounds

∏   What do you understand by the term 'heterocyclic'? Give some examples of heterocyclic compounds.

Heterocyclic compounds are aromatic compounds in which one or more of the $=C-H$ groups have been replaced with heteroatoms such as $=N$, $-NH-$, $-O-$, $-S-$. Fig. 9.10a gives some examples. Did some of these appear on your list?

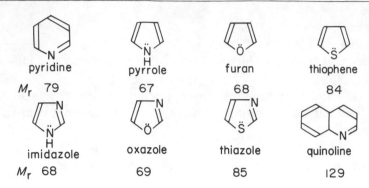

Fig. 9.10a.   *Some common heterocyclic compounds*

Note that they all have aromatic sextets or, as in the case of quino-line, dectets, of delocalised electrons and so possess aromatic sta-bility. Such systems, and their benzo-analogues such as quinoline are very common in biological chemistry. Since they form impor-tant components of living organisms and many drugs their analysis is important.

∏   We have already considered one heterocycle in Section 9.6 because it is also classed as an amine. Do you remember what it was and how it fragmented?

It was pyridine, C_5H_5N. Pyridine has an intense $M^{\dot{+}}$ and loses HCN to form m/z 52 (75%). There is also present m/z 51 but no other important fragments. We also added that alkylpyridines lose H to give azatropylium ions, which also lose HCN. If there are γ-H atoms in the alkyl chain a 'McL' rearrangement occurs with loss of alkene and formation of m/z 93, isomeric with the $M^{\dot{+}}$ of the methylpyridines. If you have forgotten these points have another look at Section 9.6.

We can now expand on and extend this pattern to other heterocycles. For the most part they give intense $M^{\dot{+}}$, and the alkyl analogues lose H (if CH_3-substituted), or cleave the β-bond to lose $R^{\cdot}$ and give a tropylium species, usually as base peak. The rings themselves break down by loss of small, stable, neutral molecules, analogous to HCN such as CH=NH, CO, HC=O, HC≡S as well as HC≡CH. In the substituted heterocycles, ArX types show varying amounts of Ar^+, and ArCOX types almost always have ArCO$^+$ as intense peaks (or ArCOOH$^+$ if X has a γ-H for 'McL' rearrangement to occur).

Fig. 9.10b summarisies the main features of the mass spectra of the heterocycles of Fig 9.10a.

System	Neutrals Lost (amu)		Ar$^+$	Ions Formed Tropylium	'McL'	(m/z) ArCO$^+$
Pyridines	HCN	(27)	78	92	93	106
Quinolines	HCN	(27)	128	142	143	156
	HC≡CH	(26)				
Pyrroles	HCN	(27)	66	80	–[3]	94
	CH=NH	(28)				
Furans	HC≡CH	(26)	67	81	–[3]	95
	CHO	(29)[1]				
Thiophenes	HC≡CH	(26)	83	97	–[3]	111
	HC≡S	(45)				
Imidazoles	HCN	(27)	67	81	–[3]	95
Oxazoles	HCN	(27)	68	82	83	96
	CO	(28)				
	CHO	(29)[1]				
Thiazoles	HCN	(27)[4]	84	112[2]	99	112
	HCN + H	(28)				

[1] Major loss of those shown

[2] Peak shown by dialkylthiazoles, methylthiazoles do not form a tropylium ion.

[3] β-cleavage to give the tropylium ion much predominates.

[4] HCN loss occurs first, followed by H loss. The ions formed are similar in intensity.

Fig. 9.10b. *Main features of the mass spectra of some heterocyclic systems*

You will note from Fig. 9.10b that apart from furans and imidazoles all the typical ions diagnostic of these heterocycles have different m/z values. It should be possible to spot which system you have from these, and remember too that sulphur has a significant (M + 2) isotope which distinguishes sulphur heterocyclic ions. Furans are distinguished by the loss of CHO (29 amu). Imidazoles behave like pyridines rather than pyrroles in only losing HCN (27 amu). This points to a common feature of the mass spectra of nitrogen-containing heterocycles.

∏   Can you see what this common factor is?

Whenever a heterocycle contains the structural unit $-N=CH-$, losses of HCN feature strongly in its spectrum. Imidazoles and thiazoles lose HCN almost exclusively (not $CH=NH$ or $HC\equiv S$, respectively) and oxazoles lose both HCN and CHO, though the latter does tend to be more favoured. There is another common feature involving this group and that is 'McL' rearrangement from 2-substitutents having γ-H atoms. Furans, pyrroles and thiophenes form tropylium ions by straight-forward β-cleavage.

An interesting question concerning 2-substituents is whether compounds containing them can be distinguished by mass spectrometry. In heterocycles containing $-N=CH$, and γ-H atoms in the substituent, the 'McL' rearrangement ions will be found and the answer is yes. 2-Methyl substituents can nearly always be distinguished because they will lose an extra 14 amu in the neutral fragment, eg 2-methylpyridine loses CH_3CN, 41 amu, 2-methylfurans lose CH_3CO 43 amu, 2-methylthiophenes CH_3CS 59 amu, and so on for the other systems. Furans and thiophenes also show ions corresponding to the neutral species, eg HCO^+ m/z 29, CH_3CO m/z 43, HCS^+ m/z 45, CH_3CS^+ m/z 59 which support this distinction. Once the side-chain at the 2- position is *longer than CH_3, however*, the usual mode of breakdown is β-cleavage to the corresponding tropylium ion and *this distinction is lost*.

You will also find that ortho-effects between two neighbouring substituents in a disubstituted heterocycle occur in a similar fashion to those found in benzene compounds, ie not involving the heteroatom. For example, 2-methyl-3-carboxyethylthiazole eliminates ethanol while 2-methyl-5-carboxyethylthiazole does not.

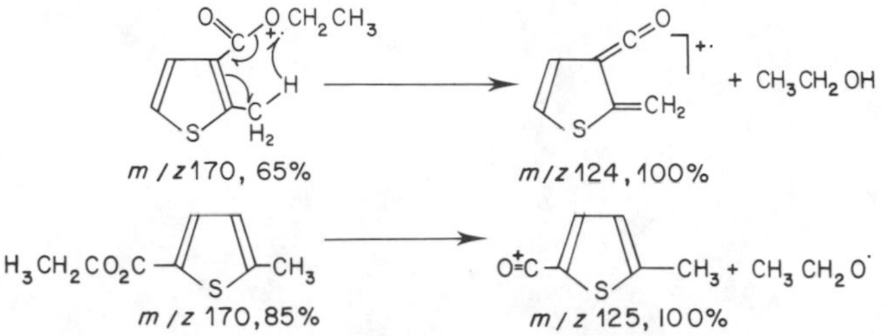
m/z 170, 65%
m/z 124, 100%
+ CH₃CH₂OH
H₃CH₂CO₂C
m/z 170, 85%
m/z 125, 100%
+ CH₃CH₂O·

SAQ 9.10a

Compound *V* has $M^{+\cdot}$ *m/z* 126 (82%), 111 (100%), 83 (36%), 57 (16%), 45 (18%) and 39 (25%) with m* 62.1 and 39.1. All the ions listed here have (M + 2) peaks of about 5%.

V could be:

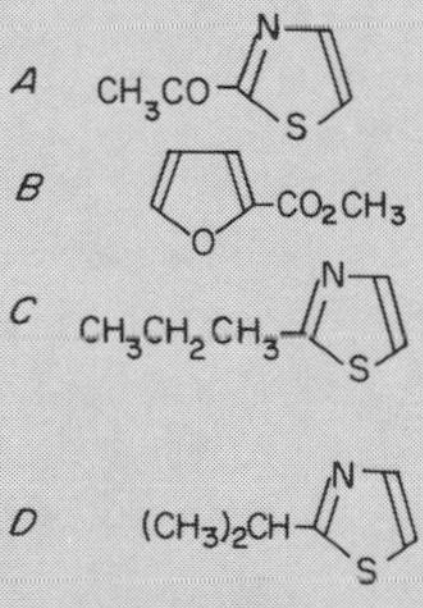
A CH₃CO
B CO₂CH₃
C CH₃CH₂CH₃
D (CH₃)₂CH

Which do you think it is?

SAQ 9.10a

SAQ 9.10b   *Compound W* has M$^+$ *m/z* 95 (100%), 67
(21%), 41 (23%), 40 (26%), 39 (50%) with m*
47.3 and 22.7. *W* could be: $\longrightarrow$

**SAQ 9.10b
(cont.)**

E

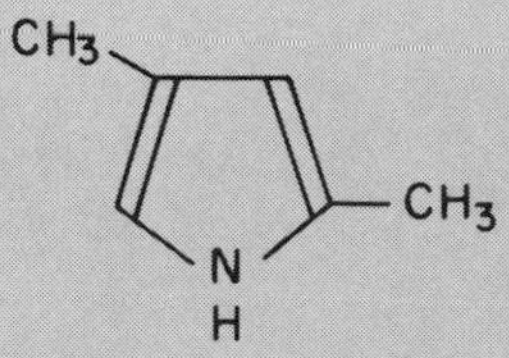

F

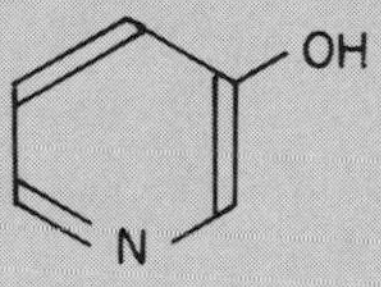

G

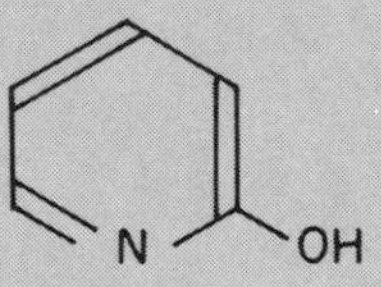

H Either *F* or *G*

Suggest which of the options *E–H* you think is most likely.

SAQ 9.10c

Compound X has $M^{\ddagger}$ m/z 110 (44%), 95 (100%), 67 (5%), 43 (13%), 39 (14%), 38 (4%), m^* 82.1, 47.3, 22.7. X could be:

I

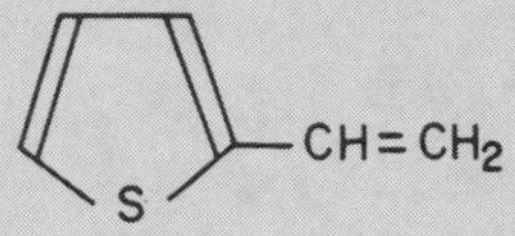

J

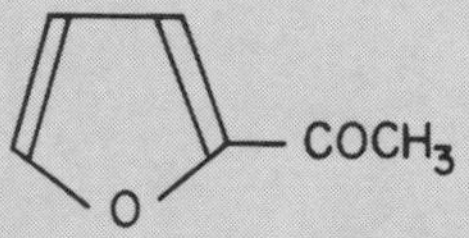

K

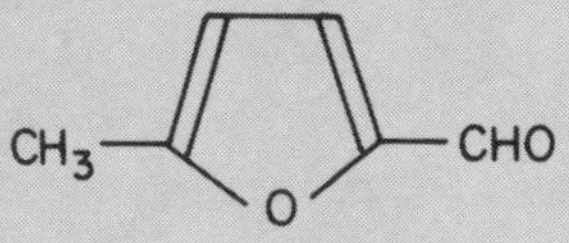

L

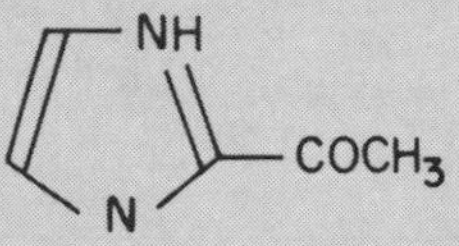

Which do you think it is?

9.10.2. Thiols

When comparing sulphur compounds with the analogous oxygen compounds, apart from the presence of the 4% (M + 2) isotope (Section 7), the M^{+} peaks are in general more intense. In other respects their behaviour is similar to the oxygen analogue with just a few marked differences.

Π   What features would you expect to see in the mass spectrum of a thiol, by analogy with those of alcohols?

Alcohols show α-cleavage ions in the series m/z 31, 45, 59, 73 ... etc. ($CH_2=\overset{+}{O}H$ etc) so thiols should show $CH_2=\overset{+}{S}H$, m/z 47, then m/z 61, 75, 89 ... etc. Alcohols also show (M-H$_2$O) and M-(H$_2$O + CH$_2$=CH$_2$) so thiols would be expected to show (M-H$_2$S) and M-(H$_2$S + CH$_2$=H$_2$), ie (M-34) and (M-62) peaks. If you did not anticipate these fragments, have another look at Section 9.2.

All these ions are in fact found, along with hydrocarbon ions formed by cleavages at each point in the alkyl chain. In chains of five carbons or more, δ-cleavage is particularly favoured in thiols as the ion formed, m/z 89, is cyclic:

$$m/z\ 89$$

This ion is analogous to the cyclic oxonium and halonium ions given by alcohols and chloro- and bromalkanes.

9.10.3. Thioethers (Sulphides)

∏   What features would you expect to see in the mass spectra
       of a thioether, by analogy with those of ethers?

You would expect to see α-cleavage of the longest chain giving ions
such as $R\overset{+}{S}{=}CH_2$, ie in the series m/z 61, 75, 89 ... etc though some
loss of both possible radicals would be expected. If the R group
contained a β-H, transfer to the $S^+{=}CH_2$ would occur next with
loss of a neutral alkene, hence $H\overset{+}{S}{=}CH_2$ m/z 47 would be expected.
If you did not anticipate these sort of fragments, have another look
at Section 9.3.

However, thioethers show two differences from ethers. Because RS^+
ions are more stable than RO^+, cleavage occurs at the $S{-}R$ links
and RS^+ ions of fairly high RA appear in the series m/z 47, 61,
75, 89 ... etc. Thus, a thioether $R{-}S{-}R'$ gives RS^+ and $R'S^+$ of
unusual intensity. The second difference is that one of the alkyl
groups cleaves off with transfer of a β-H to the charged sulphur
atom, giving an ion 1 amu higher than the RS^+.

$$R-\overset{+\cdot}{S}-CH_2\overset{\overset{H}{|}}{CH}{\longrightarrow}\quad RS\overset{+\cdot}{H} + CH_2{=}CH_2$$

The presence of such an even mass ion is very useful in fixing the
position of the sulphur in the chain.

SAQ 9.10d

A compound Y contains one sulphur atom and
its mass spectrum shows the following major
ions: m/z 90 (100%) [M^+]; 75 (97%); 49 (42%);
48 (75%); 47 (39%); 43 (68%); 41 (77%) and
27 (27%). m/z 90, 75, 49, 48 and 47 contain a
sulphur atom. Compound Y could be:

M $CH_3CH_2CH_2CH_2SH$

N $CH_3SCH_2CH_2CH_3$ $\longrightarrow$

SAQ 9.10d (cont.)

> O $(CH_3CH_2)_2S$
>
> P $CH_3SCH(CH_3)CH_3$
>
> Which do you think it is?

9.10.4. Arythioethers and Thiophenols

∏   What features would you expect to see in the mass spectrum of an arylthioether, by analogy with an arylether?

The $M^{\ddagger}$ would be intense and lose $R^{\cdot}$ to give ArS^+ which would further expel $C{=}S$ to form cyclopentadienyl ions. Methyl aryl ethers would expel $CH_2{=}S$ and transfer a hydrogen to the aromatic ring. $M^{\ddagger}$ would also fragment by loss of $RS^{\cdot}$ to give Ar^+ ions, and longer chain R groups would lose an alkene and transfer a hydrogen to sulphur to give $ArSH^+$. If you remembered all these processes from Section 9.3, well done; if not have a quick look at that section again. Fig. 9.10c shows the mass spectrum of methylthiobenzene and its fragmentation pathways for comparison with these predictions.

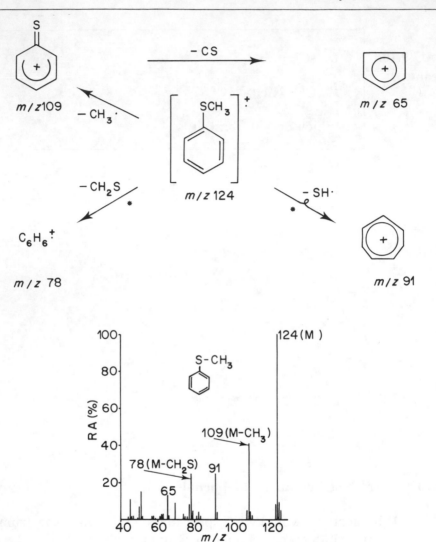

Fig. 9.10c. *Fragmentation pattern and mass spectrum of methylthiobenzene*

∏   There is one fragmentation shown in Fig. 9.10c. which is *not* as you predicted. Can you spot which it is?

It is the formation of the tropylium ion, m/z 91, by expulsion of HS·. In higher alkyl thioethers this is not so important, as formation of

$ArS^+ = CH_2$ and $ArSH^{+\cdot}$ ions tends to supersede $HS^\cdot$ loss. Hence, (M-SH) is typical of $ArSCH_3$ compounds.

∏ By comparison with phenols, what features would you expect to see in the mass spectrum of a thiophenol?

An intense $M^{+\cdot}$, followed by loss of C=S to give cyclopentadienyl ions would be expected for thiophenol itself and derivatives other than alkyl ones. Methyl and alkylthiophenols would be expected to lose either $H^\cdot$ or cleave off the β-$R^\cdot$ group of the alkyl chain to give sulphhydryltropylium ions analogous to the hydroxytropylium ions. Refer back to Section 9.4 if you have forgotten these phenol fragmentations.

In practice thiophenols differ significantly from these predictions. Take a look at Fig. 9.10d which shows the mass spectrum of thiophenol.

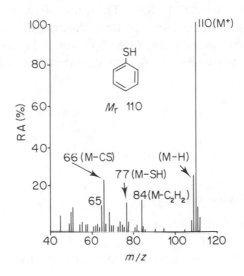

Fig. 9.10d. *Mass spectrum of thiophenol*

The two expected features *are* shown by thiophenol, eg an intense $M^{+\cdot}$ and (M-CS), m/z 66. But there are also significant ions at (M-H), m/z 109, (M-C_2H_2), m/z 84 and (M-SH), m/z 77 which have no analogies in the phenol mass spectrum. Presumably, (M-H) is a thiotropylium ion which can then lose CS to give m/z 65.

$$m/z \ 109 \qquad \qquad \qquad m/z \ 65$$

The losses of SH and C_2H_2 are quite reasonable fragmentations for a substituted benzene compound with activation energies more favourable here than in phenols.

Alkyl thiophenols also show different behaviour. They lose SH much more readily than a group from the α-carbon and give alkyl tropylium ions in the series m/z 91, 105, 119 ... etc. This is, presumably, because the ˙SH radical is more stable than ˙OH and so is more easily lost than H˙. As the alkyl chains become longer α-cleavage competes with ˙SH loss. Thiophenols also show the same sort of ortho effects that 2-substituted phenols do, eg thiosalicyclic acid and its methyl ester eliminate H_2O and CH_3OH, respectively, from M^{+}, whereas the 3- and 4-substituted isomers do not.

$$m/z \ 136 \qquad \qquad \qquad m/z \ 108$$

R = H thiosalicyclic acid

R = CH_3 methyl thiosalicylate

It is interesting that the second step in this pathway is loss of CO, not CS.

SAQ 9.10e

A compound Z contains one sulphur atom and its mass spectrum contains the following major ions: m/z 138 (100%) [M‡]; 123 (65%); 110 (66%); 109 (24%); 65 (21%); 51 (23%); 45 (32%). m/z 138, 123, 110, 109, 45 contain the sulphur atom.

Z could be:

Q $C_6H_5CH_2SCH_3$;

R 4-$CH_3C_6H_4SCH_3$;

S 4-$CH_3CH_2C_6H_4SH$;

T $C_6H_5SCH_2CH_3$;

Which do you think it is?

Summary

In Section 9.10 we have looked at the fragmentations of some common heterocyclic and sulphur compounds. They are compared with their oxygen analogues.

Objectives

Now that you have finished this section, you should be able to:

- recognise the main features of the mass spectra of pyridines, quinolines, pyrroles, furans, thiophenes, imidazoles, oxazoles and thiazoles;

- predict the main features of the mass spectrum of a given heterocyclic compound;

- distinguish between isomeric furans and imidazoles by noting the losses of CHO and HCN, respectively;

- distinguish between 2-substituted pyridines in general, and 2-methyl substituted heterocycles and their other positional isomers;

- recognise ortho effects where they occur between adjacent substituents in heterocycles;

- recognise that a given unknown spectrum is that of a thiocompound by the presence of a 4% $(M + 2)^{+ \cdot}$ isotope peak;

- recognise the typical fragmentation patterns of thiols, thioethers and thiophenols and the distinctions they show from the analogous oxygen compounds;

- suggest structures for unknown thiols, thioethers and thiophenols from mass spectral data.

Summary to Part 9

Part 9 has introduced you to a system of analysing an unknown mass spectrum in order to gain as much information as possible about the structure of the compound. In Section 9.1 two correlation tables are presented, an (M-X) table and a Mass-Composition table, which will help you to decide what class or classes an unknown compound might belong to, and likely structures for the main ions in the spectrum.

Once that has been decided, the typical fragmentations of the indicated class(es) of organic compounds can be referred to in Sections 9.2–9.10, as needed. It is not intended that you should learn all this detail at first, as it is far better in practice to build up your experience and confidence with interpretation over a period of time.

It is also important to realise that the other branches of spectroscopy, especially ir and nmr are often of great use, if not essential, in completing the final structure determination. Thus, in Section 9 we have emphasised where mass spectra cannot distinguish positional, structural and stereoisomers very well. In these instances use of other types of spectra and reference collections of mass spectra are vital and should be used as confirmatory evidence.

Objectives

Now that you have studied Part 9, you should be able to:

- analyse a mass spectrum by a logical procedure applying (M-X) and Mass Composition Tables as appropriate in order to assign the compound to its class or a limited range of classes;

- identify simple compounds from their mass spectra;

- recognise when the mass spectral information is inadequate to assign a unique structure to an unknown compound either because it does not fit any of the classes in Section 9.2–9.10, or the fragmentation pattern does not permit a clear distinction to be made;

- present your conclusions in a clear and concise manner indicating likely structures for the important ions in the mass spectrum.

10. Chromatography – Mass Spectrometry

Since the early 1960's the use of mass spectrometry has been closely associated with the development of gas chromatography (gc) in the analysis of complex volatile organic mixtures. By about the mid-1970's the linked gc-ms had matured into a routine method of analysis. It was about then that the use of high performance liquid chromatography (hplc) for the analysis of involatile organic compounds coincided with the development of new ionisation methods, suited to such compounds eg field desorption and ionisation and fast atom bombardment. This led to a growing interest in linking liquid chromatographs to mass spectrometers. Now lc-ms is expanding rapidly as research is done into efficient ways of linking the two instruments.

This section aims to introduce the two techniques gc-ms (Section 10.1) and lc-ms (Section 10.1), concentrating on the methods of efficiently linking the two types of instrument, and give some examples of their uses and the methodology of linked chromatographic-mass spectroscopic analyses. Section 10.2 takes you to the present (1986) frontiers of lc-ms. It attempts to provide a review of the state of the art in what is a fast-developing field, so that you will get a feel for the potential of the method and why it is currently attracting interest.

10.1. GAS CHROMATOGRAHPY – MASS SPECTROMETRY

It is possible, of course, to collect (trap) compounds as they elute from gas (or liquid) chromatographs and afterwards obtain their mass spectra by probe or AGHIS techniques. This is cumbersome, but it can have some advantages.

∏ Can you suggest one or two of these?

If a sample is trapped, part of it can be used for other spectro-scopic analyses, eg ir, nmr, in order to identify it conclusively. From our study of fragmentations in Section 9 you should be aware that isomers in particular, can give very similar mass spectra that may not distinguish between them. Other spectra can be very helpful, but obviously if the whole sample has been dispatched to the mass spectrometer it is not possible to recover any for such examination. Another reason might be to examine what appears to be a homoge-neous chromatographic peak on another gc column with different separating polarity to see if it really is one component.

∏ Can you suggest two or three disadvantages of this approach to mixture analysis?

Three come readily to mind: (*i*) difficulty of efficient trapping, par-ticularly of minor components; (*ii*) it is a very time-consuming method; (*iii*) it is impossible to use with capillary columns owing to the minute amounts of sample analysed.

So, the modern approach is to link chromatographs to dedicated, single, mass spectrometers rather than to try to split the sample in some way or pass it consecutively through two types of spectrometer. Thus we have the situation where a single sample of a urine extract obtained from a patient who has taken a particular drug would be analysed in parallel by gc-ms for volatile metabolites, gc-ir to con-firm these metabolites, and lc-ms for involatile metabolites.

Discussion of the chromatographic processes themselves is outside the scope of this Unit. It will be assumed that a good separation has been obtained on a conventional gc (or lc) and a need established for the identification or quantitation of some of the components of a complex mixture. These chromatographic techniques owe their prime position in analytical chemistry to their combination of versatility, wide range of applicability, and sensitivity. Where materials are too involatile or too unstable to pass through a gc, lc takes over. It is, therefore, important to establish that the mass spectrometer also meets the criteria and that the device linking the chromatograph and the spectrometer, known as the interface, does not cause unacceptable deterioration in performance. It would be pointless to get a good gc resolution if the mass spectrometer or its interface thermally degraded the compounds, or proved insufficiently sensitive.

Let us first consider the vital question of sensitivity. A routine analytical need in gc is to analyse down to one part per million of a compound in solution. It is conventional to inject one microlitre (μl) of such a solution onto the gc column and expect to get a recognisable signal from whatever detector is in use. Regarding the mass spectrometer just as a detector for the moment, let us examine if it is possible to obtain a mass spectrum of reasonable signal/noise ratio from such a 1 ppm component, assuming it is all delivered to the ion source (we shall see later how efficiently this can be done).

∏   How much compound, in g, is delivered to the mass spectrometer from a one μl injection of a 1 ppm (by weight) solution of a volatile compound, assuming 100% transmission efficiency to the mass spectrometer?

1 ppm by weight is 10^{-6} g in 1 g of solution.

1 μl is 10^{-3} g (assuming unit density).

Hence, the amount of the solute in 1 μl is 10^{-9} g, or 1 nanogramme (1 ng). Modern mass spectrometers are well capable of determining usable mass spectra on ng quantities and, in fact, can go down to 10^{-12} g, the picogramme (pg) range. Recent manufacturers' literature speaks of sensitivity in the femtogramme (10^{-15} g) range, so it is actually possible to obtain useful data at parts per billion (10^{-9} g per g solution) levels! The conclusion is that the mass spectrometer has more than enough sensitivity to be a good gc detector.

The next parameter is temperature. Gcs operate at temperatures between ambient and about 300 °C, depending on the nature of the sample and the stationary phase used on the column. Is this compatible with the mass spectrometer?

∏   What is the operating temperature range of the mass spectrometer?

It is usual to operate mass spectrometer inlet systems and source chambers at about 200 °C. This ensures efficient transfer of samples through the system and rapid removal after ionisation, but it is not so high as to decompose most covalent materials (though there are exceptions to this, where lower temperatures are need). If you did not remember this, have a look at Section 2.2 to remind you. Hence, gc and mass spectrometer are quite compatible in this respect so long as the interface between them is also heated to at least the temperature of the gc oven.

∏   Why is it important to heat the interface?

For the same reason that AGHIS are heated (Section 2.2). If there were 'cold spots' in the interface, separated components would condense out in the transfer lines and might not reach the source chamber. If they partially condensed, the next component to emerge from the gc column would catch up with them and the separation would be degraded. It is usual to heat all transfer lines and interface devices to 20–30 °C above the oven temperature used in the gc, and line them as far as possible with inert glass surfaces to avoid metal catalysed thermal decompositions and rearrangements.

Let us next consider the time factor. It is vital that the mass spectrometer can obtain a good spectrum on each and every component which emerges from the gc column before it has passed through the system and been lost forever. You can't stop the gc while the spectrum is scanned, it just keeps on eluting more peaks! Ideally, we would like to obtain about ten scans of the mass spectrum across a gc peak while it is eluting.

∏   Why is this? Surely one good spectrum is all we need? Why
         not wait until the component reaches its maximum and then
         make one scan of the mass spectrometer?

In principle, one good representative mass spectrum *is* all we need.
The problem is getting it when the compound is flowing through the
source chamber and its concentration is changing quite rapidly over
what might be several orders of magnitude as it elutes off the column. Taking the one scan when it has reached a maximum sounds
fine in theory but is difficult in practice. You could attempt it manu-
ally by waiting until the gc peak reached a maximum and then press
the mass spectrometer 'scan' button, but this is very tedious in a long
separation with say 40 components. It is not easy to achieve with a
computer-controlled mass spectrometer, either. This may surprise
you, as you might think that a threshold could be set, and scans
automatically start when it was exceeded, but the difficulty is that
in many samples the concentrations of the components vary over a
wide range, say 4 orders of magnitude in essential oil and flavour
studies, or metabolic studies. Setting a suitable threshold for all of
them is impossible.

Another reason for wanting as many scans as possible across a gc
peak is to minimise the effect of the changing concentration of the
sample on the appearance of the mass spectrum obtained, in other
words, to obtain at least one scan where the concentration is changing as little as possible.

∏   What would the effect be on the relative concentration of
         the ions in a mass spectrum if the scan went from high mass
         to low mass starting when the sample concentration was (*i*)
         increasing rapidly (gc peak just entering the source); (*ii*) de-
         creasing rapidly (gc peak clearing the source)?

(*i*)   In this case, when the sample concentration is increasing
         rapidly, the M^+ and closely related ions would be recorded
         when the concentration was small and their intensity would be
         reduced relative to the lower mass ions in the spectrum.

(*ii*)   In this case, when the sample concentration is decreasing
         rapidly, the M^+ and the higher mass ions would be more in-
         tense relative to the lower mass ions.

I hope you got these the right way round!

In either case the mass spectrum recorded would differ from that obtained under steady-state conditions using a direct probe or AGHIS. Comparison of the spectrum with reference collections and computer data bases such as the EPA National Institutes of Health or National Bureaux of Standards files (both Washington, DC), and the Eight Peak Index Collection (UK Chemical Information Service, Nottingham University) would give poor matches and possibly mis-identification of the component. This effect is called 'mass discrimination' and is worse for lower scanning speeds.

Returning to the point about scan time, this is related to the time of elution of a gc peak. For a packed column this could be as long as one minute, but for a capillary column could be as little as one second. In order to get ten scans across a peak taking one minute to elute we would require a mass spectrometer scanning rate of about six seconds per scan. This is no problem because even the older conventional electromagnets can scan a decade of mass (say 500 to 50 amu) in one second and reset in two seconds, thus giving a repetitive scan time of three seconds. So almost any mass spectrometer can scan fast enough to meet the basic requirement for packed column gc-ms and minimise mass discrimination.

∏   What scan rate is required for the avoidance of mass discrimination in capillary gc-ms?

If ten scans are required across a gc peak width of one second, the scan rate must be 10^{-1} s. This implies a scan speed of 5×10^{-2} s decade^{-1}, allowing for 'fly-back' time before the next scan.

∏   Which types of mass spectrometer can achieve such scanning rates? Is it feasible to use a mass spectrometer for capillary gc-ms?

Quadrupole and time-of-flight mass spectrometers can easily achieve this sort of scan rate. Recently Fourier-transform instruments have been introduced which separate ions by a resonance high frequency method and they also scan very rapidly. This is why from about 1970 onwards there was a surge of interest in quadrupole mass spectrometers in particular because they made good gc detectors. There has been strong rivalry between 'quad' and 'mag' mass spectrometer manufacturers ever since to produce machines for this market. The magnetic focusing instruments have been improved by the introduction of faster scanning laminated magnets which reduce eddy currents in the core to allow faster scans and importantly, reset times.

So, we can conclude that there are several mass spectrometer designs which adequately meet the scan time requirement for good reproducible mass spectra of separated components from present day capillary gc.

Another requirement for a gc detector is that it produces a chromatogram. Seems obvious, but so far we have only considered the production of mass spectra of the components of a mixture. Can we use the mass spectrometer to produce a chromatogram of time of elution versus concentration comparable with conventional gc detectors? This is vital because we need to establish retention indices and concentrations as well as identify the components. This is, fortunately, very easy. All we have to do is place a small collector plate in the edge of the positive ion beam emerging from the source chamber, give it a negative potential, and tap off a small proportion of the positive ion stream. The potential produced from this total ion current (TIC) detector, as it is called, can be amplified in the usual way and outputed as a TIC trace on chart paper. This will look very like a conventional gc chromatogram produced by a flame ionisation detector (FID) for example and is just as sensitive in most cases.

An alternative method often used when a data system is employed (regarded as almost essential in view of the hundreds of scans which will be recorded during a typical gc-ms run of say 30 min) is to use the number-crunching ability of the computer to sum up the intensities of all the ions recorded by the electron multiplier at the end of the analyser system for each scan made. The total intensity is then outputed as a voltage–time plot. Such a TIC trace will be stepped as it is updated after every scan but with the high scanning speeds of modern instruments you have to look closely to see the steps.

In some systems provision is made for the diversion of some of the gc effluent stream to a FID or other conventional gc detector while the rest goes to the mass spectrometer. This gives a gas chromatogram as per normal. This system is satisfactory for packed column work as there is plenty of sample for both the FID and mass spectrometer, but for capillary columns the reduced sample size makes it less desirable and it is not used.

From the above discussion you will have gathered that gas chromatographs and mass spectrometers are very compatible instruments indeed! But there is one factor we have not discussed so far, and this is the one which provides the chief difficulty in linking the two.

∏   Can you think what this factor is?

It is pressure. A gc operates at an outlet pressure of one atmosphere and a mass spectrometer ionisation chamber at about 10^{-6} torr. Since 1 torr = 1 mmHg pressure this is a difference of nearly nine orders of magnitude. This means that the interface should be capable of sustaining a pressure drop of this magnitude without excessive loss of sample or degradation of the chromatographic separation. It is also desirable to separate the carrier gas, most often helium, from the sample in order to concentrate it as much as possible.

A number of separators for the gc-ms interface have been developed over the last twenty years but one has shown itself to be the most generally useful by virtue of its robust construction, simplicity, cheapness and reasonable efficiency and is in most common use today. This is the *jet separator*, originally devised by Becker and Ryhage in Sweden, and is shown in Fig. 10.1a.

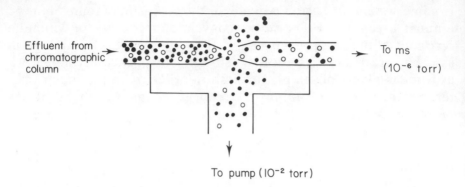

Fig. 10.1a. *Jet separator showing enrichment of the sample* (o) *by removal of carrier gas* (●) *molecules*

The jet separator relies on the differential diffusion of the lighter carrier gas molecules away from a jet created by passing the effluent stream from the gc into a small vacuum chamber. The gas stream expands as it enters the vacuum chamber from a small orifice and, according to Graham's Law, the lighter carrier gas molecules diffuse at a rate inversely proportional to their density, so as the jet expands into a cone the outer areas are richer in helium. The gas stream is directed towards a small orifice which leads to the mass spectrometer, about 1 mm away. When it has reached this, most of the helium has diffused outwards and misses the hole, while the heavier solute molecules have deviated little from the straight path between the two jets and so enter the mass spectrometer, though clearly some will be lost. This design of separator, operated at oil-pump pressure, $\approx 10^{-2}$ torr, removes about 90% of the carrier gas. About 60% of the sample reaches the mass spectrometer with virtually no delay. There is also a very small surface for the passing sample.

∏   Why is it important that there should be a small contact surface for the passing sample?

This minimises decomposition and rearrangement of the components caused by contact with hot surfaces. It also reduces sample dead volume which maintains chromatographic quality.

You will realise, of course, that a jet separator can only be optimised for one particular ratio of densities of carrier gas to sample, and one particular flow-rate if the gap between the jets is held constant. This means that its transmission efficiency is not constant for all the components of a mixture. In particular, much of the sample will be lost if it is of a low relative molecular mass, ie jet separators become more efficient as the relative molecular mass difference between the carrier gas and sample molecules rises. This is one good reason why helium is widely used for gc-ms work. Hydrogen would be even better because it is lighter than helium but it presents other handling problems!

∏   Can you suggest a way these disadvantages of the fixed jet separator could be overcome?

The clue was in the word 'fixed'. If one of the jets was moveable, particularly from outside the vacuum system, by a micrometer say, you could adjust the jet separation to allow for different flow rates. In practice no manufacturer has produced an externally adjustable jet separator, though some are available which can be set up with different separations to suit particular flow-rates and/or samples. Once set, however, these are fixed and take quite a time to dismantle and reassemble and pump down. It is usual to adjust the gc conditions to suit the separator if at all possible.

A further useful refinement of the jet separator is to introduce a 'dump valve'. This is a thin strip of tungsten in a holder which can be slid in between the jets by operating a magnetic link outside the vacuum system. This deflects the gc effluent stream completely down the inlet backing system pump to waste.

∏   What portion of the gc effluent stream might we want to deflect from the ms?

Gc samples are often made up as dilute solutions in a solvent. Obviously, we know what this is, so there is no need to determine its spectrum. As it is in large amount, it might cause the ms to be overloaded, so it is best dumped. Some samples contain large known components which are best selectively deflected to avoid excessive pressure rise in the ion source and possible filament damage.

Having removed most of the carrier gas, the sample reaching the ionisation chamber from the jet separator is not much different from that entering from the conventional inlet system, so the usual methods of ionisation can be used. We can do gc-ms in EI, CI and field ionisation modes just as easily as with single samples. Indeed, it is quite possible now to switch from EI to CI and back to EI in consecutive scans on the same gc component peak so as to get information both on its relative molecular mass from the MH^+ ion seen in CI and structural information and identity from the fragments produced in the EI spectrum. However, it is probably true to say that the 'hybrid' sources fitted with these devices are not as efficient in any one of them as a source designed and optimised for one particular inlet and ionisation method.

There is one part of the interface – source system we have not said much about – the pumps. These tend to be taken for granted but mass spectrometrists owe a great deal to the expertise of the pump manufacturers and the rapid developments of recent years. The removal of carrier gas from a packed column gc requires the exhausting of 30–40 cm^3 min^{-1} through the inlet system backing pump, which is a lot of gas at 10^{-2} torr, and the maintenance of 10^{-6} to 10^{-7} torr in the source housing by the main diffusion or turbomolecular pump of the mass spectrometer.

This implies flow-rates of the order of 1000 1 min^{-1} through these pumps. It is a credit to the manufacturers that they have provided reliable pumps which can achieve this 24 hours per day for months on end. There is another implication of this high pumping capacity, apart from the ability to remove large volumes of CI gas at the same time, which is very important to gc-ms.

Π   Can you suggest what this might be? If you are not experienced in modern gc analysis do not spend too much time trying to puzzle this out, but read on.

The high pumping capacity of the pumps used makes it feasible to couple a capillary gc column to the source chamber *directly*, without needing a separator. The gas flows used in capillary columns are about 10% of those in packed columns, so the pumping system designed to cope with the 2–4 cm^3 min^{-1} from a jet separator has no difficulty coping with the flow from a capillary column which typically will be $1 \text{ cm}^3 \text{ min}^{-1}$ or less. This means that no sample is lost in the interface and full advantage can be taken of the tremendous separations achievable using capillary columns. Most gc-ms work today is being done on capillary columns threaded directly through the glass-lined heated steel tubing of the interface into the ionisation chamber.

There are limitations on the kinds of gc column we can use for gc-ms, and the column temperatures. This is because the stationary phases, being organic polymers, will depolymerise as the temperature is raised and give characteristic decomposition products known as 'bleed'. These continuously elute from the column through the interface into the mass spectrometer where they contaminate the source and can give rise to ions as intense as the sample ions.

∏   Can you suggest two or three practical measures we can take to minimise 'bleed' problems?

The following measures are useful:

(*i*)   use longer columns with low stationary phase loadings in packed column work, say 5% stationary phase or less;

(*ii*)   be very careful not to exceed the decomposition temperature of the stationary phase;

(*iii*)   in capillary work use the recently introduced bonded phase columns which are much more stable.

Bonded phase capillary columns are particularly useful in analysis of mixtures of components of widely differing volatilities where temperature programming of the column is essential if the less volatile ones are ever to emerge in a finite time. Since the 'bleed' increases in a non-linear way with rise in temperature it is very difficult to correct for. It is sensible to use the expensive bonded phase columns which produce little 'bleed' up to really high temperatures.

The two commonly used gc stationary phases are polyethyleneglycol (PEG, Carbowax) and dimethylsiloxane (OV1) polymers. These are available as bonded phases called BP20 and BP1. BP20 is useful for natural oil and flavour work as it is a polar phase, while BP1 is non-polar and widely useful for polar molecules such as biochemicals, drugs and metabolities. BP20 is relatively bleed-free, but BP1 and other silicone phases give silicon-containing 'bleed' peaks at m/z 75, 147, 207 and 281 which you should be aware of. You should always try to establish and eliminate 'bleed' peaks from your gc-mass spectra. Many data systems include routines which will subtract one scan from another and these can be used to eliminate 'bleed' peaks by selecting a scan from a region just before or after a component of interest and subtracting this from the component mass spectrum.

∏   It is particularly important to use this routine if the gc col-
       umn is being temperature programmed. Can you suggest a
       reason for this?

It is because the rate of decomposition of the polymer phase will rise significantly during the run giving a non-linear increase in the 'bleed' peaks. So the background spectrum should be taken from a region of the chromatogram as close as possible to the component of interest.

Many types of compound need to be derivatised before gc separation because of involatility. It is important to choose derivatives which are suitable for mass spectrometric identification of the compounds in gc-ms work. We have already mentioned trimethylsilyl (TMS) derivatives in Part 9. Acids give TMS esters and amines TMS amides with trimethylsilylating reagents. *Tert*-butyldimethylsilyl is also used as a derivatising group:

$$ROH \rightarrow R-OSi(CH_3)_2 t\text{-Bu}$$

but such silyl derivatives do not always yield good structural infor-mation from their mass spectra, so acetyl and trifluoroacetyl esters and methyl esters are also used for alcohols and amines.

$$ROH + (CH_3CO)_2O \rightarrow R-OCOCH_3 + CH_3CO_2H$$

$$ROH + (CF_3CO)_2O \rightarrow R-OCOCF_3 + CF_3CO_2H$$

$$ROH + CH_2N_2 \qquad \rightarrow R-OCH_3 \qquad + N_2$$

Acids are frequently converted to their methyl esters, as we have already mentioned (Section 9.5.5). Aldehydes and ketones may be converted into their O-methyloximes which yield stable M^+ and structurally informative fragment ions:

$$\text{>\!=\!O} + NH_2OCH_3 \rightarrow \text{>\!=\!NHOCH}_3$$

Vicinal and 1,3-diols, and related systems with neighbouring active hydrogens, eg aminoalcohols, hydroxyamides, are easy to derivatise with alkyl boronic acids such as butyl-, *t*-butyl- and phenyl-boronic acids to give volatile cyclic boronates.

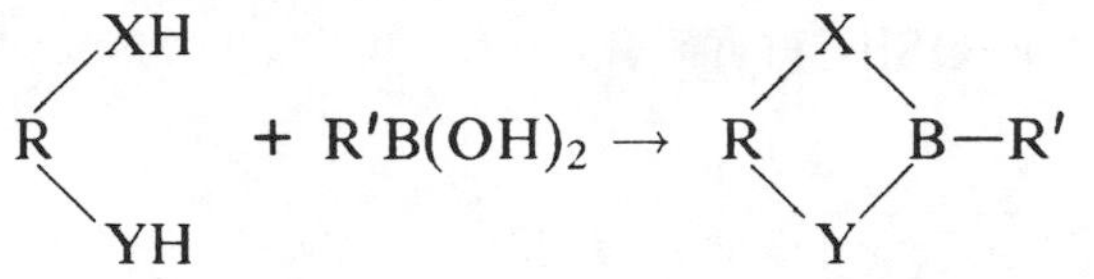

$$R\begin{array}{c} XH \\ \\ YH \end{array} + R'B(OH)_2 \rightarrow R\begin{array}{c} X \\ \diamond \\ Y \end{array}B-R' \qquad \begin{array}{l} R' = Bu, \; t\text{-Bu,Ph} \\ \\ X,Y = O, N \end{array}$$

These derivatives are formed only if 5- or 6-membered rings are possible, and are widely used in the study of natural products related to carbohydrates.

Π What do you think is the most important factor in the choice of a derivative for gc-ms analysis?

The over-riding factor is normally a chromatographic one. It is vital that an efficient gc separation is achieved before mass spectrometric identification is possible. However, all other things being equal, it should be remembered that from the mass spectrometric point of view the derivative group should have a small mass increment and should not undergo too easy fragmentations, ie should give abundant high mass ions if it is to be of any use in identifying the components. Occasionally though, a larger group is chosen deliberately to move the M^+ of the compound away from the region of a persistent contaminant, eg a 'bleed' peak, cholesterol in blood plasma metabolite extracts. Fig. 10.1b. summarises some common derivatives and the mass increment they cause per derivatised group.

Typical Substrate	Derivative	Mass Increment (amu)
Alcohols	$-OCH_3$	14
Phenols	$-OSi(CH_3)_3$	72
Enols	$-OSi(CH_3)_2 t\text{-Bu}$	114
Thiols	$-OCOCH_3$	42
Amines	$-OCOCF_3$	96
Acids	$-OCH_3$	14
	$-OSi(CH_3)_3$	72
Aldehydes / Ketones	$=NOCH_3$	29

OH OH
 | |
>C — C<

OH   OH
 | |
>C — C — C<
         |

NH$_2$OH
 | |
>C — C<

NH$_2$ OH
 | |
>C — C — C<
         |

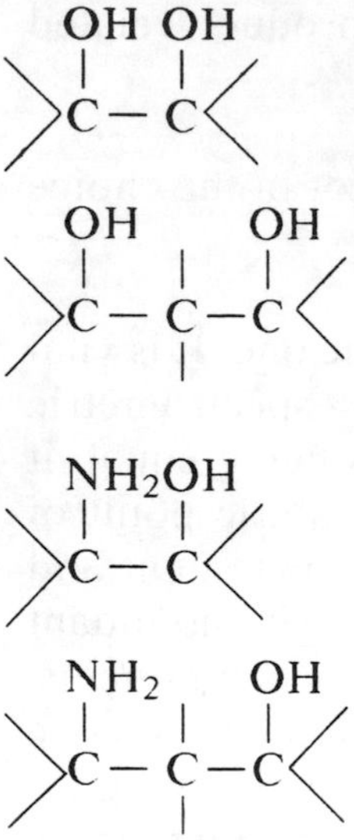	BuB-	66
	t-BuB-	66
	PhB-	86
	$-CH(CH_3)_2$	40

Fig. 10.1b.   *Commonly used derivatives and their associated mass increments*

SAQ 10.1a

Suggest derivatives which will be suitable for the gc-ms analysis of each of the following classes of compound:

(*i*)   fatty acids;

(*ii*) amino acids and peptides;

(*iii*) carbohydrates;

(*iv*) phenols;

(*v*)   2-hydroxyacids;

(*vi*) steroid hormones (contain C=O and OH groups);

(*vii*) fruit essences (contain alcohols, carbonyl compounds and esters).

In some cases there is a choice – select which derivative you think would be best from the ms point of view.

Quantitative work can be performed by gc-ms at very low levels, eg ng cm^{-3} of blood plasma for drugs and metabolites if suitable internal standards are available. Fig. 10.1c shows the methodology employed in quantitative mass spectrometry by gc-ms. Purification of the crude biological extract may not always be necessary. Mass spectrometric sensitivity and selectivity are greatly aided by a data system capable of monitoring selected ions from the mass spectra, a technique known as selected ion monitoring (SIM).

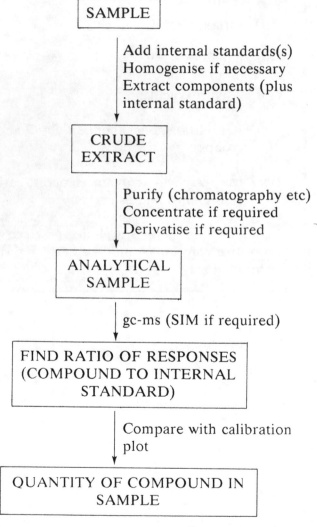

Fig 10.1c.   *Typical analytical methodology for quantitative gc-ms*

The standard has to be a compound closely similar to the compound analysed so that its gc and mass spectrometric behaviour are very similar. Deuterated drugs are particularly useful as they are generally quite safe to administer to patients with the drug itself, but extract and chromatograph very similarly. Trideutromethyl groups can be introduced fairly easily at the synthetic stage, or perhaps added later before derivatising the mixture. By using SIM on the M^+ (or other prominent ion) of the drug and its deuterated analogue, at say $(M + 3)^+$ if it contains a CD_3 group, the ratio of the drug and the standard can be obtained *without the necessity for chromatographic separation to be achieved*. This is one of the great advantages of the gc-ms technique because we can use the mass spectrometer to pick out specific ions which we know are characteristic of particular molecules, in this case the drug and the deuterated standard, and ignore all the other background peaks in the spectrum.

∏   (*i*)   Why would we need at least three deuterium atoms in the standard compound?

      (*ii*)   What other advantages does SIM have over ordinary mass scanning?

      (*iii*) Can radio-isotopes be used in gc-ms standards?

(*i*) $(M + 1)$ and $(M + 2)$ peaks due to ^{13}C, ^{2}H, ^{15}N, ^{18}O and other isotopes would seriously interfere with the determination of the ratio of drug and standard if only one or two deuterium atoms were incorporated. An increment of 3 amu is the minimum required to shift the standard's mass away from the common isotope region.

(*ii*)   Using SIM *sensitivity* is much increased, down to the pg level, because we are only trying to measure two m/z values. The dwell time on each ion can be 100 times greater than that possible in a full scan of the mass spectrum, so an increase in sensitivity of two orders of magnitude may be achieved.

(*iii*) There are a number of reasons why radio-isotopes cannot be used. The trace amounts of radio-isotopes usually available are diluted too much by the natural isotopes to be accurately detected in the spectra. The use of high specific activities would make them measurable but they could not be given to human subjects. Lastly, the pump effluent might be vented into the laboratory which is inadvisable, if not illegal.

SAQ 10.1b   In what ways are gas chromatography and mass spectrometry compatible analytical techniques?

SAQ 10.1c   In what ways are gas chromatography and mass spectrometry incompatible techniques?

SAQ 10.1d

How have the difficulties been overcome to marry the two instruments successfully?

SAQ 10.1e

A number of factors are considered important for choosing a column for gc. Which do you think is the most important for gc-ms?

SAQ 10.1f   Think back to how a magnetic mass spectrome-
                 ter is scanned. What do you think is the best way
                 of carrying out SIM? Are quadrupole analysers
                 suitable for SIM work?

An example of the power of gc-ms analysis is shown by a recently
completed study of a natural oil used in perfumery, ylang-ylang.
Ylang-ylang is extracted from the flowers of a tree of the same name
which is cultivated in the Philippines and Indian Ocean islands. The
oil is very complex, as its gas chromatogram on an OV 101 capillary
column shows (Fig. 10.1d). The major component, peak 49, is 19%
of the total oil but minor components down to 0.03%, eg peak 2,
ethanoic acid were identified by gc-ms. The TIC trace obtained on
a BP1 capillary column (very similar in properties to OV 101) is
shown in Fig. 10.1e.

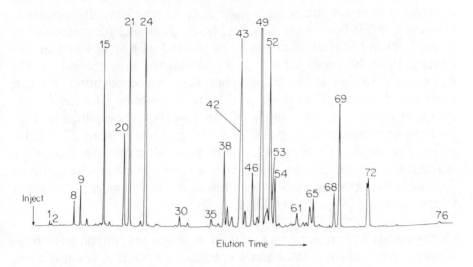

Fig. 10.1d. *Standard gas chromatogram of ylang-ylang oil*
(OV101, 60–220 °C, 4 °C min⁻¹)

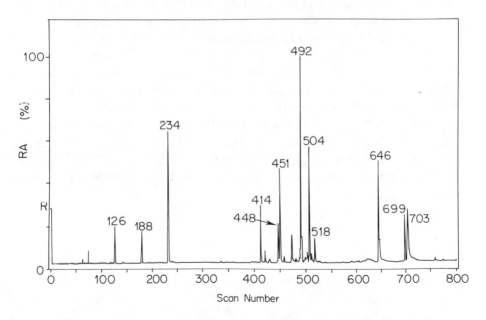

Fig. 10.1e. *Gc-ms Chromatogram of ylang-ylang oil*
(BP1, 60–220 °C, 4 °C min⁻¹)

At first sight, Fig. 10.1e does not look too impressive as many of the components seen in the gc chromatogram seem to have disappeared. However, this is because in this TIC trace, peak 492 (corresponding to peak 49 in the gcs trace) has been plotted as 100%, while in the gc trace Peak 49 is well off scale, so allowing the minor components to be seen. In fact if the mass spectral data are outputed on the same relative scale, the minor peaks *are* detectable. The important thing, of course, is the quality of the spectra, represented in Fig. 10.1e as scan numbers, eg peak 24 in the gc trace is Scan 234 in the gc-ms run, peak 49 is Scan 492, peak 52 is Scan 504, and so on. It is interesting, for example, that peak 72 in the gc trace which shows a shoulder is resolved into two components in the gc-ms TIC trace (Scans 699 and 703, respectively).

Of the nearly 80 components resolved in the gc-ms run, 50 were positively identified from their mass spectra and relative retention indices. Of these, 17 had not previously been known in ylang-ylang oil. Of the remaining 30, the mass spectra showed two were monoterpenes (M^{+} 136), three were sesquiterpenes (M^{+} 204) and five were sesquiterpene alcohols (M^{+} 222) but no reference mass spectra were available to help to identify them. The identified components ranged from propanone (peak 1), esters, such as methylbenzoate (peak 20), alcohols, such as geraniol (peak 30), ethers, such as 4-methylphenyl methyl ether (peak 15), phenols, such as 4-methylphenol (peak 19) to sesquiterpenes, such as germacrene-d (peak 49). This last identification was very interesting as previous workers on ylang-ylang had identified its major component as α-farnesene or α-humulene (Fig. 10.1f). Both these compounds were in fact present in the sample of ylang-ylang but in quite small amounts (9% and 2.7%, respectively). The mass spectrum of germacrene-d (Scan 491) is shown in Fig. 10.1g.

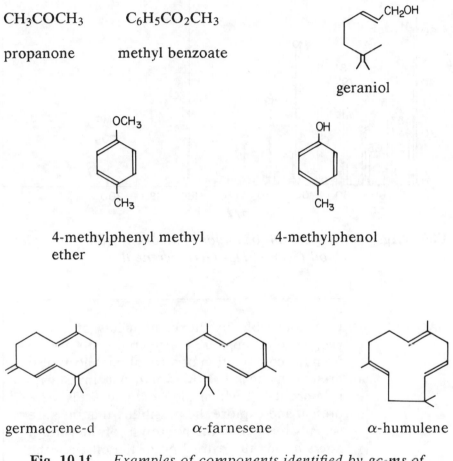

CH₃COCH₃   C₆H₅CO₂CH₃

propanone   methyl benzoate

geraniol

4-methylphenyl methyl ether

4-methylphenol

germacrene-d   α-farnesene   α-humulene

Fig. 10.1f. *Examples of components identified by gc-ms of ylang-ylang oil*

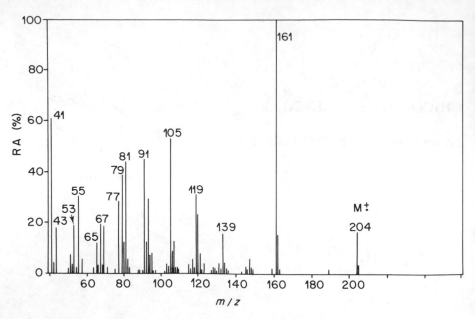

Fig. 10.1g.   *Mass spectrum of major component from ylang-ylang oil (Scan 491) – Germacrene-d*

SAQ 10.1g   It is noticeable in the components of ylang-ylang oil that some are esters, apparently arising from the combination of alcohols and acids also present in the oil. How would you investigate in more detail which alcohols and acids were present and explore the possibility that they are actually formed by the hydrolysis of the natural esters during the extraction of the oil from the flowers by steam distillation?

SAQ 10.1g

Within days of the suspiscion that diethyleneglycol, (DEG), $HOCH_2CH_2OCH_2CH_2OH$ had been added to certain Austrian wines to sweeten them, two food laboratories in Britain and Germany had developed gc-ms methods to monitor DEG in wine down to the 5 ppm (5 mg dm^{-3}) level which is the maximum permitted in foods under EEC regulations (see Fig. 10.1h). Initially hplc with refractive-index detection was used but it is only reliable down to 50 ppm and does not actually prove that the peak measured is DEG. Gc proved sensitive enough with FID detection but did not provide the necessary legal evidence of identity, so gc-ms was used to do both.

Hmm ... An impertinent little diethylene glycol, south side of the fractionating column—but what's this wine doing in it?

Fig. 10.1h. *This cartoon appeared in Chemistry in Britain September 1985 (reproduced by permission)*

Aliquots of wines were taken, all the OH compounds were derivatised with a standard TMS reagent then run on a polar capillary column of the BP20 type.

$$\underset{M_r\ 106}{HOCH_2CH_2OCH_2CH_2OH} + CF_3\overset{\overset{\displaystyle OTMS}{|}}{C}=NTMS \rightarrow$$

$$\underset{M_r\ 250}{TMSOCH_2CH_2OCH_2CH_2OTMS}$$

$$TMS = (CH_3)_3Si$$

This gave very large peaks due to $CH_3CH_2OSi(CH_3)_3$ and other alcohol TMS derivatives, but (bis-TMS)DEG could be easily resolved, then identified from its mass spectrum by comparison with a pure sample. The method required about ten minutes on the column so enabled about 50 samples per day to be checked.

A German team, faced with a huge sample backlog (the whole range of wines on sale was suspect in Germany rather than just Austrian wines as in Britain) had to find a faster procedure. They used a low resolution quadrupole gc-ms system equipped with the NBS library and a polar capillary column *without* derivatisation of the wine samples (bonded phase columns will stand injection of aqueous solutions). An automatic sampler provided fast, accurate injection with dwell times of less than 100 ms. Fig. 10.1i shows a typical TIC trace of wine containing 6.4 mg dm^{-3} DEG, eluting after 15.268 min. Even at this level it was possible to obtain an excellent match for DEG using the NBS library search at 15.27 min (Fig 10.1j.). You will see that the base peak of DEG is m/z 45, $HOCH_2CH_2^+$, with no $M^{\ddot{+}}$ visible. This explains why the other team preferred to use a TMS derivative to move the base peak to m/z 103, $CH_2=\overset{+}{O} Si(CH_3)_3$. This is far less subject to background interference when low levels of detection by SIM are required.

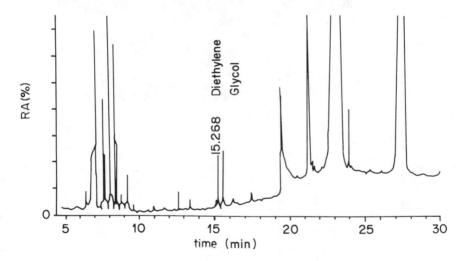

Fig. 10.1i. *TIC chromatogram of wine containing 6.4 mg dm^{-3} DEG*

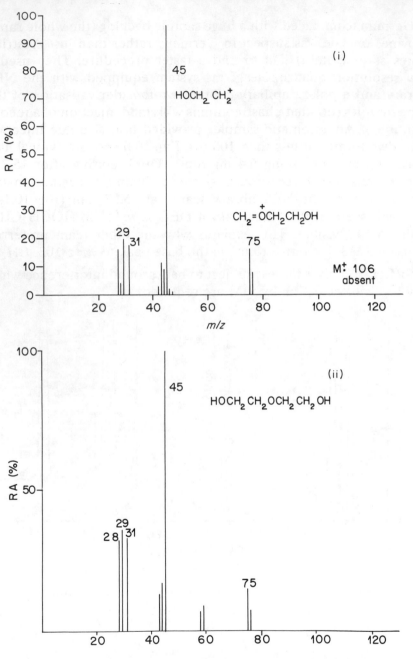

Fig. 10.1j. *Comparison of a mass spectrum of DEG found in wine containing 6.4 mg dm^{-3} DEG (i) with library spectrum of DEG (ii)*

10.2. LIQUID CHROMATOGRAPHY – MASS SPECTROMETRY

It is quite feasible to collect fractions from an hplc separation after they have passed through a non-destructive detector, evaporate them and subject them to mass spectrometry using probe insertion and EI, CI, FD or FAB methods of ionisation. Until about five years ago this was all that was routinely available. Several manufacturers and laboratories were well aware of the growing importance of hplc analysis in many fields, but particularly in the biomedical world, and the tremendous advantages which would follow from a successful on-line coupling of the two instruments. The criteria for this development included the capability for handling aqueous buffers at flow-rates of at least 1 cm^3 min^{-1} without degrading the performance of either the lc or the ms, and the ability to obtain M_r and structural information on large, non-volatile and thermally labile molecules.

Gc-ms works for neutral molecules up to about 1500 M_r. Lc-ms works for charged substrates at M_r up to around 3000 at present, and M_r in the mass range 10 000–20 000 are claimed for the bigger magnetic analysers. Without the more modern soft ionisation methods such as FD and FAB the linking of lc and ms would be much less attractive. The development of these two techniques has coincided with lc development and stimulated much research directed into making the best of both analytical methods. There can be little doubt that the next five years will see not only the perfection of the lc-ms interface but the discovery of new components of living systems which are presently unknown because they lie in the M_r 5 000–20 000 range hitherto impossible to separate and characterise.

Combination of lc and ms presents a major challenge. The mobile liquid phases used in lc range from low boiling organic solvents to aqueous mixtures, modified with a variety of acids, bases and organic and inorganic salts to buffer them and improve chromatographic performance. The problem of removing the high gas volumes produced by on-line evaporation of the mobile phases in order to reach the vacuum required for good mass spectra is much more acute than in gc-ms. Typical flow-rates for lc are 0.5–5 cm^3 min^{-1} which translates into gas flow-rates in the range 100–3000 cm^3 min^{-1}. Interfaces for lc-ms can be broadly divided into those which use the mobile phase to assist in ionisation, ie as a CI reagent gas, and those where the mobile phase is removed before ionisation.

In the direct liquid introduction approach (DLI), one of the first tried by Fred McLafferty's group in 1974, a small portion of the eluent from the lc is fed into the mass spectrometer ion source *via* a capillary inlet and the vaporised solvent becomes a CI reactant gas, giving MH$^+$ ions. This approach can be extremely effective, and various refinements have resulted in interfaces capable of solving a wide range of problems, but there are several disadvantages. Only certain solvents can be used. Structural information from the CI spectra obtained is limited. Flow-rates of only 10–20 μl min^{-1} can be tolerated. Only a limited range of compounds give MH$^+$ readily.

An alternative approach that overcomes some of these problems was developed after the chance observation that the ion beam of a lc-ms persisted even after the source had accidently burned out. At first sight it may seem surprising that ionisation could occur without the use of an electron beam or external electrical fields. Normally, FD ionisation requires fields of 10^8–10^9 V m^{-1}, but it was clear that the spectra produced were typical of that type of soft ionisation, rather than CI.

Charge exchange takes place between salt ions and the eluent when the liquid flow from the lc is heated to about 150 °C in a capillary then sprayed into a vacuum chamber. The solvent contains buffer ions, eg ammonium ethanoate and as the solvent evaporates charged solid particles containing a core of the solute molecule form in the gas phase. Because of their very small size the field gradient created across them is about 10^7 V m^{-1} and increases rapidly as the particles dissociate in the vacuum, releasing the substrate usually as a MH$^+$ or MX$^+$ species where X$=$NH$_4$, Na or K depending on the nature of the buffer, Fig. 10.2a. This is now known as thermospray (TS).

It appears that direct ion evaporation may be applicable to virtually any molecule (no matter how large) present as either *a positive*

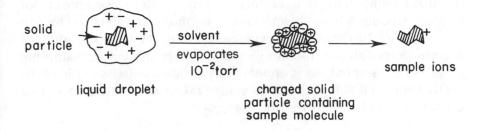

Fig. 10.2a. *Schematic representation of ionisation during thermospray*

or negative ion in aqueous solution. Solvents such as methanol and acetonitrile do not give rise to ions even when dissolved salts (ca 0.1 mol dm^{-3}) are used, but provided at least 10% water is present in reversed phase lc efficient direct ion evaporation occurs. The ionised substrate molecules need to be at concentrations of about 10^{-3} mol dm^{-3}, which is clearly a significant limitation on sensitivity. Ammonium ethanoate buffers which are commonly used in lc solvents give rise to ammonium ions. These will bring about ammonia CI, but many samples will have proton affinities less than ammonia and so will not be observed as MH^+, although polar samples may yield $(M + NH_4)^+$ ions. Relatively non-polar molecules may produce no detectable positive ions under these conditions.

To turn TS into a universal lc-ms interface an electron beam is placed between the end of the capillary and the ion sampling aperture into the ms analyser. The filament burnout problem is solved by replacing the usual tungsten or rhenium filament with a thoriated irridium filament. It is then possible to obtain ammonia CI, using the NH_4^+ ions produced in the spray of an ammonium ethanoate buffered lc effluent, on those non-polar molecules which do not respond to direct ion evaporation. The whole interface can be designed to insert into an ion source through the direct probe insertion lock, and is applicable to any mass spectrometer capable of CI operation. Fig. 10.2b shows a typical design.

This is essentially similar to a CI source, with the addition of a pumping line from the ion source connected through a cold trap to an oil pump. This is necessary to remove the large amount of vapour produced from 1 cm^3 min^{-1} or more of solvent. The ion exit aperture to the mass spectrometer analyser is at the apex of a cone perpendicular to the TS jet. The use of such a 'sampling cone' is not essential but it appears to provide a better sample of directly vaporised and ionised molecules rather than those deposited on surfaces and revaporised or pyrolysed.

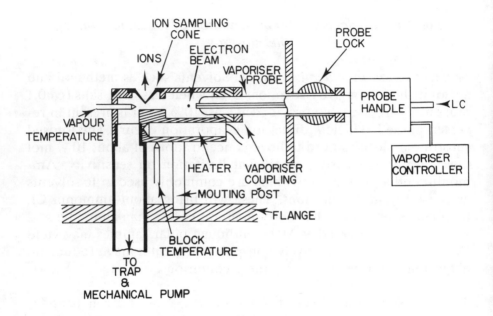

Fig. 10.2b.   *Design of a thermospray (TS) lc-ms interface*

Cyanocobalamin (vitamin B$_{12}$, M_r 1354) has been used as a test compound for evaluating 'soft' ionisation techniques. A thermospray spectrum obtained by Vestal is shown in Fig. 10.2c, using 0.1 mol dm^{-3} ammonium ethanoate as solvent.

In this case, the intact M$^{\ddot{+}}$ or MH$^+$ is not obtained, but rather the cation C$^+$ produced by loss of CN$^-$ is the only peak in the molecular ion region. The major peaks in the spectrum are the doubly charged cation C^{+2} (m/z 664) and satellites corresponding to the replacement of a proton by an alkali ion. At the low mass end some

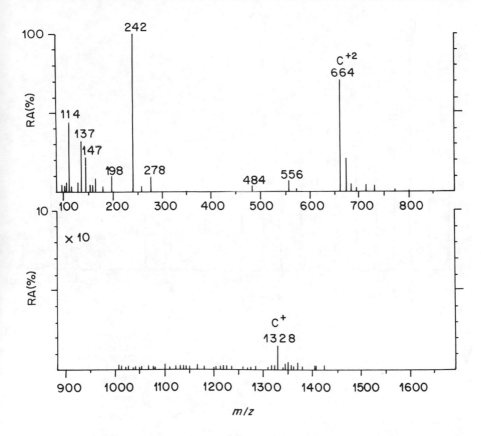

Fig. 10.2c. *Thermospray mass spectrum of vitamin B*$_{12}$
(cyanocobalamin), 10 nmol

structurally significant fragment ions are also produced. Multiply protonated molecular ions are not infrequently found when complex polar molecules undergo TS, particularly in buffers of low pH. For example, the peptide glucagon (M_r 3483) has been detected as the triply and quadruply protonated molecular ions using a quadrupole ms having a mass range of only 1300 amu.

The analysis of a series of alkaloids and drugs is shown in Fig. 10.2d (*i*) and (*ii*). In this study, aliquots containing 5 µg of each of the listed compounds were injected at 1.5 min intervals onto an lc column eluted with $CH_3OH(0.2$ mol $dm^{-3})$, NH_4OCOCH_3 (80:20) flowing at 0.8 cm^3 min^{-1}. A quadrupole mass spectrometer was used recording over the range 40–450 amu.

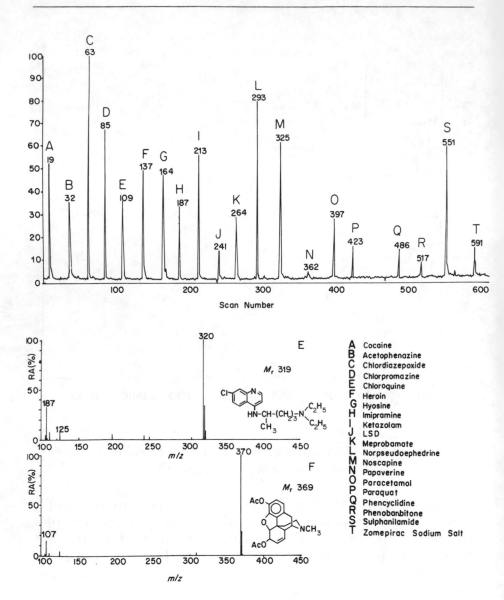

Fig. 10.2d.   *(i) lc-ms TIC trace of compounds A-T (ii) mass spectra of chloroquine* (E) *and Heroin* (F)

Fig. 10.2d shows that all the drugs were detected but there were considerable differences in sensitivity, over a 20-fold range, papaverine, (N) being the least sensitive. With three exceptions, TS ionisation yields MH$^+$ ions which were quite intense, thus providing M_r information, as shown for chloroquine and heroin, E and F, Fig. 10.2d (*ii*). Whether structure-related ions were obtained depended on the intrinsic stability of the system. Molecules that had fused rings gave few fragments but those which were linear or monocyclic usually gave structurally informative ions, perhaps by thermal cleavages.

An alternative approach to the lc-ms interface is to remove the solvent prior to entry into the mass spectrometer. The most popular method is the moving belt interface, Fig. 10.2e.

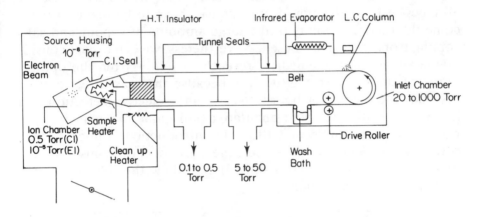

Fig. 10.2e. *Moving belt interface for lc-ms*

In this system, the lc eluent is sprayed onto an endless moving belt in a low vacuum chamber. The solvent evaporates rapidly and this is helped by means of a small infrared heater. The sample deposits are carried into the ion source through a system of seals and vacuum chambers where any remaining solvent is progressively removed. Transfer efficiencies of 40–80% can be achieved. EI, CI and, most importantly, FAB methods of ionisation can be used straight off the belt surface, which means that large biomolecules can be examined. A washbath is placed at the exit from the vacuum system to clean the belt. This also has the advantage that the matrix required for FAB can be dissolved in the washbath and renewed each cycle.

It has recently been found that the use of microbore columns combined with angled spray deposition onto the moving belt improves sensitivity, which had not been impressive either with larger flow rates, or just allowing the eluent to drip onto the belt. Belt interfaces in comparative tests show sensitivity up to ten times those of thermosprays. There have also been recent improvements in the polyimides used to make the belts, giving much longer working lives though they are less reliable than TS interfaces in the long-term.

Lc-ms has now advanced to the point where it has become relatively routine. It is applicable to a wide range of compounds and phases. On the other hand the sensitivity obtained is still variable and invariably less than gc-ms. There is no universal interface suitable for all types of lc application. Compromises have to be made to overcome the incompatability of the large amounts of vapour and salts coming from the solvents used. Nevertheless, we can expect considerable advances in biomedical reseach to follow from the optimisation of these interfacing devices because the mass spectrometer provides the means for both universal and selective detection and it is very sensitive for both qualitative and quantitative analysis, as we saw for gc-ms in Section 10.1. The introduction of microbore lc columns may well have the same beneficial effect on lc-ms that capillary columns did on gc-ms since they reduce the solvent volume to manageable proportions.

SAQ 10.2a	What are the fundamental requirements for a lc-ms system?

SAQ 10.2b

> What coincidental developments have aided lc-ms?

SAQ 10.2c

> What is the chief problem in coupling liquid chromatographs directly to mass spectrometers?

SAQ 10.2d	What methods are used to overcome the problem of direct coupling of lc with ms and to what extent are they successful?

Summary

Section 10 has considered the linking together of chromatography (gas and liquid) with mass spectrometers. Mass spectrometers, although expensive, provide the sensitivity, generality and, by use of fragmentation patterns and selected ion monitoring, the specificity to make attractive gc and lc detectors.

Interfacing is a much easier proposition in gc-ms than lc-ms and, by means of the jet separator and direct introduction of the capillary column into the ion source, is effectively solved. Sensitivity down to the pg range of components injected onto columns can be achieved. A wide range of derivatising reagents are available which enhance volatility of polar samples and also their mass spectrometric behaviour. Relative molecular mass values in the 1000–2000 amu range are routinely measurable, but ionic compounds are not directly analysed by this method.

Interfacing for lc-ms is under active development. Direct liquid introduction, thermospray and belt interfaces are all giving good results, but no universally applicable interface has yet emerged. Sensitivities down to the ng range are being achieved, with the belt interface giving the better sensitivities. Ionisation of charged substrates is readily achieved in thermospray inlets, with some useful fragmentation often occurring, while belts are quite compatible with all the current ionisation methods, especially FAB. This makes lc-ms and gc-ms into complementary techniques, because lc is more suited to ionic compounds, such as acid salts, base hydrochlorides of biomolecules. These can be separated by lc using buffered eluents but are unsuited to gc without extensive extraction and derivatisation to render them volatile.

Rapid advances in biochemical and metabolic studies may be expected to follow from extensive use of lc-ms, and may well be aided as gc-ms was by miniaturisation of the columns and hence improved flow-rates.

Objectives

Now that you have completed Part 10, you should be able to:

- appreciate the practical problems involved in coupling gas and liquid chomatographs to mass spectrometers;

- appreciate the reasons why mass spectrometers are good detectors for gc and lc;

- describe the design and principle of operation of jet separators for gc-ms;

- describe the design and principle of, operation of direct liquid introduction, thermospray and belt interfaces for lc-ms;

- appreciate the range of samples and concentrations which can be examined by each of the separators;

- appreciate that before many compounds can be successfully analysed by gc-ms they need to be derivatised to reduce their polarity;

- suggest derivatives for use with given functional groups for gc-ms;

- understand the methodology involved in quantitative analysis in gc-ms and lc-ms and the importance of selecting a suitable internal standard;

- suggest suitable methodologies for the gc-ms or lc-ms analysis of mixtures of various sorts of compounds, including the basic choice of a gc or lc approach.

Self Assessment
Questions and Responses

SAQ 1.2a

> Given that 1 atomic mass unit (amu) = 1.661 × 10^{-27} kg and that the mass of the electron = 9.110 × 10^{-31} kg, express the mass of the electron in atomic mass units and use these data to calculate the mass difference between: (*i*) CH_4 and CH_4^+, and (*ii*) $C_{20} H_{42}$ and $C_{20} H_{42}^+$, if each mass is required to the nearest whole number.

Response

The mass of the electron is 9.110 × 10^{-31}/1.661 × 10^{-27} = 1.823 × 10^{-5} amu (0.00001823 amu). Relative molecular mass of CH_4 (to nearest whole number) = 16.

Mass of CH_4^+ (to nearest whole number) = 16 − 0.000018 = 16.

Relative molecular mass of $C_{20} H_{42}$ (to nearest whole number) = 282.

Mass of $C_{20} H_{42}^+$ (to nearest whole number) = 282 − 0.000018 = 282.

****

SAQ 1.2b   Calculate the mass difference, to five places of
                decimal, between the following pairs:

                (*i*) CH_4 and CH_4^+;

                (*ii*) $C_{20} H_{42}$ and $C_{20} H_{42}^+$.

Response

Relative molecular mass of CH_4 (to five places of decimals $=$ 16.03130.

Mass of CH_4^+ (to five places of decimals) $=$ 16.03128.

Relative molecular mass of $C_{20} H_{42}$ (to five places of decimals) $=$ 282.32863.

Mass of $C_{20} H_{42}^+$ (to five places of decimals) $=$ 282.32863.

****

SAQ 1.3a   On a qualitative energy level diagram show the
                order in which the above mentioned molecular
                orbitals appear. Having done this, use the dia-
                gram to decide the order of ionisation energies
                for alkanes, alkenes and compounds containing
                a carbonyl group.

Response

Your diagram should look as follows

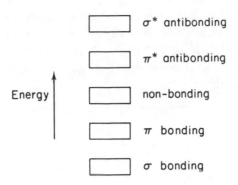

If the electron is removed from the highest occupied molecular orbital for each of alkanes, alkenes and carbonyl-containing compounds, the order of ionisation energies will probably be carbonyl containing < alkenes < alkanes.

The word probably is included, because depending on the detailed structure of the molecule, the energy of non-bonding orbitals in one molecule may be slightly lower than that of π bonding orbitals in another. This feature is illustrated by the following examples.

Ionisation Potentials (eV) for $CH_3CH_2CH_2X$:

X	IP	X	IP
H	11.07	CHO	9.86
CH_3	10.63	$CH=CH_2$	9.5
C_2H_5	10.34	$COCH_3$	9.34
OH	10.17	NH_2	8.78

The rule of thumb given above is broken by $CH_3CH_2CH_2CH=CH_2$, which would have predicted it to have an ionisation potential between that of $CH_3CH_2CH_2C_2H_5$ and $CH_3CH_2CH_2OH$.

****

SAQ 1.3b   For the molecule, C_2H_6 represent three fragmentation processes for the molecular ion, each of which involves only C—H bond cleavage but which lead to the production of a neutral atom, a neutral radical and a neutral molecule respectively.

Response

One can in fact write four processes, one produces an atom, another produces a radical and the others both produce a molecule. These are:

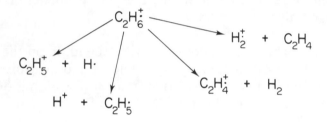

****

SAQ 1.4a   Draw up a fragmentation pattern for methanol based on the above assignments and underline the most intense ion in the spectrum.

Response

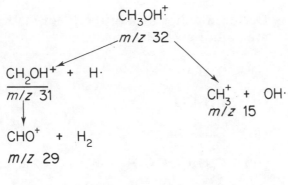

****

SAQ 1.4b   By measuring the intensities of the four major peaks in Fig. 1.4a, setting the most intense to 100% relative abundance and normalising the others to it, construct a bar diagram for methanol.

Response

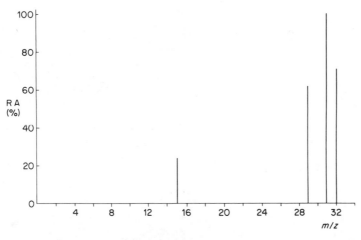

Bar diagram of methanol

****

SAQ 1.4c   Decide whether ions of the following formulae
                 are odd or even electron.

   (*i*) C_3H_8

   (*ii*) CH_3CO

   (*iii*) C_6H_4

   (*iv*) $C_6H_5COOC_2H_5$

   (*v*)   HCl

   (*vi*) C_7H_{15}

   (*vii*) C_7H_7

   (*viii*) $C_2H_3OC_3H_7$

   (*ix*) C_7H_{13}

   (*x*) $C_6H_5NO_2$

Response

(*i*) $C_3H_8^{+\cdot}$

(*ii*) CH_3CO^+

(*iii*) $C_6H_4^{+\cdot}$

(*iv*) $C_6H_5COOC_2H_5^{+\cdot}$

(*v*) $HCl^{+\cdot}$

(*vi*) $C_7H_{15}^+$

(*vii*) $C_7H_7^+$

(*viii*) $C_2H_3OC_3H_7^+$

(*ix*) $C_7H_{13}^+$

(*x*) $C_6H_5NO_2^+$

Note that (*i*), (*iv*), (*v*), (*viii*) and (*x*) are the formulae of stable neutral molecules, (*iii*) is the formula of one stable molecule (C_6H_6) minus another (H_2) and (*ii*), (*vi*), (*vii*) and (*ix*) are the formulae of stable molecules minus an atom or radical.

****

SAQ 1.4d

Calculate m/z values (to the nearest whole number) for the following molecular ions:

(*i*) $C_2H_6^+$

(*ii*) $(CH_3)_2CO^+$

(*iii*) $C_6H_6^+$

(*iv*) $CH_3NH_2^+$

(*v*) CH_3COOH^+

(*vi*) $C_6H_5CH_2CN^+$

(*vii*) $C_4H_9SH^+$

(*viii*) $C_6H_4(NH_2)_2^+$

(*ix*) $C_6H_3(NH_2)_3^+$

(*x*) $(CH_3O)_3P^+$

(*xi*) $C_6H_6^{2+}$

Response

(*i*)   30

(*ii*)   58

(*iii*)   78

(*iv*)   31

(*v*)   60

(*vi*)   117

(*vii*)   90

(*viii*) 108

(*ix*)   123

(*x*)   124

(*xi*)   39

****

SAQ 1.4e   Calculate mass to charge ratios for singly charged ions of the following formulae:

(*i*) C_4H_{10};

(*ii*) C_6H_5;

(*iii*) CH_2COOH;

(*iv*) $C_6H_5NH_2$; $\longrightarrow$

SAQ 1.4e **(cont.)**	(*v*) Cl_2;
	(*vi*) $(C_6H_5)_3P$;
	(*vii*) $H_2NOCCONH_2$;
	(*viii*) C_7H_7.

Response

(*i*)   58

(*ii*)   77

(*iii*) 59

(*iv*)   93

(*v*)   266

(*vi*)   262

(*vii*) 88

(*viii*) 91

****

SAQ 1.4f	For each of the ions in SAQ 1.4e decide whether it should be classified as an even or an odd electron ion.

Response

(*i*) $C_4H_{10}^{+\cdot}$

(*ii*) $C_6H_5^+$

(*iii*) CH_2COOH^+

(*iv*) $C_6H_5NH_2^{+\cdot}$

(*v*) $CI_2^{+\cdot}$

(*vi*) $(C_6H_5)_3P^{+\cdot}$

(*vii*) $(CONH_2)_2^{+\cdot}$

(*viii*) $C_7H_7^+$

Note that (*i*), (*iv*), (*vi*) and (*vii*) are the formulae of stable neutral molecules, (*ii*), (*iii*) and (*viii*) are the formulae of stable neutral molecules minus an atom or radical and that (*v*) is the formula of one stable molecule (CH_2I_2) minus another (H_2).

****

SAQ 1.4g   Examine the following fragmentation processes. Fill in the missing pluses ($+$) and dots (.) indicating ionic and radical character respectively. The ionic product is written first on the right hand side of each equation. (*i*) serves as an example.

(*i*) $C_4H_{10}^{+\cdot}$ $\rightarrow$ $C_4H_8^{+\cdot} + H_2$

(*ii*) $C_5H_{11}CONH_2$ $\rightarrow$ $C_5H_{11} + CONH_2$
                                                      $\longrightarrow$

SAQ 1.4g
(cont.)

(*iii*) C_7H_7 $\rightarrow$ $C_5H_5 + C_2H_2$

(*iv*) $(CH_3)_2CHOH$ $\rightarrow$ $CH_3CHOH + CH_3$

(*v*) C_2H_5 $\rightarrow$ $C_2H_3 + H_2$

(*vi*) C_2H_5SH $\rightarrow$ $C_2H_4 + H_2S$

(*vii*) $C_{10}H_{22}$ $\rightarrow$ $C_7H_{15} + C_3H_7$

(*viii*) C_4H_9 $\rightarrow$ $C_3H_5 + CH_4$

(*ix*) $C_6H_5NO_2$ $\rightarrow$ $C_6H_5 + NO_2$

(*x*) $C_6H_5SO_2NH_2$ $\rightarrow$ $C_6H_5NH_2 + SO_2$

Response

(*i*) $C_4H_{10}^{+\cdot}$ $\rightarrow$ $C_4H_8^{+\cdot} + H_2$

(*ii*) $C_5H_{11}CONH_2^{+\cdot}$ $\rightarrow$ $C_5H_{11}^+ + CONH_2^\cdot$

(*iii*) $C_7H_7^+$ $\rightarrow$ $C_5H_5^+ + C_2H_2$

(*iv*) $(CH_3)_2CHOH^{+\cdot}$ $\rightarrow$ $CH_3CHOH^+ + CH_3^\cdot$

(*v*) $C_2H_5^+$ $\rightarrow$ $C_2H_3^+ + H_2$

(*vi*) $C_2H_5SH^{+\cdot}$ $\rightarrow$ $C_2H_4^+ + H_2S$

(*vii*) $C_{10}H_{22}^{+\cdot}$ $\rightarrow$ $C_7H_{15}^+ + C_3H_7^\cdot$

(*viii*) $C_4H_9^+$ $\rightarrow$ $C_3H_5^+ + CH_4$

(*ix*) $C_6H_5NO_2^{+\cdot}$ $\rightarrow$ $C_6H_5^+ + NO_2^\cdot$

(*x*) $C_6H_5SO_2NH_2^{+\cdot}$ $\rightarrow$ $C_6H_5NH_2^{+\cdot} + SO_2$

****

SAQ 1.4h

Given in Fig. 1.4b is the mass spectrum of ethanol, C_2H_5OH.

(*i*)   Write down the m/z of the molecular ion.

(*ii*) Write down the m/z of the base peak.

(*iii*) The base peak is formed from the molecular ion in a single step. Write down an equation for this fragmentation process (nb refer back to the discussion of methanol for the formula of the base peak).

(*iv*) Suggest two fragmentation processes for production of ions of m/z 29 which may have different formulae (refer back to discussion of the methanol spectrum if you are stuck).

(*v*)   Write down fragmentation processes for the stepwise decompositions,

$$m/z\ 46 \rightarrow m/z\ 45 \rightarrow m/z\ 43.$$

(*vi*) Assign a formula to m/z 18 and give a fragmentation process for its direct formation from the molecular ion.

(*vii*) Combine the fragmentation pathways discussed above to give the fragmentation pattern for ethanol. $\longrightarrow$

SAQ 1.4h
(cont.)

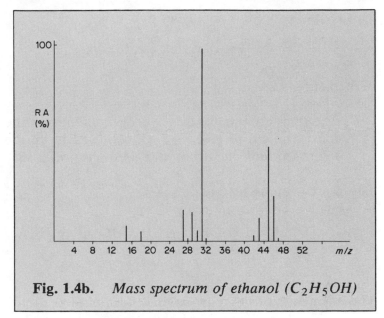

Fig. 1.4b. *Mass spectrum of ethanol (C_2H_5OH)*

Response

(*i*) m/z 46 (molecular ion),

(*ii*) m/z 31 (base peak)

(*iii*) $CH_3CH_2OH^{+\cdot} \rightarrow CH_2OH^+ + CH_3^{\cdot}$

The base peak has the same formula as that in the methanol spectrum. It is formed by C—C bond cleavage in the case of ethanol.

(*iv*) In methanol m/z 29 was formulated as CHO^+, which arose by the fragmentation

$$CH_2OH^+ \rightarrow CHO^+ + H_2$$

This may also happen with ethanol, but the m/z 29 ion may also be formulated as $C_2H_5^+$ and this may be formed by the fragmentation

$$C_2H_5OH^{+\cdot} \rightarrow C_2H_5^+ + OH^{\cdot}$$

This second process is analogous to

$$CH_3OH^{+\cdot} \rightarrow CH_3^+ + OH^{\cdot}$$

in methanol.

(v)   The steps are the loss of a hydrogen atom from the molecular ion (m/z 46) to give m/z 45, followed by further loss of a hydrogen molecule from m/z 45 to give m/z 43.

This can be represented as

$$CH_3CH_2OH^{+\cdot} \rightarrow CH_3CHOH^+ + H^{\cdot}$$
$$\downarrow$$
$$CH_3CO^+ + H_2$$

The fragmentations exactly parallel those of methanol.

(vi) m/z 18 is $H_2O^{+\cdot}$

$$C_2H_5OH^{+\cdot} \rightarrow H_2O^{+\cdot} + C_2H_4$$

(vii)

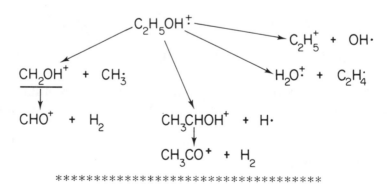

****

SAQ 2.1a   Can you recall from Part 1 why such a high value is used when most molecules are ionised at energies of 8–15 eV?

Response

High values of electron beam energy are used to induce fragmentation of ions.

****

SAQ 2.1b   Underline the correct answer from the options given in the sentences below.

(*i*)   Ionising electrons are emitted from the trap/filament/repeller plate.

(*ii*) The collimating magnets affect the ionising electrons/ions/electrons expelled from molecules on ionisation.

(*iii*) In positive ion mass spectrometry, the repeller plate is held at a positive/negative potential with respect to the first accelerating plate.

(*iv*) The trap is held at a positive/negative potential with respect to the filament.

Response

(*i*)   The correct answer is *filament*. The function of the trap is to collect the ionising electrons and that of the repeller plate is to cause the ions to move out of the ion source.

(*ii*) *The purpose of the magnets is to collimate the beam of ionising electrons.* There will also be some effect on both expelled electrons and ions, as any charged particle is affected by a magnetic field (we will make use of this fact later), but the principle effect is on the electron beam.

(*iii*) The correct answer is *positive*. Since the ion repeller causes the positively charged ions to leave the source and as like charges repel, the repeller plate must be positively charged.

(*iv*) The correct answer is also *positive*. The trap attracts the electrons from the filament and since opposite charges attract, negatively charge electrons are attracted to a positively charged trap.

****

SAQ 2.1c   As far as you can, sequence the following components of the single focussing magnetic sector mass spectrometer in the order that an ion would experience their effect, after its formation to its detection.

Magnet;
Accelerating plates;
Vacuum system;
Electron multiplier;
Repeller plate.

Response

The order is:

(*i*) Repeller plate (the positive potential on this plate causes the ions to leave the source).

(*ii*) Accelerating plates (passage through the high potential difference between these plates give the ions their kinetic energy).

(*iii*) Magnet (this performs the mass analysis).

(*iv*) Electron multiplier (this is the detecting device).

You may have been puzzled where to place the vacuum system in this sequence. The answer is that it could be placed anywhere. As the whole instrument is under vacuum, the vacuum pumping system can be connected at almost any part of the instrument.

Note that the above order (*i*)–(*iv*) is always the same and is the only correct answer. If you did not get it right, go back and study Fig. 2.1b again.

****

SAQ 2.1d

In Fig. 2.1d are shown the trajectories of two types of ions that do not follow the correct flight path to reach the detector. These are shown as:

ions of type (*i*) which are following a circular flight path with a radius larger than that required, and

ions of type (*ii*) which are following a circular flight path with a radius smaller than that required.

Which of the two types of ions is the heavier?

Response

Ions of the type (*i*) are heavier.

If you recall Eq. 2.1d

$$m/z = B^2 r^2 / 2V \qquad (2.1d)$$

If B, V and z are constant, then ions following a path of larger radius must be heavier and those following a path of smaller radius must be lighter than the ions reaching the detector.

Another way of looking at this is to consider the centrifugal force associated with the ions and given by mv^2/r.

The heavier the ion the larger the centrifugal force it develops and the more likely it is to follow a flight path tangential to the circle unless the magnitude of the magnetic force is increased. This is the case for ions of type (i).

For lighter ions, the magnetic force is greater than the centrifugal force and thus these ions are directed in towards the centre of the circle, as shown by ions of type (ii).

****

SAQ 2.2a   Typically, a mass spectrometer will have a number of inlet systems. What systems are required to meet the following requirements for mass spectrometric analysis?

(i) A laboratory performing daily gas analysis with occasional requirements for the analysis of solids. The instrument is not linked to a computer.

(ii) A laboratory exclusively dedicated to mixture analysis using a computerised mass spectrometer.

(iii) A laboratory performing analysis of single substances that are invariably solids or liquids, using a computerised spectrometer.

Response

(*i*) This laboratory would required only a cold inlet system and a direct insertion probe.

(*ii*) This laboratory would require the special inlet systems appropriate to gc-ms and/or lc-ms and a septum inlet or an AGHIS for the computer calibrant.

(*iii*) This laboratory would required an AGHIS and a direct insertion probe. They would probably find a septum inlet useful as well, as this is the quickest method of admitting the computer calibrant.

****

SAQ 3.1a
Can you think of other groups or atoms commonly encountered in organic chemistry which would also be likely to dominate the negative ion spectra of molecules of which they are part? On what basis did you make your choice?

Response

F, Cl, Br, I and OH are prime examples of such groups and atoms. The reason why is that they are very electronegative and very prone to electron attachment.

****

SAQ 3.2a
In that part of Section 2.1, a number of reasons were advanced for using such low pressures. Can you list them?

Response

The use of low pressures:

(*i*)   avoids the filament burning out;

(*ii*) helps to vaporise the sample;

(*iii*) prevents ions, once formed, being lost by collisions with other
        molecules;

(*iv*) helps to remove sample from the instrument after analysis.

If you had difficulty with this question, revise the appropriate part
of Section 2.1 before proceeding.

****

SAQ 3.2b	What is the consequence of this difference in internal energies likely to be?

Response

The quasi-molecular ion produced by chemical ionisation will be
less likely to fragment than the molecular ion produced by electron
impact, since the former has a much lower internal energy.

****

| SAQ 3.2c | The major fragment ion in both the EI and CI spectra of proline occurs at m/z 70. This represents the loss of 45 mass units from M^{+} in the EI spectrum and the loss of 46 mass units from $(MH)^{+}$ in the CI spectrum. Suggest a formula for the neutrals produced in each spectrum. |

Response

45 mass units corresponds to $-COOH$ and 46 mass units corresponds to $C(OH)_2$ or an isomer of it.

****

| SAQ 3.2d | Are quasi-molecular ions odd- or even-electron ions? Can you rationalise the fragmentations of the quasi-molecular ion of proline in terms of its electron configuration? |

Response

Quasi-molecular ions must always be even-electron ions, since they are formed by adding only a proton to an even-electron molecule.

In Part 1 we discussed the fact that even-electron ions are usually more stable than odd-electron ions. Thus fragmentation of an even-electron quasi-molecular ion might be expected to form even-electron fragment ions by the loss of even-electron neutrals. This is what happens in the case of proline.

****

> **SAQ 3.2e** We have now spent a considerable time discussing CI mass spectrometry. Before we move on to the next ionisation method, list the advantages and limitations of the CI method, paying particular attention to those problems we outlined for EI at the end of Section 3.1.

Response

The problems outlined for EI mass spectrometry were that:

(i) in some cases molecular ions were very weak or not observed because of fragmentation;

(ii) isomers could not be distinguished;

(iii) thermally unstable molecules do not give useful mass spectra;

(iv) mass spectra cannot be obtained for involatile molecules.

In a large number of cases CI mass spectrometry can help with problem (a), in that quasi-molecular ions usually have a high relative intensity. There is a problem with some molecules in knowing what type of quasi-molecular ion they are going to form:

$(MH)^+$, $(M - H)^+$ or $(M + C_2H_5)^+$ etc and this may introduce some uncertainty in interpreting the results.

Because fragmentation is reduced, the structural information given by the fragments is reduced or completely lost. However, for some molecules fragmentation can reflect fine structural differences allowing differentiation between isomers.

Chemical ionisation still requires volatilisation of the sample since the ion-molecule reactions on which it depends are gas phase reactions. Thus, if the sample is thermally unstable or involatile, CI is not the answer.

****

| SAQ 3.3a | Using the figures given above, calculate the distance between the electrodes. |

Response

$$\text{Distance} = \frac{\text{voltage}}{\text{voltage/cm}} = \frac{10 \times 10^3}{10^8} = 10^{-4} \text{ cm.}$$

****

| SAQ 3.3b | These ions will have passed through a high potential difference. What effect will this have on them? |

Response

The ions will gain kinetic energy. If you did not answer this correctly go back to Section 2.1 where this is discussed in relation to the accelerating potential of an EI ion source.

****

| SAQ 3.3c | What must be done to these ions so that the mass spectrometer can analyse them? |

Response

The ions must be slowed down somewhat.

****

| SAQ 3.3d | What is the consequence of this as far as the mass spectrum is concerned? |

Response

Lower internal energies mean less fragmentation of the molecular ion.

****

| SAQ 3.3e | List any advantages and disadvantages of FI over EI or CI and any limitations of the technique. |

Response

As with CI, because FI is a soft ionisation technique, $M^{\ddot{+}}$ or $(MH)^+$ ions of considerable relative abundance are usually formed. Thus the method is useful for relative molecular mass determinations.

Its value is, however, lowered somewhat by the uncertainty about whether $M^{\ddot{+}}$ or $(MH)^+$ ions will be formed. Unlike CI, no higher adduct ions such as $(M + C_2H_5)^+$ are observed. The amount of fragmentation is less than for EI mass spectrometry, but some structurally informative ions may be formed.

The sample requires vaporisation prior to ionisation and the method is, therefore, not suitable for thermally unstable or involatile samples.

****

SAQ 3.4a   What is the consequence of not producing fragment ions?

Response

FD can be used for relative molecular mass determinations, but is no help in elucidating the structures of uncharacterised materials.

****

SAQ 3.4b   List the advantages and shortcomings of the FD technique.

Response

The advantages are:

(*i*) no need to vaporise the sample;

(*ii*) intense M^{+} or $(MH)^{+}$ ions aid relative molecular mass determination.

The disadvantages are:

(*i*)   no structural information is produced as very little, if any, fragmentation occurs;

(*ii*) the compound must be soluble in a suitable volatile solvent to be applied to the emitter;

(*iii*) preparation of the emitter is a specialised technique;

(*iv*) ion beams may not be persistant, thus not giving sufficient time for tuning the spectrometer.

****

SAQ 3.5a	What will be the result of passing ions such as Xe^{+} through the electric field?

Response

The ions will gain kinetic energy. We have now encountered this technique of accelerating charged particles several times.

****

SAQ 3.5b List the advantages and disadvantages of FAB.

Response

The advantages of FAB are:

(*i*)   abundant M^+ or $(MH)^+$ ions;

(*ii*) some fragment ions produced which may be structurally informative;

(*iii*) no need to vaporise the sample;

(*iv*) persistant ion beams.

The disadvantages are:

(*i*)   use of the matrix material produces extra ions in the spectrum that may complicate its interpretation;

(*ii*) the sample must be soluble in the matrix material.

****

SAQ 3.6a Why is spark source mass spectrometry of virtually no use to the organic chemist?

Response

The organic chemist is usually using mass spectrometry to establish relative molecular masses and molecular structures. Spark source mass spectrometry will only give information on elemental composition and there are very well established inexpensive methods of doing this without recourse to mass spectrometry.

****

SAQ 3.7a

Electron Impact Ionisation

(*i*)   EI ionisation can produce both positive and negative ions. Why is it more difficult to produce negative molecular ions than their positive counterparts?

(*ii*) Draw an ionisation efficiency curve and explain why this leads mass spectrometrists to use 70 eV bombarding electrons.

(*iii*) What might be the advantage of recording a low voltage spectrum?

(*iv*) List the major shortcomings of the EI method.

Response

(*i*)   Negative molecular ions are formed with high internal energies because they retain the energy of the bombarding electron. They are thus very prone to fragmentation.

(*ii*)

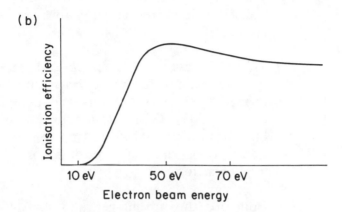

(b)

At 70 eV the ionisation efficiency is close to the maximum value, but is relatively unaffected by small changes in beam energy. This gives greatest reproducibility and sensitivity.

(*iii*) Low voltage spectra show reduced fragmentation. This often allows important fragment ions to be picked out more easily. It also increases the relative abundance of the molecular ion, making its identification easier.

(*iv*) Some molecules show very weak molecular ions or do not show them at all. Many groups of isomeric compounds show very similar spectra. Samples have to be vaporised, making the technique unsuitable for thermally unstable or involatile compounds.

****

SAQ 3.7b

Chemical Ionisation

(*i*) Ion-molecule reactions of methane lead to two predominant species being formed. Write equations for their formation. $\longrightarrow$

SAQ 3.7b
(cont.)

　　(*ii*)　What properties govern the ease of proton transfer from a reactant ion to a sample molecule?

　　(*iii*)　The CI mass spectra of methyl esters of carboxylic acids show fragment ions formed by loss of CH_3OH and the elements of CH_3COOH. Rationalise this behaviour in terms of the electron configuration and possible structures of the pseudo-molecular ion.

　　(*iv*)　Compare the advantages and disadvantages of the EI and CI methods.

Response

(*i*)　$CH_4 \rightarrow CH_4^{+\cdot} + e$

$CH_4^{+\cdot} + CH_4 \rightarrow CH_5^+ + CH_3^{\cdot}$

$CH_4^{+\cdot} \rightarrow CH_3^+ + H^{\cdot}$

$CH_3^+ + CH_4 \rightarrow C_2H_5^+ + H_2$

(*ii*)　Proton transfer reactions are governed by the proton affinity of the sample and the acidity of the reactant ion.

(*iii*)　Quasi-molecular ions of methyl esters will be likely to have one of two structures.

A　　　　　　　　　　　　　　　　　　　　B

Both CH_3OH and CH_3COOH are even-electron neutrals. Loss of such neutrals would be expected from an even-electron quasi-molecular ion giving an even-electron fragment ion. Structure A could easily lose both CH_3OH and $CO.O(CH_3)H$. Structure B could equally easily lose $C(OH)OCH_3$ and a hydrogen transfer is all that is required to give loss of CH_3OH.

(*iv*) EI gives identifiable M^+ ions for most but not all types of molecule, so in some cases relative molecular masses are not observed. CI gives $(MH)^+$ ions for most type of molecules, usually with an enhanced relative abundance compared to M^+ in the CI spectrum. In some cases $(M - H)^+$ or $(M + C_2H_5)^+$ ions are formed which complicates identification. EI usually gives a wide range of structurally informative fragment ions. In CI the fragmentation is less extensive, but structurally informative ions are sometimes formed. Both techniques require sample vaporisation and are therefore unsuitable for thermally unstable or involatile samples.

****

SAQ 3.7c

Field Ionisation and Field Desorption

(*i*) These are two techniques based on the same principle. Describe this and state how FD differs from FI.

(*ii*) Explain what is meant by 'whiskering'.

(*iii*) The EI, FD and FI spectra of D-glucose, ($M_r = 180$) are shown in Fig. 3.7a. Use these to highlight the comparative features of the techniques.

(*iv*) Explain the origin of the ions at m/z 163, 145 and 127 in the FI spectrum. $\longrightarrow$

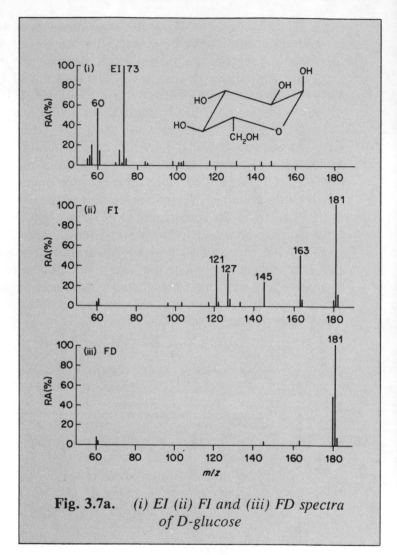

Fig. 3.7a.   *(i) EI (ii) FI and (iii) FD spectra*
of D-glucose

Response

(*i*)   In both FI and FD, ionisation occurs when a molecule is subjected to a high potential gradient while close to an anode which can accept electrons. The positive ions are drawn towards a cathode and then into the mass analyser. In FI the sample is vaporised and molecules come very close to or impinge upon the anode (emitter) where they are ionised. In FD the sample is coated onto the emitter and the ions are desorbed from the solid state.

(*ii*) Whiskering is coating the emitter with small needles or whiskers of carbon prior to analysis. This both increases the surface area and aids ionisation efficiency for both techniques.

(*iii*) The EI spectrum shows no molecular ion and no fragment ions above m/z 73. It would be almost impossible to identify the molecule from this spectrum. Glucose has a relative molecular mass of 180 and thus both the FI and FD spectra are giving intense $(MH)^+$ ions. In the FD spectrum $M^{\ddot{+}}$ ions are also formed. In the FI spectrum there are a number of fragment ions, but none are observed in the FD spectrum with a relative abundance of greater than 10%.

(*iv*) The fragment ions at m/z 163, 145 and 127 represent successive losses of H_2O from $(MH)^+$. I think you might expect this in glucose.

****

SAQ 3.7d	*Fast Atom Bombardment*
	(*i*) How are fast atoms generated?
	(*ii*) What is meant by the term 'matrix material'?

Response

(*i*) Fast atoms (A) are generated by the following sequence of reactions:

$$A + e \rightarrow A^+ + 2e$$

$$A^+ \xrightarrow{\text{acceleration}} A^+ \text{ (fast)}$$

$$A^+ \text{ (fast)} + A(\text{thermal}) \rightarrow A(\text{fast}) + A^+ \text{ (thermal)}$$

(*ii*) The matrix material is the involatile liquid in which the sample is dissolved for FAB ionisation.

****

SAQ 3.7e   On the basis of the discussion of ion sources, describe the features you would seek in a perfect ionisation method.

Response

The perfect ionisation method would give identifiable molecular ions (M^+), not pseudo-molecular ions, for all compounds. It would also give sufficient fragmentation for structural analysis. Samples would not require vaporisation so that samples, both covalent and ionic, including thermally unstable and involatile materials, could be analysed. Ion beams should be persistant for at least 15 minutes. No complicating additives, such as matrix materials, should be required. The source should be simple to operate and cheap to build and run. If you can design such a source, and patent it, you'll make a fortune!

****

SAQ 4.1a	By way of revision, can you describe how a magnetic analyser functions and write down the equation governing the passage of an ion (mass, m; charge, z; kinetic energy, $\frac{1}{2}mv^2$) through a magnetic field of strength, B?

Response

An ion enters the magnetic field with a velocity, v, which is dependant on its m/z value and the accelerating voltage through which it has passed. The ions are deflected into a circular path by the magnetic field. The magnetic field imposes a force on the ion which is directed towards the centre of the circle of radius r and which has a magnitude given by the expression Bzv. This is counteracted by the centrifugal force the ions develop, which is given by mv^2/r. When these two forces are exactly balanced, the ions travel around the circumference of the circle and $Bzv = mv^2/r$.

****

SAQ 4.1b	Can you use the relationship derived in SAQ 4.1a, together with the relationship between accelerating voltage and kinetic energy of the ion, to derive an equation relating m/z to field strength and accelerating voltage?

Responses

The relationship between accelerating voltage and kinetic energy is

$$zV = \tfrac{1}{2} mv^2$$

We also know the relationship $Bzv = mv^2/r$.

By expressing both these equations in terms of v^2 and equating them, we obtain

$$v^2 = 2zV/m = B^2r^2z^2/m^2$$

$$\therefore \qquad m/z = B^2r^2/2V$$

If you found difficulty in answering these two SAQ's, go back and revise the appropriate part of Section 2.1 before continuing.

****

SAQ 4.1c   Let's suppose we are using an instrument with a resolving power (rp) of 5000, but that the mass region we are interested in is 500. How accurately could we measure such masses?

Response

$$\mathrm{rp} = m/\Delta m$$

$$5000 = \frac{500}{\Delta m}$$

$$\Delta m = 0.1$$

This is, with an instrument of resolving power 5000, we can measure masses of around 500 to an accuracy of 0.1 rmm unit. Similarly, we can measure masses of around 50 to an accuracy of 0.01 rmm unit.

****

SAQ 4.1d	What is the mass of the methane molecular ion?

Response

16 amu.

****

SAQ 4.1e	Assuming we had the right kind of mass spectrometer for the job, would you expect the accurately measured mass of CH_4^+ to be a whole number? Explain your answer.

Response

No, the mass of CH_4^+ would not be a whole number. The relative mass of an atom, molecule or ion can be measured more accurately than the nearest whole number. This is because relative atomic masses of the elements are not exactly whole numbers when measured with respect to the relative atomic mass of ^{12}C (12.000000). Thus masses can be measured to an accuracy of several places of decimals. The relative atomic mass (A_r) of the four most common elements in organic chemistry are:

$$^{12}C \ = \ 12.00000$$
$$^{1}H \ = \ 1.00783$$
$$^{16}O \ = \ 15.99492$$
$$^{14}N \ = \ 14.00307$$

****

SAQ 4.1f	Calculate the mass of CH_3OH^{+} to an accuracy of five decimal places using the A_r values given in the response to SAQ 4.1e

Response

32.02624 amu.

****

SAQ 4.1g	As an example of the identification of molecular formulae by means of accurate mass measurement of molecular ions, imagine we are presented with a colourless liquid and a colourless gas to identify by mass spectrometry. These two compounds have the same nominal mass of 32, however, the accurate mass of the molecular ion of the liquid is found to be 32.0263 $\pm$ 0.0001 and that of the gas is 31.9898 $\pm$ 0.0001. Use this information and the list of accurate A_r values in the response to SAQ 4.1e to identify the molecules.

Response

32.0263 clearly corresponds to CH_3OH^{+}, there being no other atomic combination that has this mass value. The molecular ion of mass 31.9898 must contain oxygen, as this is the only element listed in SAQ 4.1e which is mass deficient. In fact, 31.9898 is the accurate mass of O_2^{+}. These are simple, and perhaps trival examples, but the method is of vital importance for determining the molecular formulae of organic molecules.

****

SAQ 4.1h	If we had a mixture of methanol and oxygen in the mass spectrometer, what resolving power would be needed to distinguish between the molecular ions?

Response

The mass difference between $CH_3OH^{\ddot{+}}$ and $O_2^{\ddot{+}}$ is $32.02624 - 31.98984 = 0.03640$

$$rp = \frac{32}{0.0364} = 879.12$$

It should, therefore, be a relatively simple task to differentitate between these two ions with a magnetic analyser of the type discussed in Section 2.1.

****

SAQ 4.1i	Calculate the resolving power needed to distinguish between the following pairs of ions:
	(i) $C_8H_{16}^{\ddot{+}}$ and $C_7H_{12}O^{\ddot{+}}$
	(ii) $C_{24}H_{50}^{\ddot{+}}$ and $C_{23}H_{46}O^{\ddot{+}}$
	(iii) $C_{40}H_{82}^{\ddot{+}}$ and $C_{39}H_{78}O^{\ddot{+}}$

Response

(i)   The mass of $C_8H_{16}^{+}$ is 112.12528 and that of $C_7H_{12}O^{\ddot{+}}$ is 112.08888. The resolving power required is

$$\frac{112}{0.03640} = 3077$$

(*ii*) The mass of $C_{24}H_{50}^+$ is 338.39150 and that of $C_{23}H_{46}O^+$ is 338.35510. The resolving power required is

$$\frac{338}{0.03640} = 9286$$

(*iii*) The mass of $C_{40}H_{82}^+$ is 562.64206 and that of $C_{39}H_{78}O^+$ is 562.60566. The resolving power required is

$$\frac{562}{0.03640} = 15,440$$

****

SAQ 4.2a	Examination of these three equations reveals there are three properties of an ion which are vital in mass spectrometry. What are these three properties?

Response

The properties are mass, charge and velocity. If you included any of field, voltage or radius, think again as these are not properties of the ion but instrumental parameters that can be varied to control the behaviour of the ion.

****

SAQ 4.2b	In combining Eq. 2.1b and 2.1c to generate Eq. 2.1d, what assumption do we make?

Response

We assume that the velocity of the ions, v, remains constant once they have been accelerated.

****

SAQ 4.3a
Can you recall what the maximum value is for such an instrument?

Response

About 5000.

****

SAQ 4.5a
Calculate the resolving power needed to distinguish between $C_{27}H_{56}^+$ and $C_{26}H_{52}O^+$ ($^{12}C = 12.00000$; $^{1}H = 1.00783$; $^{16}O = 15.99492$).

Response

$$\text{Mass of } C_{27}H_{56}^+ = 380.43848$$
$$\text{Mass of } C_{26}H_{52}O^+ = 380.40208$$
$$\Delta m = 0.0364$$

Resolving power $= m/\Delta m = 380/0.0364 = 10\ 440$

****

SAQ 4.5b   What is the maximum permitted overlap between a pair of peaks which are said to be resolved?

Response

The maximum permitted overlap is 10% of the peak height. This
gives rise to the '10% valley definition' of resolving power.

****

SAQ 4.5c   What is it that limits the resolving power of a
mass spectrometer?

Response

Resolving power is limited by the spread of kinetic energies of ions
of the same m/z entering the magnetic analyser.

****

SAQ 4.5d   Show that ions of the same kinetic energy, when
accelerated through a potential of V and passed
through an electric sector with a potential of E,
all follow a curved flight path of radius R while
in the electric sector, independent of their m/z
values.

Response

If ions of mass m and charge z are passed through an accelerating potential of V they gain kinetic energy as shown by the equation:

$$\text{ke} = \frac{1}{2} mv^2 = zV$$

When the ions enter the electric sector they follow a curved flight path of radius R, provided that

$$zE = mv^2 R$$

$$\therefore \qquad mv^2 = 2zV = zER$$

$$\therefore \qquad R = \frac{2V}{E}$$

Thus, the radius of the flight path depends only on V and E not on m or z.

****

SAQ 4.5e Fill in the missing words or phrases in the following description of a quadrupole analyser.

A quadrupole analyser consists of .... parallel rods with adjacent .... linked to a .... consisting of a .... component and an .... component. The m/z range is scanned by altering both the .... and the ...., while the ratio between them is ....

Response

A quadrupole analyser consists of *four* parallel rods with adjacent *pairs* linked to a *voltage* consisting of a dc component and an rf component. The m/z range is scanned by altering both the *dc voltage* and the *rf field*, while the ratio between them is *kept constant*.

****

SAQ 4.5f

Suppose you are asked to recommend the design of mass spectrometer to be included in an unmanned space probe to Mars, what type of ion source and analyser would you suggest for the instrument, given that the following points are of importance.

(*i*) The spectrometer will be analysing for relatively simple organic molecules of low M_r (up to 150), but needs positive identification.

(*ii*) The instrument must be small and not too heavy to fit into the probe, but it must be robust to withstand take-off and landing.

(*iii*) It has to be reliable – there are no service engineers on Mars.

(*iv*) It has to be computer controlled – this is an unmanned flight.

Response

Most organic molecules of low M_r are volatile, so as long as you can get the sample into the instrument, one should not need an ion source designed for involatile materials. To obtain positive identification of compounds by mass spectrometry one ideally requires a high resolution measurement of the mass of the molecular ion to yield a molecular formula and fragmentation data to confirm the structure. All of this suggests an electron impact ion source would be best. Possibly a chemical ionisation ion source would be more suitable for some molecules, but the increased complexity of this type of source may rule it out.

High resolution mass measurements require a double focussing magnetic sector instrument, but these are often large and complex. All the other requirements of size, weight, robustness, reliability and the need for computer control are available in the quadrupole analyser, but this is of limited resolution. Here one must weigh up the positive and negative features of a double focussing magnetic sector instrument against one with a quadrupole analyser.

In fact the choice was the double focussing instrument with an electron impact ion source for the Viking landing craft in 1975. Don't be too concerned if you didn't make the right choices in answering this question, as it does really require a considerable amount of experience to confidently make such judgements. It does, however, highlight some of the factors that must be considered in choosing an instrument for any laboratory, except that in this example, the scientists had a very considerable budget. Because of this they were able to develop a double focussing instrument that occupied a space of 10 × 10 × 20 cm (incredibly small compared to the normal laboratory instrument) and although this was very costly, it was deemed necessary to have the certainty of accurate mass measurements for positive identification of compounds. It would have been much more expensive to launch another probe, if the first one hadn't given quite the desired information!

****

SAQ 5.1a   If you recall our general discussion in Section
                1.3 concerning the energetics of the formation
                of molecular and fragment ions, you will remem-
                ber that molecular ions with less than a certain
                threshold energy do not have enough energy to
                decompose. These are analysed and recorded as
                M^+ ions. Those with internal energies over the
                threshold decompose in the ion source and are
                analysed and recorded as fragment ions.

                Can you represent this on a plot of internal en-
                ergy distribution? You were shown this earlier.

Response

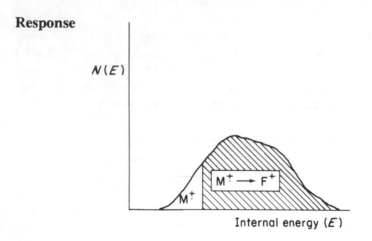

You may remember discussion of this aspect in Section 1.3. In fact
the figure shown here is Fig. 1.3a. If you had difficulty with this
question revise this earlier section.

****

| SAQ 5.1b | Can you recall the relationship between accelerating potential, V and kinetic energy of the ion being accelerated? |

Response

$zV = \frac{1}{2} m_1 v_1^2$. If you need to remind yourself of this equation reread Section 2.1.

****

| SAQ 5.2a | All the metastable ions observed in the mass spectrum of benzene, arise from the molecular ion (m/z 78). Using the formula |

$$m^* = m_2^2/m_1$$

assign the fragment ions (M_2^+). As an example the fragmentation

$$C_6H_6^+ \quad \rightarrow \quad C_6H_5^+ \quad + \ H^{\cdot}$$

$$m/z \ 78 \qquad m/z \ 77$$

shows a metastable ion at $77^2/78 = 76.0 \ m/z$

Response

$C_6H_6^+ \rightarrow C_6H_4^+ + H_2$

$\quad$ 78 $\qquad\quad$ 76

$m^* = 76^2/78 = 74.1$

$$C_6H_6^+ \rightarrow C_4H_4^+ + C_2H_2$$
$$\quad 78 \qquad\qquad 52$$

$$m^* = 52^2/78 = 34.7$$

$$C_6H_6^+ \rightarrow C_3H_3^+ + C_3H_3^+$$
$$\quad 78 \qquad\quad 39$$

$$m^* = 39^2/78 = 19.5$$

****

SAQ 5.2b   Other metastable ions are observed in the benzene spectrum:

M^*	M_1^+	$\rightarrow$	M_2^+
50.0	52	$\rightarrow$	51
48.1	52	$\rightarrow$	50
33.8	77	$\rightarrow$	51
32.9	76	$\rightarrow$	50

Use these metastable ions and those used in SAQ 5.2a for the $C_6H_6^+$ ion, to draw up a complete fragmentation pattern for benzene.

Response

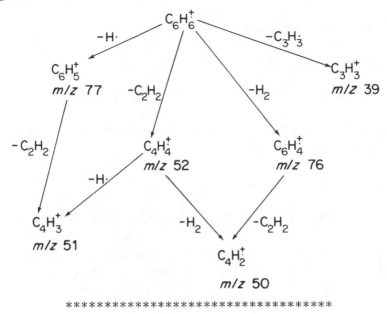

****

SAQ 5.2c Fig. 5.1a is actually the mass spectrum of aniline, $C_6H_5NH_2$. The two metastable ions shown at m/z 46.8 and 45.9 correspond to fragmentations of the molecular ion ($M^{\ddagger}$) and the fragment ion $(M-H)^+$ respectively. What are the daughter ions formed in these two fragmentation processes?

Can you suggest a formula for the neutral species lost?

Response

The metastable ion at m/z 46.8 arises from the molecular ion of aniline (m/z 93). Therefore as the mass of daughter ion of this fragmentation is given by

$$m_2^2 = m^* . m_1,$$

we calculate that the daughter ion has m/z of 66.

The fragmentation is thus

$$C_6H_5NH_2^{+\cdot} \ (m/z \ 93) \rightarrow m/z \ 66 + 27 \ amu$$

Similarly, m/z 45.9 arises from $(M-H)^{+\cdot}$ $(m/z$ 92) and here the mass of the daughter ion is m/z 65.

The fragmentation is thus

$$(M-H)^+ \ (m/z \ 92) \rightarrow m/z \ 65 + 27 \ amu$$

At first sight fragmentations resulting in the loss of a neutral of mass 27 seems unlikely, since we have just seen benzene, a closely related molecule, loses C_2H_2 (26 amu) or C_3H_3 (39 amu) and aniline might be expected to behave similarly and not lose C_2H_3. Alternatively, one might expect aniline to lose $\cdot NH_2$ (16 amu).

The fact that the neutral has an *odd* mass is a vital clue. It means it may well contain nitrogen and thus a likely formulation is HCN. The formulae of the fragment ions at m/z 66 and 65 must therefore be $C_5H_6^{+\cdot}$ and $C_5H_5^+$ respectively.

****

SAQ 6.1a Can you remember what the most commonly employed type of detector is called and how it works?

Response

It is called the electron multiplier and it consists of a detector plate and a series of dynodes. When the ions hit the plate, a small electrical current is generated and this is amplified by the cascade effect across the dynodes. If you had forgotten this, revise Section 2.1.

****

SAQ 6.1b   How long would it take to scan from m/z 10 to m/z 1000 at 10 s decade^{-1}?

Response

20 seconds. That is, 10 seconds for scanning from m/z 10 to m/z 100 and a further 10 seconds for scanning from m/z 100 to m/z 1000. If you had forgotten this, revise Section 4.4.

****

SAQ 6.1c   Can you recall how these traces differ and why we need more than one?

Response

The traces are recorded with different sensitivities. The trace (i) is ten times the sensitivity of the trace (iii) and trace (ii) is three times more sensitive than trace (iii). This allows both very intense and very weak peaks to be recorded in one scan.

****

| SAQ 6.1d | How are spectra usually presented? |

Response

They are usually presented in tabulated form (m/z values *vs* relative abundances) or in the form of bar diagrams.

****

| SAQ 6.2a | Briefly list the steps involved in the acquisition of mass spectral data. |

Response

(*i*)   A mass spectrum of a reference compound is fed to the computer *via* the ADC in the form of voltage reading with respect to time, rejecting values below a preset threshold.

(*ii*)   The computer calculates a peak centroid value from this data for each peak and stores the time value corresponding to the centroid.

(*iii*) The computer calculates the peak areas by summing all voltage values for each peak and stores these for using in giving relative abundances.

(*iv*) The time values for the peaks in the reference compound are 'plotted' against the known mass values to give the scan law of the instrument.

(*v*)   The procedure is repeated for the sample to be examined to obtain peak times and abundances.

(*vi*) The peak times of the unknown are converted to mass/charge values using the scan law.

(*vii*) The mass spectrum of the unknown is printed in tabular form or plotted as a bar diagram.

****

SAQ 7.2a   Can you think of an example of a monoisotopic element? (If you have access to a 'Rubber Book' have a glance through the Table of the Relative Atomic Masses of the Elements, but don't spend too much time on this).

Response

If you thought of one or more of the elements in the table below, well done. If not, don't worry too much, a knowledge of the distribution of isotopes is rarely required outside mass spectrometry.

Common Elements Having Single Isotopes

		Name	A_r
(*i*)	Metals	Beryllium	9
		Sodium	23
		Scandium	45
		Cobalt	59
		Niobium	93
		Rhodium	103
		Caesium	133
		Gold	197
		Bismuth	209
(*ii*)	Non-metals	Fluorine	19
		Phosphorus	31
		Arsenic	75
		Iodine	127

****

SAQ 7.2b	Taking the data for lead in Fig. 7.2b, plot the normalised spectrum of the Pb^+ ion cluster.

Response

Step 1

For the most intense isotope (m/z 208) determine the factor by which its relative abundance would be converted to 100% ie $100/51.7 = 1.93$.

Step 2

Now all the other RAs for m/z 204, 206 and 207 may be converted to the appropriate value on the same 0 to 100 scale by multiplying by 1.93.

For plotting purposes it is sufficiently accurate to round all figures off the the nearest whole %. You should have obtained the following data:

At. No.	m/z	Rel.Ab.	%
82	204	1.4	3
	206	25.2	49
	207	21.7	42
	208	51.7	100

and plotted a graph like this:

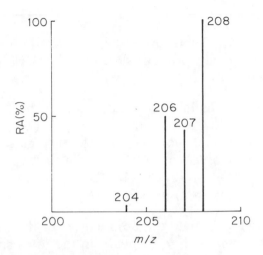

Normalised Spectrum of Pb^{+}.

****

SAQ 7.2c   A normalised spectrum of a molecular ion cluster is shown in Fig. 7.2d. Calculate the isotopic abundances of the element concerned. Given that the species has the formula $(CH_3)_2X$, what is the element X?

Response

The base peak here is m/z 93, and m/z 95 is 44.5% (or thereabouts as near as the graph can be read). Hence of the total positive ion intensity of 144.5%, the most abundant isotope is 100/144.5 or 69.2% and the less abundant is 44.5/144.5 or 30.8%. Since the mass of two CH_3 groups is 30 atomic mass units (amu), the atomic weight of the isotopes must be 63 (ie 93–30) and 65 (ie 95–30). Reference to mass and abundance tables of the isotopes of the elements shows that this is *copper*, ^{63}Cu, 69.1%; ^{65}Cu, 30.9%. You probably guessed this from your residual memory of common atomic weights even if you didn't known that copper had these two isotopes.

****

SAQ 7.2d

(*i*)   Calculate the abundance of the $(M + 1)$ peaks of CO, N_2, and $CH_2=CH_2$ ($M^+ = 28$).

(*ii*) Calculate the abundance of the $(M + 1)$ peak of naphthalene, $C_{10}H_8$ ($M^+ = 128$). Could you expect to distinguish naphthalene from cyclohexanecarboxylic acid, $C_7H_{12}O_2$ (also M_r 128) by the height of its $(M + 1)$ peak alone?

(*iii*) Calculate the abundances of the $(M + 1)$ and $(M + 2)$ peaks in the mass spectrum of cholesterol, $C_{27}H_{45}OH$.

(*iv*) Calculate the abundance of the $(M + 2)$ peak in the mass spectrum of trimethyl phosphate, $(CH_3O)_3PO$.

(*v*)   At what carbon number, approximately, would you expect the $(M + 2)$ peak of an organic compound containing C, H, N, F, P or I and 4 oxygen atoms to appear above a general background level of 2% (relative to $M^+ = 100\%$)?

Response

(*i*)   For this question, you need to use Eq. 7.2b and the data in Fig. 7.2c

$$(M + 1)/M = (1.1 \times 1)\% = 1.1\% \text{ for CO,}$$

which has one carbon atom and no hydrogens or nitrogen. The $(M + 1)$ isotope of O, ^{17}O, is 0.04% (quite negligible). For N_2 there is no carbon or hydrogen so only the third term in Eq. 7.2b matters, hence

$$(M + 1)/M = (0.36 \times 2)\% = 0.72\%.$$

For ethene ($CH_2{=}CH_2$), there is no nitrogen so we need only calculate the first two terms, thus:

$$\frac{(M + 1)}{M} = (1.1 \times 2)\% + (0.016 \times 4)\% = 2.68\%$$

These calculations show an interesting point. If you could measure the height of the $(M + 1)$ peak (m/z 29) accurately, you could distinguish the isomeric compounds CO, N_2 and $CH_2{=}CH_2$ by the RA of their $(M + 1)$ peaks alone. In practice the background in the mass spectrometer would have to be very low in order to achieve this, say 10% of the intensity of the smallest m/z 29 (that of N_2) ie 0.07% relative to the N_2 molecular ion at m/z 28.

(*ii*) Taking naphthalene first, for $C_{10}H_8$ Eq. 7.2b gives:

$$\frac{(M + 1)}{M} = (1.1 \times 10)\% + (0.016 \times 8)\% =$$
$$(11.0 + 0.13)\% = 11.13\%$$

Notice that the bulk of the $(M + 1)$ peak consists of $^{12}C_9{}^{13}C^1H_8$ with $^{12}C_{10}{}^1H_7{}^2H$ making a very minor contribution (0.13%). In practice, bearing in mind the inherent inaccuracy in determining RAs in ordinary mass spectrometers you can neglect the deuterium in organic compounds until the number of hydrogen atoms reaches 20 or so. At this point $(M + 1)$ for ^2H is $(0.016 \times 20)\% = 0.32\%$.

Now considering cyclohexanecarboxylic acid, $C_7H_{12}O_2$,

$$\frac{(M + 1)}{M} = (1.1 \times 7)\% + (0.016 \times 12)\% =$$
$$(7.7 + 0.19)\% = 7.9\%$$

You can now see that the difference in RA of the $(M + 1)$ peaks for these two isomeric compounds is 3.24%, so it should be possible to distinguish them (but bear in mind the point made about background levels in (*i*)). You can also see from this example that molecules containing a large proportion of carbon ie hydrocarbons in general, or molecules having large hydrocarbon residues within them, *have relatively intense (M + 1) peaks.*

(*iii*) In this answer, both Eq. 7.2b and Eq. 7.2c are used:

$$\frac{(M + 1)}{M} = (1.1 \times 27)\% + (0.016 \times 46)\%$$

$$= (29.7 + 0.72)\%$$

$$= 30.42\%$$

$$\frac{(M + 2)}{M} = \frac{(1.1 \times 27)^2}{200}\% + (0.2 \times 1)\%$$

$$= (4.4 + 0.2)\%$$

$$= 4.6\%$$

The (M + 2) peak consists of contributions from $^{12}C_{25}{}^{13}C_2{}^1H_{45}OH$ and $^{12}C_{27}{}^1H_{45}{}^{18}OH$. Note that these larger C, H and O organic molecules have substantial (M + 1) peaks mainly due to ^{13}C, and their (M + 2) peaks are readily observable due to the finite chance that *two* ^{13}C are incorporated in the *same* molecule.

(*iv*) Here we need to check first in Fig. 7.2c whether P has a significant (M + 2) isotope. It does not – it is one of the few monoisotopic elements. Check SAQ 7.2a for this. Hence we can apply Eq. 7.2c with no special modification:

$$\frac{(M + 2)}{M} = \frac{(1.1 \times 3)^2}{200}\% + (0.2 \times 4)\%$$

$$= 0.05\% + 0.8\%$$

$$= 0.85\%$$

In this example note that the $2 \times {}^{13}C$ isotope contribution given by the first term is really negligible but the ^{18}O isotope term starts to become appreciable as the number of O atoms increases to four.

(*v*) None of the atoms mentioned except C and O have isotopes which contribute significantly to (M + 2). The answer to (*iv*) showed that the contribution to this peak from four ^{18}O atoms is 0.8%. This problem, then, boils down to finding how large *n* is for the term $(1.1 \times n)^2/200$ to reach $(2 - 0.8)\%$ ie 1.2%.

When this happens the (M + 2) peak will exceed 2% and appear above the background.

So $$\frac{(1.1 \times n)^2}{200} = 1.2$$

$$1.21n^2 = 240$$

$$n^2 = 198$$

$$n = 14.08$$

Hence a compound would have to contain *14* carbon atoms to satisfy this criterion. A couple of benzene rings plus a small side-chain would do the trick.

****

SAQ 7.2e

What are the molecular ions of:

(*i*)   1-chloronaphthalene, $C_{10}H_7Cl$;
(*ii*) diethylmercury, $(C_2H_5)_2Hg$;
(*iii*) tetraethyl lead, $(C_2H_5)_4Pb$ and
(*iv*) ferrocene, dicyclopentadienyliron $(C_5H_5)_2Fe$?

Response

In all these calculations, the A_r of the most abundant isotope is used.

(*i*) $C_{10}H_7Cl$ $M_r = (12 \times 10) + (1 \times 7) + (35 \times 1) = 162$

   $M^{+}_{.} = m/z$ 162

(*ii*) $(C_2H_5)_2Hg$ $M_r = (12 \times 4) + (1 \times 10) + (202 \times 1) = 260$

   $M^{+}_{.} = m/z$ 260

If you obtained $M^{+}_{.} = m/z$ 258 here, have another look at Fig. 7.2b, where you will see 200 = 23.1%, 202 = 29.8% for Hg, hence 202 is the correct A_r to use for calculating the M_r of Hg compounds.

(*iii*) $(C_2H_5)_4$ Pb $M_r = (12 \times 8) + (1 \times 20) + (208 \times 1) = 324$

   $M^{+}_{.} = m/z$ 324

You should not have used A_r 206 or 207 here for Pb because these are not the most abundant Pb isotopes (see Fig. 7.2b).

(*iv*) $(C_5H_5)_2$ Fe $M_r = (12 \times 10) + (1 \times 10) + (56 \times 1) = 186$

   $M^{+}_{.} = m/z$ 186

The principal isotope of Fe is ^{56}Fe, 91.52%

****

SAQ 7.3a	Use Eq. 7.3b to calculate the ratios of the isotope peaks expected for a Br_2-containing ion. Compare your prediction with Fig. 7.3e (*ii*).

Response

The ratios of the Br_2 isotope peaks are obtained from $a^2 + 2ab + b^2$ where $a = b = 1$. Hence three peaks are predicted $1^2 + 2 \times 1 \times 1 + 1^2$ which becomes $1:2:1$. Examination of Fig. 7.3d (*ii*) shows that $M:(M + 2):(M + 4)$ is very nearly $1:2:1$.

****

SAQ 7.3b	Use Eq. 7.3b to calculate the ratios of the isotope peaks expected for a Br_3-containing ion. Compare your prediction with Fig. 7.3e (*iii*).

Response

The ratios of the Br_3 isotope peaks are $1:3:3:1$. calculated from Eq. 7.3c as follows:

$$a^3 + 3a^2b + 3ab^2 + b^3 \text{ becomes } 1^3 + 3 \times 1^2 \times 1 + 3 \times 1 \times 1^2 + 1^3, \text{ or } 1:3:3:1.$$

Inspection of Fig. 7.3e (*iii*) shows that $M:(M + 2):(M + 4):(M + 6)$ is very nearly $1:3:3:1$.

****

SAQ 7.3c

(*i*) Expand the expression $(a + b)^5$ and use it to calculate the relative abundances of the ions in the Cl_5 and Br_5 clusters.

(*ii*) Expand the expression $(a + b)^6$ and use it to calculate the relative abundances of the ions in the Cl_6 and Br_6 clusters. (Optional)

Response

(*i*) $(a + b)^4 = a^4 + 4a^3b + 6a^2b^2 + 4ab^3 + b^4$

$(a + b)^5 = (a + b)(a + b)^4$

$$= a^5 + 4a^4b + 6a^3b^2 + 4a^2b^3 + ab^4 +$$
$$a^4b + 4a^3b^2 + 6a^2b^3 + 4ab^4 + b^5$$

Collecting together like terms we have:

$(a + b)^5 = a^5 + 5a^4b + 10a^3b^2 + 10a^2b^3 + 5ab^4 + b^5$

Use this equation to calculate the relative abundances of Cl_5 and Br_5.

For Cl_5 substituting $a = 3$, $b = 1$; the cluster will have the relative abundances:

$$243 : 405 : 270 : 90 : 15 : 1$$

For Br_5 substituting $a = 1$, $b = 1$; the cluster will have the relative abundances:

$$1 : 5 : 10 : 10 : 5 : 1$$

(*ii*) $(a + b)^6 = (a + b)(a + b)^5$

$$= a^6 + 5a^5b + 10a^4b^2 + 10a^3b^3 + 5a^2b^4 +$$
$$ab^5 + a^5b + 5a^4b^2 + 10a^3b^3 + 10a^2b^4 +$$
$$5ab^5 + b^6$$

Collecting like terms together we have:

$$(a + b)^6 = a^6 + 6a^5b + 15a^4b^2 + 20a^3b^3 + 15a^2b^4 + 6ab^5 + b^6$$

Substituting into this equation $a = 3$, $b = 1$, the Cl_6 cluster will have the relative abundances:

729 : 1458 : 405 : 540 : 135 : 18 : 1 which simplifies to
40 : 81 : 23 : 30 : 8 : 1 : 0.1 very nearly.

Similarly for Br_6, where $a = 1$, $b = 1$, the Br_6 cluster will have the relative abundances:

1 : 6 : 15 : 20 : 15 : 6 : 1

****

SAQ 7.3d Fig. 7.3f shows the un-normalised mass spectrum of 4-bromochlorobenzene. Are the molecular ions in the predicted 3 : 4 : 1 ratio? What other ions containing halogens are present in the spectrum?

Response

First draw a baseline across the molecular ion region under the m/z 190–194 peaks then measure the peak heights. In mm the ratios are m/z 190 : 192 : 194 = 56 : 71 : 17 or 3.3 : 4.2 : 1. This is not too good agreement with the expected 3 : 4 : 1 ratio but ions occur at m/z 111 and 113, and 36 and 38, in the 3 : 1 ratio expected of chlorine, and 80 and 81 in the $1:1$ ratio expected of bromine. $(190 - 79)$ is 111, and $(192 - 79)$ is 113 showing that $M^{\ddot{+}}$ decomposes by loss of bromine to give an ion containing *one* chlorine. Hence the sample is confirmed as bromochlorobenzene. m/z 111/113 is $C_6H_4Cl^+$, m/z 80/82 is $HBr^{\ddot{+}}$ and m/z 36/38 is $HCl^{\ddot{+}}$.

****

SAQ 7.3e Calculate the relative abundances of the ions in the following clusters:

(*i*) Cl_3Br;

(*ii*) Cl_2Br_2;

(*iii*) $ClBr_3$.

Would mass spectrometry enable these to be distinguished from one another, and also from Cl_4 and Br_4 clusters?

Response

(*i*)   The expression for Cl_3Br is

$$(a + b)^3(c + d)$$

which expands partially to

$$(a^3 + 3a^2b + 3ab^2 + b^3)(c + d)$$

and fully to

$$a^3c + 3a^2bc + 3ab^2c + b^3c + a^3d + 3a^2bd + 3ab^2d + b^3d$$

Grouping ions of common nominal mass this becomes

$$a^3c + (3a^2bc + a^3d) + (3ab^2c + 3a^2bd) + (b^3c + 3ab^2d) + b^3d$$

Substituting in this equation; $a = 3, b = c = d = 1$, we get

$$M^{\ddagger} : (M + 2)^{\ddagger} : (M + 4)^{\ddagger} : (M + 6)^{\ddagger} : (M + 8)^{\ddagger}$$

$$= 27:54:36:10:1$$

or

$1:5:3:1:0.1$ approximately dividing through by 10.

(*ii*) The expression for Cl_2Br_2 is

$$(a + b)^2(c + d)^2$$

which expands partially to $(a^2 + 2ab + b^2)(c^2 + 2cd + d^2)$ and fully to

$$a^2c^2 + 2(a^2cd + abc^2) + (a^2d^2 + b^2c^2 + 4abcd) + 2(abd^2 + b^2cd) + b^2d^2$$

grouping ions of common nominal mass. Substituting as usual for a, b, c and d we get:

$$M^{\ddagger} : (M + 2)^{\ddagger} : (M + 4)^{\ddagger} : (M + 6)^{\ddagger} : (M + 8)^{\ddagger}$$

$$= (9:24:22:8:1).$$

(*iii*) The expression for $ClBr_3$ is $(a + b)(c + d)^3$

which expands partially to

$$(a + b)(c^3 + 3c^2d + 3cd^2 + d^3)$$

and fully to:

$$ac^3 + (3ac^2d + bc^3) + 3(acd^2 + bc^2d) +$$
$$(3bcd^2 + ad^3) + bd^3$$

grouping ions of common nominal mass. Substituting as usual for a, b, c and d we get

$$M^{+} : (M + 2)^{+} : (M + 4)^{+} : (M + 6)^{+} : (M + 8)^{+}$$

$$= 3 : 10 : 12 : 6 : 1$$

Summarising then as before, the relative abundances are as follows:

	M^{+}	$(M + 2)^{+}$	$(M + 4)^{+}$	$(M + 6)^{+}$	$(M + 8)^{+}$
Cl_4	81	108	54	12	1
Cl_3Br	27	54	36	10	1
Cl_2Br_2	9	24	22	8	1
$ClBr_3$	3	10	12	6	1
Br_4	1	4	6	4	1

In principle, it should be possible to distinguish all five of these clusters.

$$************************************$$

SAQ 7.3f

Calculate the relative abundances, using the quick calculation method, of the major isotope peaks in ions containing:

(*i*) S_2Cl_2;
(*ii*) $CuBr_2$;
(*iii*) $SClBr$;

Response

(i) S_2Cl_2

For sulphur, $^{32}S : {}^{34}S = 95:4$, see Fig. 7.2c.

For S_2, $(a + b)^n = (95 + 4)^2 = 9025:760:16$

or $564:48:1$

For Cl_2, $(a + b)^n = (3 + 1)^2 = 9:6:1$

Combining these relative abundances gives,

$$(564:48:1)\ (9:6:1)\ (564:48:1) \times 9 = 5076:\ \ 432:\ \ 9$$
$$(564:48:1) \times 6 = \ \ \ \ \ \ 3384:288:\ \ 6$$
$$(564:48:1) \times 1 = \ \ \ \ \ \ \ \ \ \ \ \ \ \ \ \ 564:48:1$$
$$\text{Total} = 5076:3816:861:54:1$$

Hence $M^+ : (M + 2)^+ : (M + 4)^+ : (M + 6)^+$ is $94:71:16:1$ approximately with $(M + 8)^+$ being negligible.

(ii) $CuBr_2$

For copper $^{63}Cu : {}^{65}Cu = 69.1:30.9$

or $2.24:1$ (SAQ 7.2c)

For Br_2 $(a + b)^2 = (1 + 1)^2 = 1:2:1$

Combining these relative abundances gives,

$$(2.24:1)\ (1:2:1)\ (2.24:1) \times 1 = 2.24:\ \ \ 1$$
$$(2.24:1) \times 2 = \ \ \ \ \ 4.48:\ \ \ 2$$
$$(2.24:1) \times 1 = \ \ \ \ \ \ \ \ \ \ \ \ 2.24:1$$
$$\text{Total} = 2.24:5.48:4.24:1$$

Hence $M^+ : (M + 2)^+ : (M + 4)^+ : (M + 6)^+$ is $2.24:5.48:4.24:1$

(*iii*) SClBr

Combining the three relative abundances we have, $(95:4)(3:1)(1:1)$ $= (95:4)(3:4:1)$ when Cl and Br are combined as in Section 7.3.2., Fig. 7.3g

Hence:
$$
\begin{aligned}
(95:4) \times 3 &= 285: \ 12 \\
(95:4) \times 4 &= 380: \ 16 \\
(95:4) \times 1 &= 95:4 \\
\hline
\text{Total} &= 285:392:111:4
\end{aligned}
$$

Hence $M^+ : (M + 2)^{\pm} : (M + 4)^{\pm} : (M + 6)^{\pm}$ is $71:98:28:1$

Notice how close the ratio is to the $3:4:1$ expected for an ion containing ClBr, neglecting the $(M + 6)^{\pm}$ ion $(2.5:3.5:1)$.

****

SAQ 7.4a

> Use the expression derived in Eq. 7.4b to calculate the relative abundances of the first four ions in Cl_4 cluster, and compare your result with that found by expanding $(a + b)^4$ in Fig. 7.3g.

Response

According to Eq. 7.4b, the first four ions have intensities of:

$$
a^3 \ : \ na^2b \ : \ \frac{n}{2}(n - 1)ab^2 \ : \ \frac{n}{6}(n - 1)(n - 1)b^3
$$

where $a = 3$, $b = 1$, and $n = 4$ for a Cl_4 cluster. Substituting these values in Eq. 7.4b, we get $M^+ : (M + 2)^{\pm} : (M + 4)^{\pm} : (M + 6)^{\pm} = 27:36:18:4$.

The ratios for Cl_4 given in Fig. 7.3g are $81:108:54:12$ for the first four isotope peaks, which divided through by 3 also gives $27:36:18:4$, in excellent agreement.

****

SAQ 7.4b In a similar way to the previous exercise in the text, compare the observed and calculated relative abundances in the S_4^+ and S_5^+ clusters shown in Fig. 7.4b.

Response

Cluster	m/z	height (mm)	%	Predicted %
S_4	128	114	100	100
	130	21	18.4	16.8
	132	4	3.5	1.1
S_5	160	129	100	100
	162	29	22.4	21.1
	164	3.5	2.7	1.8

The predicted % are obtained using Eq. 7.4b as follows:

$$S_4 \ (95 + 4)^4 \approx 95^3 : 4 \times 95^2 \times 4 : \frac{4}{2} (3) \ 95 \times 4^2$$

$$= \ 100 : 16.8 : 1.1$$

$$S_5 \ (95 + 4)^5 \approx 95^3 : 5 \times 95^2 \times 4 : \frac{5(4)}{2} \times 95 \times 4^2$$

$$= \ 100 : 21.1 : 1.8$$

It is noticeable that the observed $(M + 2)^+$ and $(M + 4)^+$ are always slightly higher than predicted. This is because at each m/z there is an unrelated background peak. These will affect the less intense peaks proportionately more than the M^+ peak, tending to increase them.

****

SAQ 7.4c   Reference to Fig. 7.2c shows that ^{33}S has a natural abundance of 0.78% relative to $^{32}S = 100\%$.

(*i*)   What is the relative abundance of $(M + 1)^+$ in the S^{8+} cluster? (Hint – use 'I never Rang Ann')

(*ii*)   What is the contribution made by species containing two ^{33}S to the $(M + 2)^+$?

(*iii*)   Would $(M + 3)^+$ be observable in the S_8^+ cluster? In attempting (*ii*) and (*iii*) you will find the approximate binomial expansion we have just discussed most helpful (Eq. 7.4b).

Response

(*i*)   The simple formula of Eq. 7.2a applies here:

% $(M + 1)^+ = (0.78 \times 8) = 6.24\%$

(*ii*)   It is best to evaluate Eq. 7.4b here to the fourth term, as this latter will give the answer to (*iii*) also. The ratio of $^{32}S:^{33}S$ is $100:0.78$ or $128:1$. Hence $M^+:(M + 1)^+:(M + 2)^+:(M + 3)^+$ for ^{33}S is:

$$(128)^3: 8 \times 128^2 \times 1 : \frac{8}{2}(7)\, 128 \times 1^2 : \frac{8}{6}(7)(6)1^3$$

$$= 2.097 \times 10^6 : 1.31 \times 10^5 : 3584 : 56$$

$$\simeq 100 : 6.24 : 0.17 : 0.003$$

Contribution to $(M + 2)^+$ due to two ^{33}S is only 0.17%

(*iii*) The answer to (*ii*) shows that $(M + 3)^{\ddagger}$ due to ^{33}S is only 0.003%. You would not expect to be able to observe such a small peak using an ordinary mass spectrometer.

You might like to note that both the simple formula Eq. 7.2a and the more complicated Eq. 7.4b give the same answer for the intensity of $(M + 1)^{\ddagger}$ to two decimal places. We hope you find this reassuring. Moral: use Eq. 7.2a wherever possible.

****

SAQ 7.4d A common method of analysing polyhydroxy compounds such as glucose is to convert each —OH group into a —OSi(CH₃)₃ (trimethylsilyl, TMS) group. This is then used in gc-ms analysis. How intense would you expect the $(M + 1)^{\ddagger}$ ions to be for the penta-TMS derivative of glucose, $C_6H_{12}O_6$, whose formula is $C_{21}H_{47}O_6Si_5$?

Response

If five —OH are converted to —OSi(CH₃)₃, the formula of the derivative is $C_{21}H_{47}O_6Si_5$. $(M + 1)^{\ddagger}$ contains mostly ^{13}C, ^{2}H and ^{29}Si, hence Eq. 7.2b applies:

$$(M + 1)^{\ddagger} = (1.1 \times 21)\% + (0.016 \times 47)\% + (5.07 \times 5)\%$$

(natural abundances taken from Fig. 7.2c).

Hence $(M + 1)^{\ddagger} = (23.1 + 0.75 + 25.35)\%$

$$= 49.2\%$$

$(M + 2)^{\ddagger}$ will contain ^{18}O, ^{30}Si, and species containing *two* ^{13}C or ^{29}Si. Here a combination of Eq. 7.2a and 7.2c can be used:

$$(M + 2)^{\ddagger} = (0.2 \times 6)\% + (3.31 \times 5)\% + \frac{(1.1 \times 21)^2}{200}\% +$$

$$\frac{(5.07 \times 5)^2}{200}\%$$

$$= (1.2 + 2.43 + 16.55 + 3.21)\%$$

$$= 23.39\%$$

$(M + 3)^{\ddagger}$ will consist of three possible species, ions containing $^{13}C_2$ ^{29}Si, $^{13}C^{30}Si$, and $^{29}Si_3$ (neglecting any ^2H contributors which will be very small).

$\%\ ^{13}C_2\ ^{29}Si = 5 \times 2.43 \times 0.0507\% = 0.62\%$

$\%\ ^{13}C\ ^{30}Si = 21 \times 0.011 \times 3.31 \times 5\% = 3.82\%$

$\%\ ^{29}Si_3$ can be obtained from Eq. 7.4b where $^{28}Si : ^{29}Si = 20 : 1$ and $n = 5$, so

$$M^{\ddagger} : (M + 1)^{\ddagger} : (M + 2)^{\ddagger} : (M + 3)^{\ddagger}$$

$$= 20^3 : 5 \times 20^2 \times 1 : \frac{5}{2} \times 4 \times 20 \times 1^2 : \frac{5}{6} \times 4 \times 3 \times 1^3$$

$$= 8000 : 2000 : 200 : 10$$

$$= 100 : 25 : 2.5 : 0.125$$

Hence $(M + 3)^{\ddagger}$ due to $^{29}Si_3$ is 0.125%.

So the total intensity of $(M + 3)^{\ddagger}$ is 4.57%

****

SAQ 7.4e
Fig. 7.4c shows the mass spectrum of the pesticide DDT, which contains a number of atoms of a halogen. What halogen is present, and how many atoms of it are there? How do you think the cluster m/z 235/237/239 has been formed from $M^{\ddot{+}}$?

Response

The most intense peaks in the clusters in Fig. 7.4c from m/z 235 onwards are all separated by 2 amu, thus indicating chlorine or bromine rather than fluorine or iodine. Since you are told a halogen is present, the general lack of symmetry of the clusters tends to rule out bromine. Also there are mass differences of 35 between some of the clusters. It appears that DDT contains Cl_n. You should then work out the normalised intensities of the Cl_n clusters for $n = 2$ to 6 from the data in Fig. 7.3g, for comparison with the $M^{\ddot{+}}$ cluster at m/z 352 etc. We got the following table when we did this:

Cl_n	$M^{\ddot{+}}$	$(M + 2)^{\ddot{+}}$	$(M + 4)^{\ddot{+}}$	$(M + 6)^{\ddot{+}}$	$(M + 8)^{\ddot{+}}$	$(M + 10)^{\ddot{+}}$	$(M + 12)^{\ddot{+}}$
2	100	65	10				
3	100	100	33	4			
4	81	100	56	13	0.6		
5	59	100	67	22	4	0.4	
6	49	100	84	37	10	1	0.1
DDT	61	100	64	22	4	–	–

The best agreement is with $n = 5$. DDT contains 5 chlorines and has the structure:

Bis-2,2-(4-chlorophenyl)-1,1,1-trichloroethane.

Note the clusters at m/z 317–325, 282–288, and 246–250 show the intensity distributions expected for Cl_4, Cl_3 and Cl_2 respectively. These ions arise by successive losses of chlorine from $M^{\ddot{+}}$. The m/z 235/237/239 clearly contains two chlorines because it has the characteristic 100:65:10 ratio expected for Cl_2. However, it cannot have been formed by loss of three chlorines because this fragment would lead to the m/z; 246/248/250 cluster (includes the loss of an additional hydrogen atom). The loss from $M^{\ddot{+}}$ to give m/z; 235/237/239 is 117 amu. This corresponds to a loss of Cl_3C leading to:

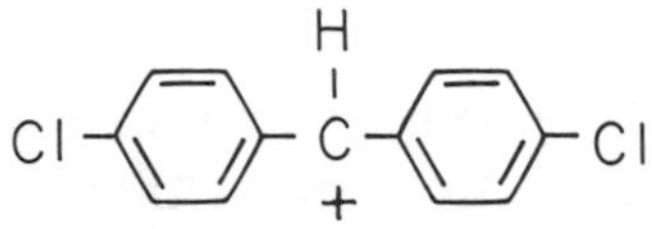

$$Cl\text{—}\hexagon\text{—}\underset{+}{\overset{\overset{\displaystyle H}{|}}{C}}\text{—}\hexagon\text{—}Cl$$

Such an ion would be specially stabilised (see Part 8). If you managed to deduce that DDT must contain a CCl_3 group, well done.

****

SAQ 8.2a

> Fig. 8.2b illustrates the mass spectra of benzene, naphthalene and anthracene. Which ions are formed from the loss of HC≡CH from (i) benzene, (ii) napththalene and (iii) anthracene.

Response

(i)   There are 26 amu in HC≡CH, so we are looking in each spectrum for mass differences of 26. In the benzene spectrum, there are two such losses, m/z 78 → 52, and m/z 77 → 51. You probably spotted the first one easily enough, but don't forget that $M^{\ddot{+}}$ of organic compounds often lose a single hydrogen, as here m/z 78 → 77, then this daughter ion also undergoes the typical loss of HC≡CH.

(ii)   There is only one clear loss of HC≡CH here, m/z 128 → 102.

(*iii*) In the case of anthracene, both m/z 178 and 177 lose 26 amu to give m/z 152 and 151 respectively.

****

SAQ 8.2b	Show how the methoxybenzene ion *B* can be converted into ion *G* (2 steps).

Response

This is best done using a combination of single and double electron shifts, as follows:

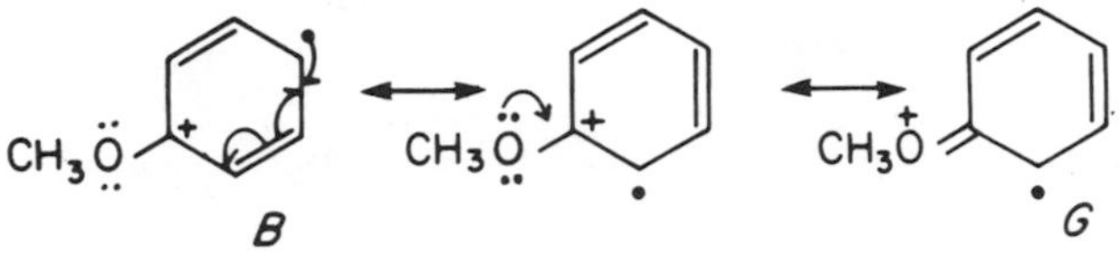

****

SAQ 8.2c	Show how the $M^{\ddagger}$ ion of phenyl ethanoate $$\text{(C}_6\text{H}_5\text{O}\overset{\displaystyle O}{\overset{\|}{\text{C}}}\text{CH}_3)$$ might be stabilised.

Response

If ionisation occurs on the oxygen next to the benzene ring, or on the benzene ring (which was shown to be equivalent in Section 8.2.2) we can stabilise $M^{\ddagger}$ as follows:

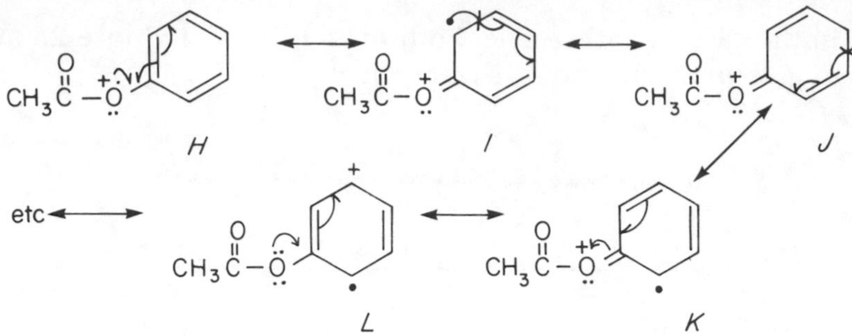

However, if ionisation occurs on the C=O group, resonance is limited to

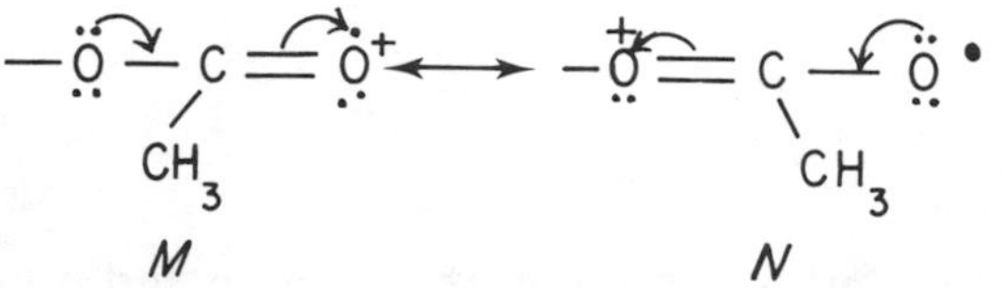

so presumably most $M^{\ddot{+}}$ would be of *H–K* type, though from the mass spectrum alone we cannot prove this.

****

SAQ 8.2d

Would you expect the $M^{\ddot{+}}$ ion of benzyl ethanoate

$$(C_6H_5CH_2O\overset{\displaystyle O}{\overset{\|}{C}}CH_3)$$ to be as stable as the

$M^{\ddot{+}}$ ion of phenyl ethanoate?

Response

$$C_6H_5-CH_2-O-\overset{\displaystyle O}{\overset{\|}{C}}-CH_3$$ could have three $M^{\ddot{+}}$ structures:

While *O* and *P* have resonance forms as already described (Section 8.2.1. and 8.2.2., SAQ 8.2c) and *Q* has the possible resonance forms

$$-CH_2-\overset{+..}{\underset{..}{O}}-\overset{\overset{\displaystyle O}{\|}}{C}-CH_3 \rightleftharpoons -CH_2-\overset{+}{\underset{..}{O}}=\overset{\overset{\displaystyle .O}{|}}{C}-CH_3$$

the presence of the $-CH_2-$group means *O*, *P* and *Q* are isolated from one another. The $M^{\ddagger}$ of benzyl ethanoate would *not* be as stable as that of phenyl ethanoate, $C_6H_5-OCOCH_3$ because when the positive charge is on the $\overset{..+}{O} -C_6H_5$ group it can be delocalised round the benzene ring, giving more resonance forms of $M^{\ddagger}$ for this compound. Remember, the more the resonance forms, the greater the stability of the ion.

* *

SAQ 8.2c

Show how the nitro group ($-\overset{\overset{\displaystyle O}{\|}}{\underset{+}{N}}-O^-$) acts to

destabilise the molecular ion of 4-nitrophenol,

Response

The key ion is R. In order to try to stabilise it, electrons are drawn out of the ring to give S:

S shows two positive charges on adjacent atoms, clearly a highly unstable arrangement. If you started from one of the resonance forms having a positive charge on the ring, you should have obtained a form such as T which also has two adjacent positive charges:

T

✳✳✳✳✳✳✳✳✳✳✳✳✳✳✳✳✳✳✳✳✳✳✳✳✳✳✳✳✳✳✳✳✳✳✳✳✳✳

SAQ 8.2f   In compounds of the type RC_6H_4COX, would the following substituents (ie R) stabilise, destabilise, or have little effect on the benzoyl ion produced:

(*i*)   3-methoxy;
(*ii*) 4-methoxy;
(*iii*) 3-cyano;
(*iv*) 4-nitro?

Try to explain your answers by means of resonance forms of the ions concerned.

Response

(*i*) Little effect

The crucial ion would be

$$\overset{+}{C}=O$$ attached to benzene ring with $\overset{\cdot\cdot}{\underset{\cdot\cdot}{O}}CH_3$ at the 3-position   *m/z* 135

Can this be stabilised by the 3-OCH$_3$ substituent? Use of curly arrows to show the $+M$ effect of the oxygen in this group shows that it could *not* stabilise the benzoyl ion:

$$\overset{+}{C}=O \quad\longleftrightarrow\quad \overset{+}{C}=O \qquad m/z\ 135$$

Hence the effect on the RA of *m/z* 135 would not be great because the 1-position where the $-\overset{+}{C}=O$ is located is by-passed in the resonance.

This is a general result for 3-substituents, whether $+M$ or $-M$.

(*ii*) Stabilising effect

When the methoxy-group is at the 4-position the benzoyl ion *is* stabilised:

$$\overset{+}{C}=O \quad\longleftrightarrow\quad \overset{O}{\underset{\parallel}{C}} \qquad m/z\ 135$$

Hence it, or fragments derived from it, would be expected to be enhanced in the spectrum.

(*iii*) Little effect

The $-C\equiv N$ group is $-M$ ((Fig. 8.2a (*ii*)) so would destabilise the benzoyl ion if it were conjugated with the $-\overset{+}{C}=O$ group. But in the 3-position it is not conjugated:

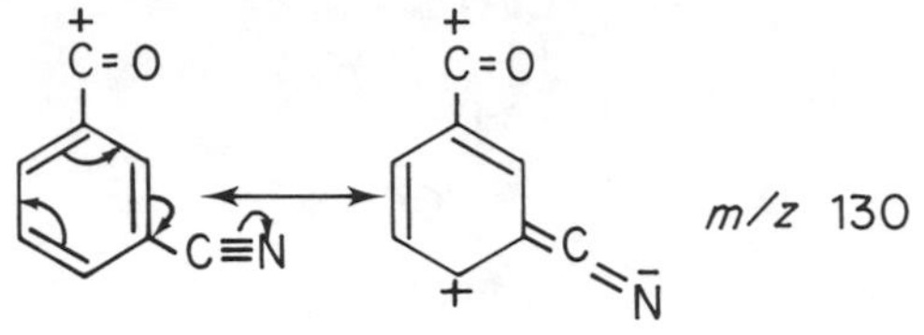

m/z 130

There might be a general destabilising effect from the withdrawal of electron density from the benzene ring, due to the $-I$ effect of the $-CN$ group at the 3-position, leading to a small decrease in *m/z* 130.

(*iv*) Destabilising effect

The $-NO_2$ group is strongly $-M$, so in the 4-position it would destabilise the benzoyl ion, *m/z* 150:

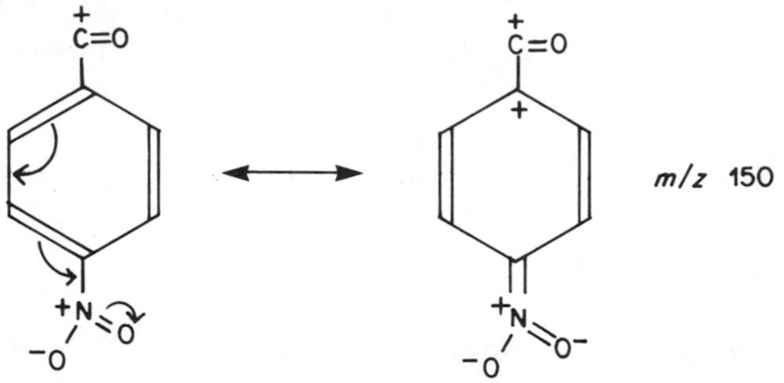

m/z 150

A less prominent *m/z* 150 would be expected. In practice the ratio of $(M-CH_3)/(M-Ar)$ in the spectrum of 4-nitrophenylethanone is 4.3/1, compared to 5.6/1 in that of 1-phenylethanone ie there is relatively less of the 4-nitrobenzoyl ion, *m/z* 150.

****

SAQ 8.2g (Optional for those who feel they need practice in use of ↷ and ↴)

Show how the following ions are stabilised:

(i) $C_6H_5\overset{+}{C}H_2$; (ii) $CH_3CH=CH-\overset{+}{C}O$; (iii) $CH_3\overset{..}{S}-CH=CH-\overset{+}{C}H_2$;

(iv) [cyclopentadienyl cation structure] (v) $H\diagdown H$ ‡ (vi) $\left(Cl-\!\!\left\langle\!\!\!\bigcirc\!\!\!\right\rangle\!-\right)_2\overset{+}{C}H$

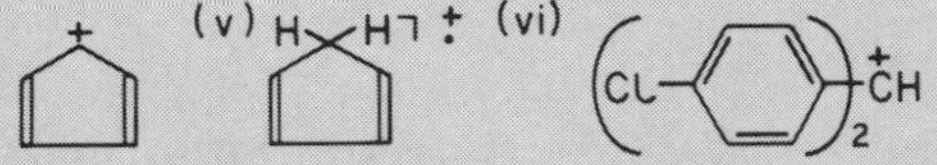

Response

(i) m/z 91:

[resonance structures of benzyl cation: $^+CH_2$ ↔ CH_2 ↔ CH_2 ↔ CH_2 ↔ $^+CH_2$]

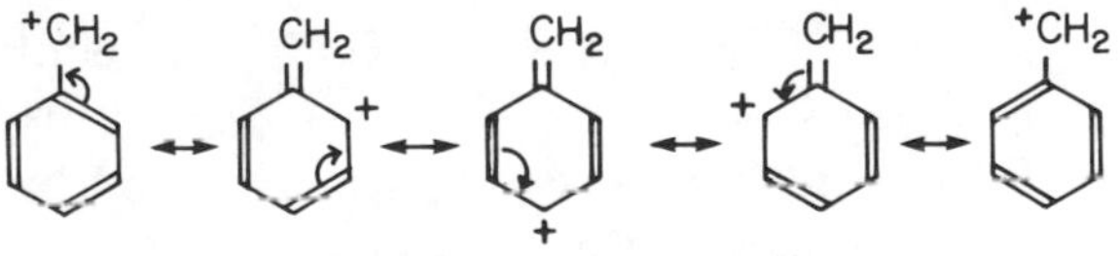

(ii) m/z 69

$CH_3CH=CH-\overset{+}{C}=O \rightleftharpoons CH_3\overset{+}{C}H-CH=C=O$

(iii) m/z 87

$CH_3\overset{..}{S}-CH=CH-\overset{+}{C}H_2 \rightleftharpoons CH_3-\overset{+}{S}=CH-CH=CH_2$

(iv) m/z 65

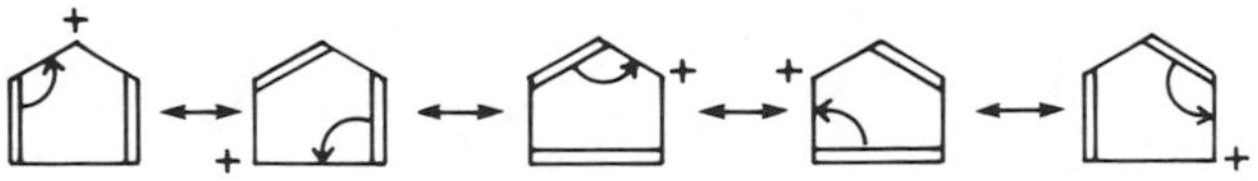

[resonance structures of cyclopentadienyl cation]

Note here all five structures are the same. Hence this must be a particularly stable ion.

(*v*)   *m/z* 66

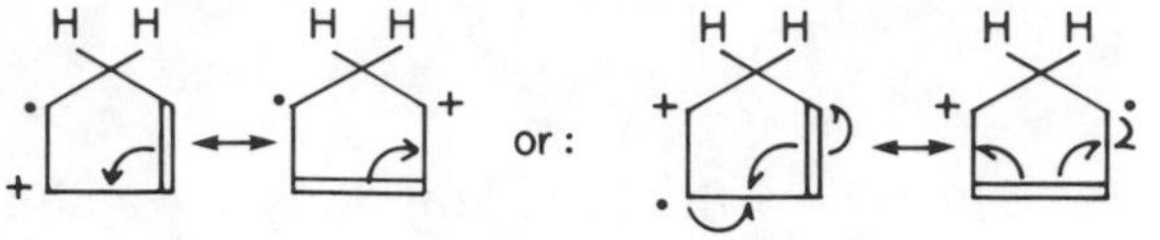

(*vi*)   *m/z* 235/237/239

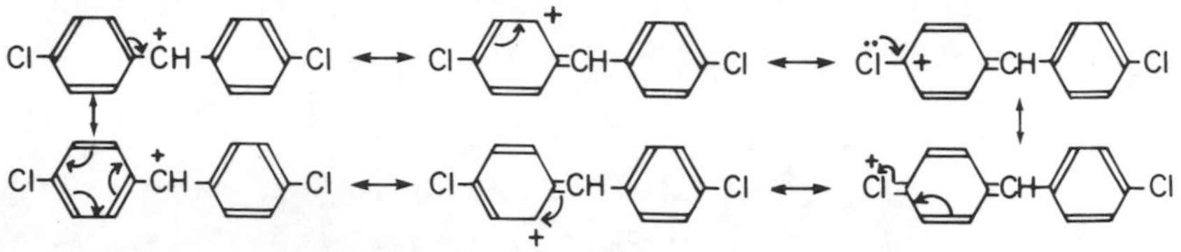

There are of course a further four similar structures possible using the right hand benzene ring making nine in all. This explains why this ion is the base peak of the DDT spectrum (Fig. 7.4c).

****

SAQ 8.3a	How would you expect the molecular ion of 1-bromo-1-phenylethane, $C_6H_5CHBrCH_3$ to fragment? Which of the ions formed by heterolytic and homolytic cleavages, α and β, would be stabilised by $+M$ and/or $+I$ effects?

Response

The four types of cleavage occur with the results opposite:

(*i*)   Heterolytic α-cleavage

$$C_6H_5-\underset{\underset{CH_3}{|}}{\overset{\overset{H}{|}}{C}}-\overset{\cdot+}{B}r \rightarrow C_6H_5-\underset{\overset{+}{|}}{\overset{\overset{H}{|}}{C}}-CH_3 + \overset{\cdot}{B}r$$

$$m/z \; 105$$

This should be an intense ion in the spectrum since the positive charge on the carbon α to the benzene ring is stabilised both by delocalisation, and the $+I$, CH_3-group.

(*ii*) Heterolytic β-cleavage

$$C_6H_5-\underset{\underset{CH_3}{|}}{\overset{\overset{H}{|}}{C}}-\overset{\cdot+}{B}r \rightarrow C_6H_5^+ + CH_3CH=\overset{\cdot}{B}r$$

$$m/z \; 77$$

This is much less likely as neither $C_6H_5^+$ nor $CH_3CH=Br$ are particularly stabilised.

(*iii*) Homolytic α-cleavage

$$C_6H_5-\underset{\underset{CH_3}{|}}{\overset{\overset{H}{|}}{C}}-\overset{+}{B}r \rightarrow C_6H_5\overset{\cdot}{C}HCH_3 + Br^+$$

$$m/z \; 79/81$$

Br^+ is not stabilised so this cleavage is not likely to be favoured.

(*iv*) Homolytic β-cleavage

$$C_6H_5-\underset{\underset{CH_3}{|}}{\overset{\overset{H}{|}}{C}}-\overset{\cdot+}{B}r \rightarrow C_6H_5-\overset{\overset{H}{|}}{C}=\overset{+}{B}r + CH_3{}^{\cdot}$$

$$m/z \; 169/171$$

This cleavage might be favoured to some extent as the $(M - CH_3)^+$ ion is conjugated through to the $+M$ phenyl group:

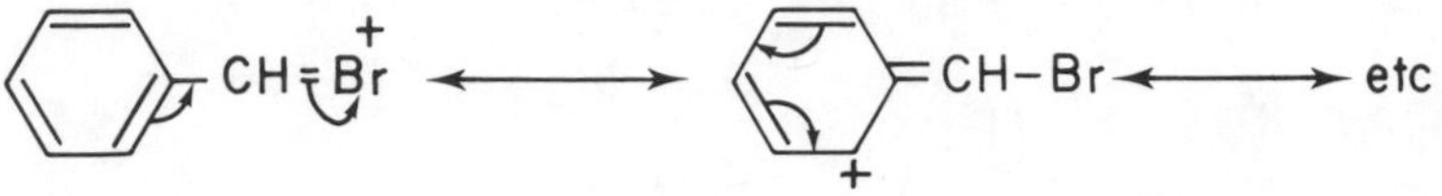

In practice, m/z 105 is the base peak of the spectrum and no other ion is greater than 16% RA, so (i) is the favoured process.

If you correctly obtained the ions described, well done, and if you decided m/z 105 and 169/171 were likely to be the most intense, you have grasped the ideas of resonance stabilisation too!

****

SAQ 8.3b

How would you expect propan-2-ol molecular ion, $CH_3-CHOH-CH_3$, to fragment? Which of the ions formed by heterolytic and homolytic cleavages, α and β, would be stabilised by $+M$ and/or $+I$ effects?

Response

The four types of cleavage occur with the results:

(i) Heterolytic α-cleavage

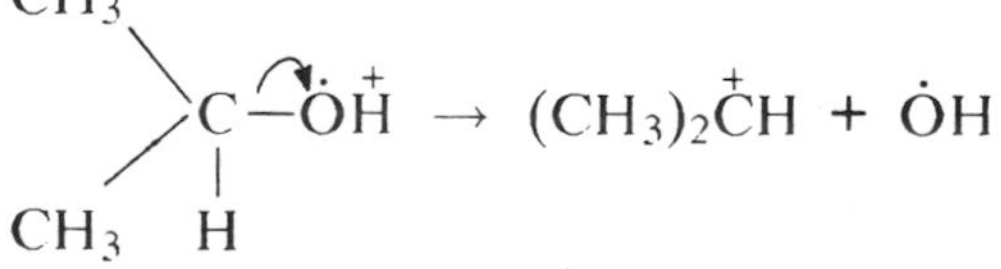

$$(CH_3)_2\overset{+}{C}H + \dot{O}H$$

$$m/z\ 43$$

m/z 43 is a secondary carbocation and would be stabilised to some extent by the $+I$ effects of the CH_3 groups. It would be reasonable to expect this ion in the spectrum.

(*ii*) Heterolytic β-cleavage

Did you spot this could occur in two ways?

$$\begin{array}{c}CH_3 \\ \diagdown \\ \diagup \\ CH_3 \quad H\end{array} C\!\!-\!\!\overset{\cdot\cdot+}{O}\,H \;\rightarrow\; (CH_3)_2C\!\!=\!\!\overset{\cdot}{O}H \;+\; H^+$$
$$m/z\ 1$$

and

$$\begin{array}{c}CH_3 \\ \diagdown \\ \diagup \\ H \qquad CH_3\end{array} C\!\!-\!\!\overset{\cdot\cdot+}{O}\,H \;\rightarrow\; CH_3CH\!\!=\!\!\overset{\cdot}{O}H \;+\; CH_3^+$$
$$m/z\ 15$$

Neither H^+ nor CH_3^+ are stabilised ions so these cleavages would not be expected to any great extent.

(*iii*) Homolytic α-cleavage

$$\begin{array}{c}CH_3 \\ \diagdown \\ \diagup \\ CH_3 \quad H\end{array} C\!\!-\!\!\overset{\cdot\cdot+}{O}\,H \;\rightarrow\; (CH_3)_2\overset{\cdot}{C}H \;+\; \overset{+}{O}H$$
$$m/z\ 17$$

This is not favoured because ^+OH is unstable.

(*iv*) Homolytic β-cleavage

Here again this can occur in two ways – did you spot this?

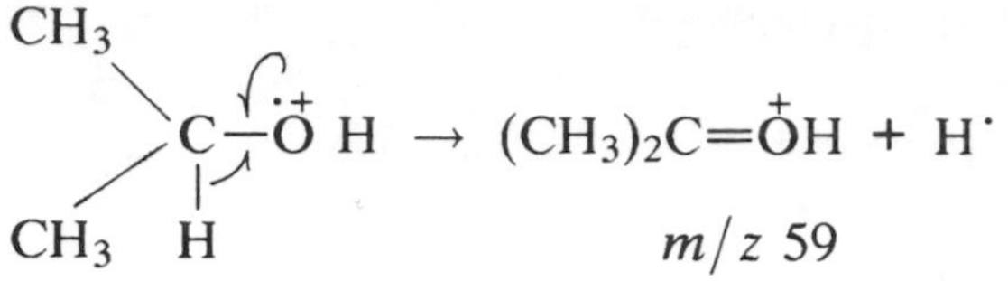

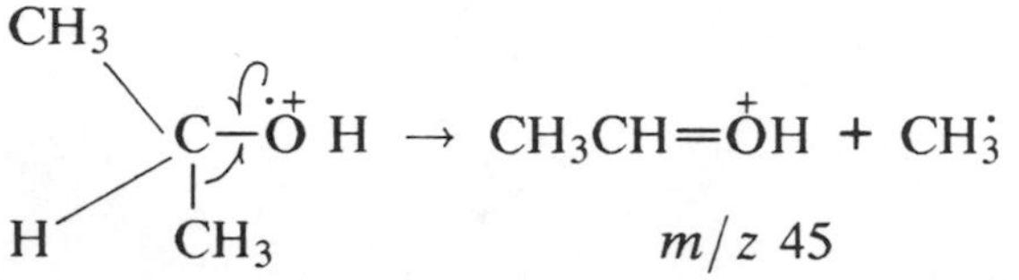

$$(CH_3)_2C=\overset{+}{O}H + H^{\cdot}$$
$$m/z\ 59$$

$$CH_3CH=\overset{+}{O}H + CH_3^{\cdot}$$
$$m/z\ 45$$

The ions m/z 59 and 45 are both resonance stabilised eg

$$(CH_3)_2C=\overset{+}{O}H \leftrightarrow (CH_3)_2\overset{+}{C}OH$$

so these would be expected to be favoured over the only other fairly stable ion, $(CH_3)_2\overset{+}{C}H$. If you correctly obtained the ions described, well done, and if you decided that m/z 43, 45 and 59 were the most likely to be formed, very well done indeed. You have certainly grasped the principles of Part 8!

****

SAQ 9.1a

Analyse the spectrum of *Unknown 4* Fig. 9.1i which contains no nitrogen. Suggest a possible structure for the compound, using the correlations in Fig. 9.1a and 9.1b (there may be more than one possible structure which is in accord with the data). $\longrightarrow$

SAQ 9.1a (cont.)

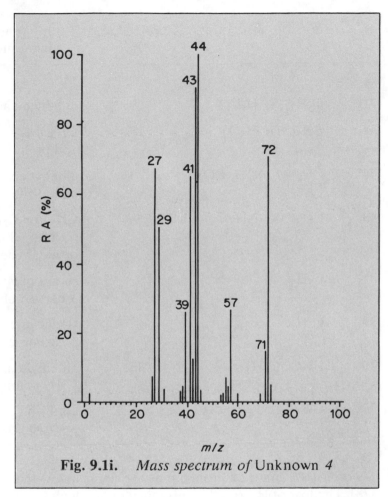

Fig. 9.1i. *Mass spectrum of* Unknown *4*

Response

Unknown 4 is butanal, $CH_3CH_2CH_2CHO$

You should have obtained an analysis table similar to the one opposite:

m/z	Possible Structures	Associated X	Inferences loss (amu)
72			$M^{\ddagger}$
71	C_5H_{11}, C_3H_7CO	1	aldehyde, acetal
57	C_4H_9, (C_3H_5O)	15	CH_3 loss – methyl compound
44	CH_2=CHOH, CO_2	28	aldehyde with γ-H, anhydride
43	CH_3CO, C_3H_7	29	CHO or C_2H_5 loss from ethyl compound or aldehyde
41	C_3H_5		shows presence of 3-carbon chain
39	C_3H_3		shows presence of 3-carbon chain
29	CHO, C_2H_5	43	aldehyde and/or C_2H_5 compound
27	C_2H_3		confirms hydrocarbon chain

Unknown 4 cannot be aromatic as the M_r is below that of benzene (78). The loss of H$^{\cdot}$ from $M^{\ddagger}$ is typical of aldehydes and acetals. The other compounds mentioned in the (M − X) table are irrelevant as they are aromatic or contain nitrogen. The presence of m/z 29 confirms an aldehyde rather than an acetal. The base peak m/z 44 is especially characteristic of aldehydes having γ-H atoms, which rearrange to give the ion CH_2=CHOH‡ (Fig. 9.1b). (M − CH_3)$^+$ and (M − CH_3CH_2)$^+$ are both present indicating an alkyl chain, along with other typical hydrocarbon ions at m/z 27, 39 and 41. Putting this together to reach the required M_r of 72 we get

$$\overset{\gamma}{CH_3}-\overset{\beta}{CH_2}-\overset{\alpha}{CH_2}-CHO$$

As shown, this has three γ-H and therefore would form m/z 44, typical of aliphatic aldehydes.

Other structures you might have thought of are $(CH_3)_2CHCHO$ (but this has no γ-H so could not form m/z 44) or $CH_3COCH_2CH_3$ (would give m/z 57, 43, 29 and 27, but not the all-important base peak m/z 44). If you thought *4* was pentane or one of its isomers because of the odd mass ions which could be hydrocarbon cations such as m/z 27, 29, 39, 41, 43, 57 this was not a bad guess because some of them *are* hydrocarbon ions, but m/z 44 cannot be obtained from pentane. Odd electron ions like m/z 44 are always very significant in mass spectra.

****

SAQ 9.1b   Analyse the spectrum of *Unknown 5*, Fig. 9.1j, which contains a nitrogen atom. Suggest a possible structure for the compound, using the correlations in Fig. 9.1a and 9.1b.

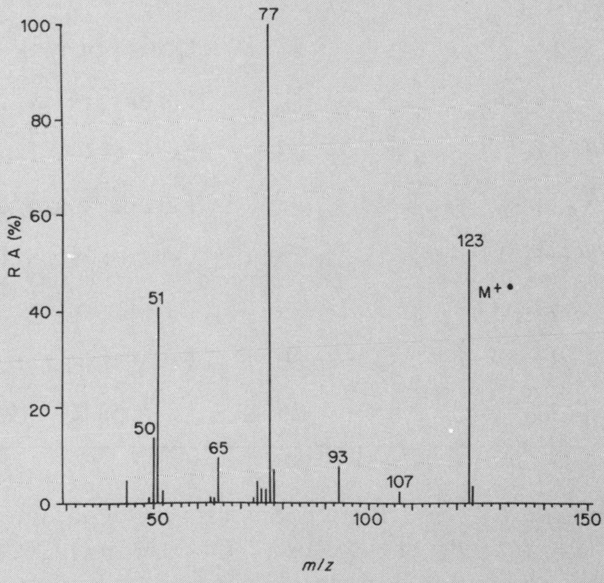

Fig. 9.1j.   *Mass spectrum of* Unknown 5

Response

Unknown 5 is nitrobenzene, $C_6H_5NO_2$

You should have obtained an analysis table similar to the one below:

m/z	Possible Structure	Associated X Loss (amu)	Inferences
123	$C_6H_5NO_2$		$M^{\ddot{+}}$ – odd mass, so contains odd number of N
107	C_6H_5NO	16	Loss of O – nitrocompound
93	C_6H_5O, C_6H_6N	30	Loss of NO from nitrocompound, or aromatic methyl ether
77	C_6H_5	46	Loss of NO_2 from nitrocompound, aliphatic alcohols?, ethyl *o*-ester?
65	C_5H_5	58	Does not make sense
93	C_6H_5O	–	Parent for m/z 66 and 65?
65	C_5H_5	28	Loss of CO from m/z 93
93	C_6H_6N	–	Parent for m/z 66 and 65?
65	C_5H_5	28	Loss of HCNH?
107	C_6H_5NO	–	Parent for m/z 77?
77	C_6H_5	30	Loss of NO from m/z 107
51	C_4H_3	26	Loss of $HC{=}CH$ from m/z 77

The intense $M^{\ddot{+}}$ and higher ions are very characteristic of an aromatic compound. The losses of O (16 amu), NO (30 amu), and NO_2 (46 amu) are characteristic of nitrocompounds. This accounts for the ions at m/z 107, 93 and 77. The presence of m/z 51 confirms that m/z 77 is $C_6H_5^+$, because this ion always gives some m/z

51 by loss of ethyne. You will not see one without the other. If m/z 77 *is* $C_6H_5^+$, then *5* must contain a single NO_2 group to get the M_r of 123. The m/z 93 ion could be C_6H_5O or C_6H_6N, but only the former can lose 28 amu easily (CO) to give $C_5H_5^+$ at m/z 65. If you said *5* was $C_6H_5NH—NO$ this was consistent with the formation of m/z 93, but this ion would be expected to lose HCN to give m/z 64. Congratulations if you said *5* was $C_6H_5O—NO$. This was a very logical structure to deduce, and in fact nitrobenzenes *do* rearrange to nitrites before they lose NO (see Section 9.9)

****

SAQ 9.2a Show how the m/z 57 ion would be formed from the M^+ of cyclopentanol.

Response

In this case the X lost is an ethyl radical.

****

SAQ 9.2b Show how the M^+ of 3-methylcyclohexanol can give rise to both m/z 57 and 71. Why is m/z 71 more intense than m/z 57 in this spectrum?

Response

The molecular ion can undergo α-cleavage in two ways, path (a) or (b), each approximately equally likely. When the ring-opened M^+ ions fragment, they each follow the mechanism of Fig. 9.2d, leading to m/z 57 or 71 respectively.

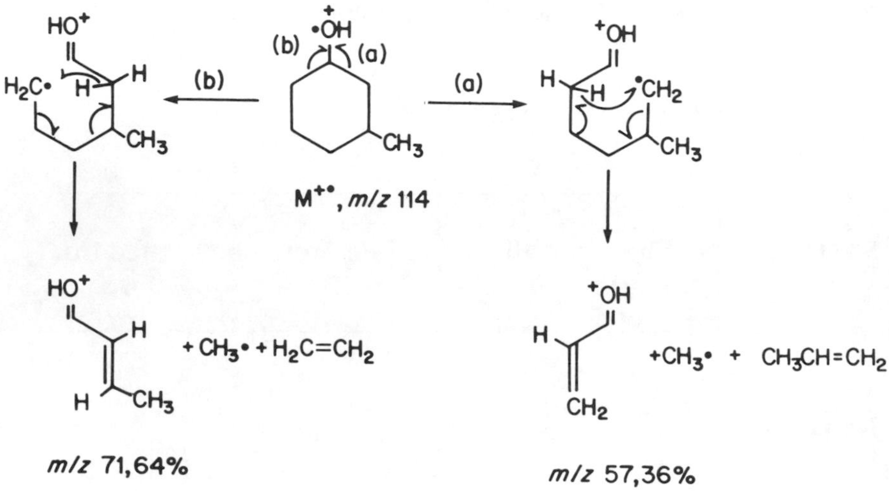

m/z 71 is more intense than m/z 57 because the double bond system is m/z 71 is further stabilised by the $+I$ CH_3 group.

****

SAQ 9.2c

Fig. 9.2i shows the mass spectrum of an alcohol, *Unknown 6*. Interpret this spectrum and identify the alcohol. Note that m/z 31, not shown in Fig. 9.2i, is actually 45%. $\longrightarrow$

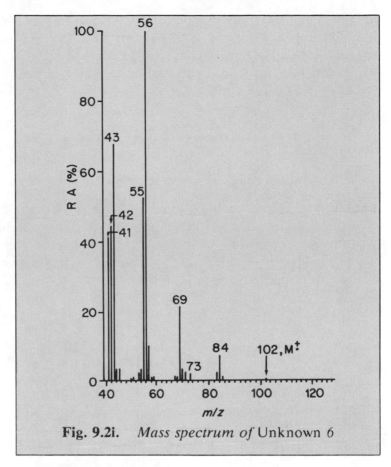

Fig. 9.2i. *Mass spectrum of* Unknown 6

Response

Unknown 6 is hexanol, $CH_3(CH_2)OH$. The M_r of 102 indicates a molecular formula of $C_6H_{14}O$, a saturated alcohol, and its low intensity is typical of primary alcohols. The base peak in this spectrum is m/z 56 ie $(M - 46)^+$. This is typical of the concerted loss of H_2O and $H_2C{=}CH_2$ shown by longer chain alcohols.

m/z 31, $CH_2\overset{+}{O}H$, is not shown in Fig. 9.2i but in fact is about 45% RA. Other peaks are $(M - H_2O)$, m/z 84; $(M - H_2O-CH_3)^+$ m/z 69; $(M - H_2O-CH_2CH_3)^+$ m/z 55; and $CH_3CH_2CH_2^+$ m/z 43. Apart from m/z 43, hydrocarbon ions are of low abundance, see eg m/z 57, 71, 85.

****

SAQ 9.2d   Fig. 9.2j shows the mass spectrum of an alcohol, *Unknown 7*. Interpret this spectrum and identify the alcohol.

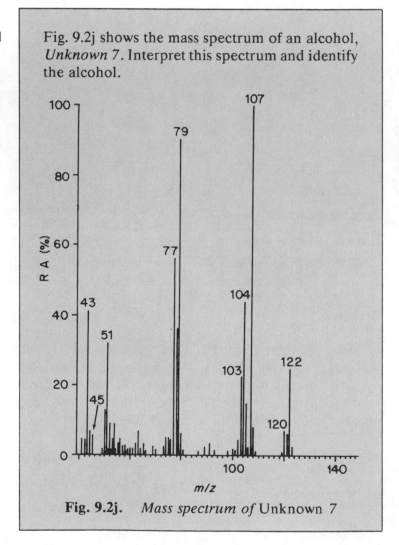

Fig. 9.2j.   *Mass spectrum of* Unknown 7

Response

Unknown 7 is 1-phenylethanol, $C_6H_5CH(OH)CH_3$. The M^+ indicates that this is an isomer of 2-phenylethanol (Fig. 9.2f) but the lack of m/z 91 shows that the $C_6H_5CH_2$ grouping is absent. The presence of m/z 107 combined with 79, 77 and 51 is very typical of the hydroxytropylium ion and its fragments so *Unknown 7* must contain a structural unit capable of forming this ion, but not containing $C_6H_5CH_2O$. The answer has to be a secondary alcohol, which loses CH_3· preferentially against the predictions of Stevenson's Rule, because of the high stability of the hydroxytropylium ion:

m/z 107

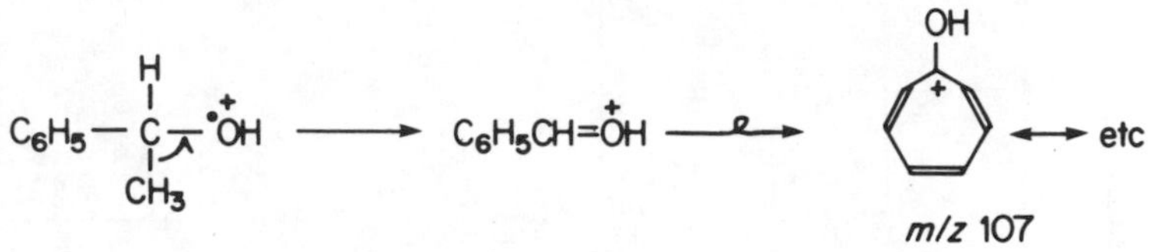

m/z 104 results from $(M - H_2O)^+$ and m/z 103 from $[(M - H)-H_2O]^+$. In this spectrum m/z 45 $CH_3CH=\overset{+}{O}H$ is replaced by m/z 43, CH_3CO^+. It appears that M^+ loses two H to give $C_6H_5COCH_3^+$ (common in alcohol mass spectra since the ketone has a more stable M^+) and CH_3CO^+ is formed by α-cleavage of this ion.

****

Fig. 9.3e shows the mass spectrum of an aliphatic ether, *Unknown 8*. Interpret this spectrum and suggest a structure for the ether.

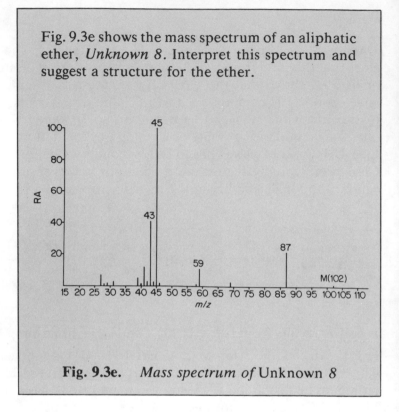

Fig. 9.3e.   *Mass spectrum of* Unknown 8

Response

Unknown 8 is di-isopropyl ether, $(CH_3)_2CHOCH(CH_3)_2$. $M^{\ddot{+}}$ has m/z 102, hence the partial formula must be $C_6H_{14}O$, a saturated ether. Various combinations of alkyl groups are possible from $CH_3OC_5H_{11}$ to $(CH_3CH_2CH_2)_2O$ and *8* itself, but the simplicity of the spectrum points to a symmetrical structure which readily loses $CH_3\cdot$ to give m/z 87, and one having at least one α-CH_3 group:

$$(M - CH_3)^+, m/z \ 87$$

m/z 87 then loses $CH_3CH{=}CH_2$ to give the base peak, $CH_3CH{=}\overset{+}{O}$.

$$CH_3CH{=}O{-}\underset{\underset{\overset{|}{H}{-}CH_2}{}}{\overset{+}{C}HCH_3} \longrightarrow CH_3CH{=}\overset{+}{O}H + CH_3CH{=}CH_2$$

$$m/z\ 45$$

The intensity of m/z 43 is quite high for the alkyl R^+ group derived from an ether R_2O so this would lead you to suspect it was a secondary isopropyl ion. m/z 59, $(CH_3)_2CHO^+$ confirms the presence of C_3H_7O, complementary to m/z 43. If you said Unknown *8* was $(CH_3CH_2CH_2)_2O$ this was a fair answer based on the mass spectrum alone.

If you suggested Unknown *8* could be $CH_3CH_2O\overset{\overset{\displaystyle CH_3}{|}}{C}HCH_2CH_3$, note that this compound would tend to lose $CH_3CH_2^\cdot$ to give m/z 73 from $M^{\ddagger}$, rather than $CH_3^\cdot$ in the first α-cleavage, though it is true that m/z 73 would then give $CH_3CH{=}\overset{+}{O}H$ as base peak by eliminating $H_2C{=}CH_2$ from the other ethyl group.

****

SAQ 9.3b

> *Unknown 9* is an aromatic ether of M_r 136. The mass spectrum shows the following peaks and intensities:
>
> | m/z | 136 | 95 | 94 | 77 | 66 | 51 | 43 | 41 | 39 |
> |-------|-----|----|-----|----|----|----|----|----|----|
> | RA(%) | 25 | 6 | 100 | 8 | 7 | 6 | 5 | 5 | 6 |
>
> Suggest a structure for *unknown 9* compatible with these data.

Response

Unknown 9 is phenyl propyl ether, $C_6H_5OCH_2CH_2CH_3$. If you suggested phenyl isopropyl ether this was a good answer, but less likely since m/z 43, $(CH_3)_2CH^+$ from this would be expected to be greater than RA 5%, and some loss of $CH_3^\cdot$ (α-cleavage) would be expected from $M^{\ddagger}$, too. The spectrum below m/z 94 is essentially that of phenol, hence a partial structure is $C_6H_5OC_3H_7$ to give $M^{\ddagger} = 136$.

C_3H_7 can only be propyl or isopropyl. (Phenol)$^{+\bullet}$ ions are formed by a four-centred mechanism.

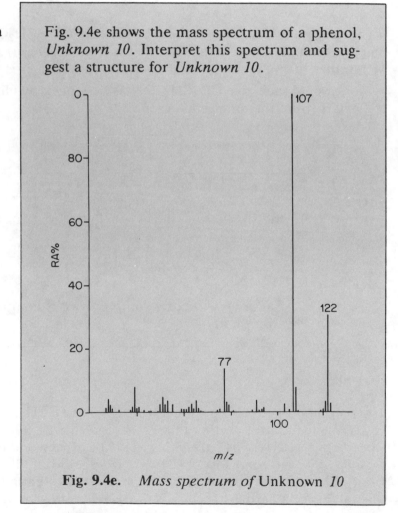

m/z 136   *m/z* 94

****

SAQ 9.4a   Fig. 9.4e shows the mass spectrum of a phenol, *Unknown 10*. Interpret this spectrum and suggest a structure for *Unknown 10*.

Fig. 9.4e.   *Mass spectrum of* Unknown *10*

Response

Unknown 10 is *p*-ethylphenol (4-ethylhydroxybenzene),

$$CH_3 \overbrace{}^{} CH_2 - C_6H_4 - OH$$

107

The spectrum shows a very intense $(M - CH_3)^+$ peak, typical of either benzylic cleavage as shown, or loss of a CH_3 from a dimethylphenol. However, a dimethyl isomer would be expected to show a significant $(M - H)^+$ at m/z 121. In this spectrum m/z 121 is very weak, leading to the conclusion that it is an ethyl substituent. If you said *Unknown 10* was *m*-ethylphenol (3-ethylhydroxybenzene), this was an acceptable answer, because mass spectra do not permit the distinction of 3- and 4- isomers. However, you should have realised that it could not be either a 2-ethyl or a 2-methyl isomer because these would have shown loss of H_2O in an ortho-effect. There is no m/z 104 $(M - H_2O)^{+\bullet}$ in this spectrum!

****

SAQ 9.5a

In which of the following compounds would you expect to see a 'McL' peak? Give its m/z and structure.

(*i*) $CH_3COCH_2CH_3$;

(*ii*) $CH_3COCH_2CH_2CH_3$;

(*iii*) $CH_3COCH(CH_3)_2$;

(*iv*) $(CH_3)_2CHCH_2CHO$;

(*v*) $CH_3CH_2COOCH_2CH_3$;

(*vi*) $CH_3CH_2CH_2CONHCH_3$;

(*vii*) $CH_3CH(CH_3)CH_2 CH(CH_3)COOH$;

(*viii*) $CH_3CH_2CH(CH_3)COC_6H_5$; $\longrightarrow$

SAQ 9.5a
(cont.)

> *(ix)* $CH_3CH=CHCH_2COCH_2CH_3$;
>
> *(x)* $$\overset{O}{\underset{\|}{(C_6H_5)_2P.SCH_2CH_3}}$$

Response

(i) $\overset{\alpha}{C}H_3\overset{}{C}O\overset{\alpha}{C}H_2\overset{\beta}{C}H_3$ has no γ-H. It has no 'McL' peak.

(ii) $\overset{\alpha}{C}H_3CO\overset{\beta}{C}H_2\overset{\gamma}{C}H_2CH_3$ has a γ-H.

Predict $CH_3\underset{\overset{|}{{}^+OH}}{C}=CH_2$, m/z 58

(iii) $CH_3CO\underset{\alpha}{\overset{\overset{\displaystyle CH_3}{|}}{C}}\underset{\beta}{H}-CH_3$ has no γ-H. It has no 'McL' peak.

(iv) $\overset{\gamma}{C}H_3-\overset{\beta}{C}H-\overset{\alpha}{C}H_2CHO$ has a γ-H.
$$\underset{CH_3}{|}$$

Predict m/z 44, $CH_2=C\overset{\overset{\displaystyle \dot{O}\overset{+}{H}}{\diagup}}{\underset{\diagdown H}{}}$

(v) $CH_3CH_2\overset{\overset{\displaystyle O}{\|}}{C}O\underset{\alpha\beta}{}CH_2\overset{\gamma}{C}H_3$ has a γ-H in the $-OCH_2CH_3$ group.

Predict m/z 74, $CH_3CH_2C\overset{\overset{\displaystyle {}^+OH}{\diagup}}{\underset{\diagdown O}{}}$ Note this is isomeric

with the 'McL' peak of a methyl ester.

(*vi*) γ   β   α
$CH_3CH_2CH_2CONHCH_3$ has a γ-H in the
$CH_3CH_2CH_2$-group.

Predict m/z 73, $CH_2{=}C\overset{\overset{\displaystyle {\overset{+}{\cdot}}OH}{\diagup}}{\underset{\diagdown}{NHCH_3}}$

(*vii*) CH_3
               |
            γ   β   α
$CH_3CHCH_2CH(CH_3)COOH$ has one γ-H.

Predict m/z 74, $CH_3CH{=}C\overset{\overset{\displaystyle {\overset{+}{\cdot}}OH}{\diagup}}{\underset{\diagdown}{OH}}$

(*viii*) γ   β   α
$CH_3CH_2CH(CH_3)COC_6H_5$ has a γ-H.

Predict m/z 150, $CH_3CH{=}C\overset{\overset{\displaystyle {\overset{\cdot}{O}\overset{+}{H}}}{\diagup}}{\underset{\diagdown}{C_6H_5}}$

(*ix*)   γ   β   α
$CH_3CH{=}CHCH_2COCH_2CH_3$ has one γ-H, but it is on a
double bond. It has no 'McL' peak. If you got

m/z 72, $CH_2{=}C\overset{\overset{\displaystyle {+}OH}{\diagup}}{\underset{\diagdown}{CH_2CH_3}}$ look at Section 9.5.1 again

(*x*)   O
               ‖   α   β   γ
$(C_6H_5)_2 P{-}S{-}CH_2{-}CH_3$ has γ-H.

Predict $(C_6H_5)_2P\overset{\overset{\displaystyle {\overset{+}{\cdot}}OH}{\diagup}}{\underset{\diagndown}{\diagdown\!\diagdown S}}$ m/z 234.

Pushing the boat out a bit here perhaps, but P=O and S=O compounds *do* behave like C=O in the 'McL' manner.

****

SAQ 9.5b   Give mechanisms for the formation of the two structures for m/z 88 in Fig. 9.5f and show that they both form the same m/z 60 in the second stage of the Double 'McL' Rearrangement.

Response

The mechanisms are:

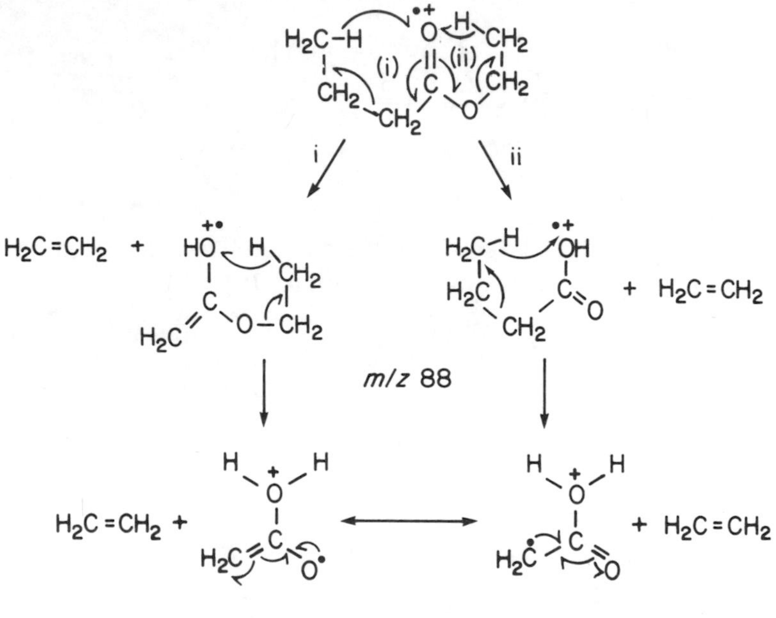

m/z 88

A m/z 60   B

The m/z 88 must obviously be different. The m/z 60 which result from the second 'McL' process appear different at first sight, but in fact they are simply related by resonance, $A \rightleftharpoons B$, so they are two forms of the same ion. You might like to note that the group $-\overset{+}{\cdot}OH$ can act as a terminus for a migrating γ-H. You saw something similar in the mass spectrum of 2-methylhydroxybenzene (Fig. 9.4b) and of 2-methylbenzylalcohol (Section 9.2).

****

SAQ 9.5c   Fig. 9.5i shows the mass spectrum of an alde-hyde, *Unknown 11*. Interpret the spectrum and identify *Unknown 11*.

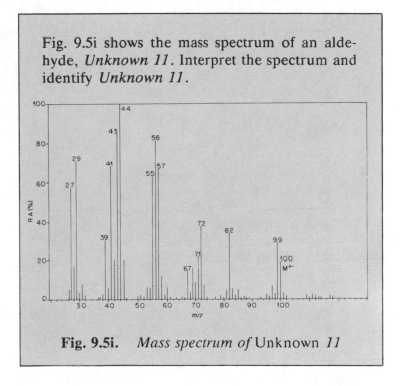

Fig. 9.5i. *Mass spectrum of* Unknown *11*

Response

Fig. 9.5i is the spectrum of *hexanal* $CH_3(CH_2)_4CHO$, *A*.

It shows most of the characteristic features of a saturated aliphatic aldehydes. $M^{\overset{+}{\cdot}}$ is present at m/z 100, and (M-1), m/z 29 and the base peak at m/z 44 indicate an aldehyde with at least three carbon atoms in the chain with a γ-H, and no substituents at C_2. The four types of β-cleavage are well-illustrated in the spectrum, they give rise to the m/z 43, 44 ('McL'), 56 and 57 peaks.

(*i*) Homolytic β-cleavage

$$CH_3(CH_2)_3 - CH_2 - C = O \longrightarrow CH_3(CH_2)_3{}^{\bullet} + CH_2 = CHO^+$$

$$\underset{\textit{m/z 43}}{}$$

(*ii*) 'McL'

$$RCH = CH_2 + CH_2 = C\underset{H}{\overset{\overset{\bullet+}{O}H}{}}$$

$$m/z\ 44$$

$$R = CH_3CH_2$$

(*iii*) Alkene cleavage

$$\underset{R = CH_3CH_2}{}$$

$${}^+CH_2 + CH_2 = CHOH$$

$$m/z\ 56$$

(*iv*) Alkyl cleavage

$$CH_3(CH_2)_3 - CH_2 - C \overset{+\bullet}{=} O \longrightarrow CH_3(CH_2)_3^+ + CH_2 = CHO^{\bullet}$$

$$m/z\ 57$$

There are in addition two features of this spectrum which sometimes occur in aliphatic aldehydes as the chains lengthen – loss of H_2O (m/z 82) and $H_2C = CH_2$ (m/z 72).

If you gave as your answer B, C or D, isomers of hexanal which all contain the $-CH_2CHO$ group and have a γ-H, well done!

$$(CH_3)_2CHCH_2CH_2CHO \quad B$$

CH₃CH₂C(CH₃)CH₂CHO   *C*

$$CH_3CH_2C(CH_3)CH_2CHO \quad C$$

$$(CH_3)_3CCH_2CHO \qquad D$$

Unknown 11 could have been any of these on the basis of the information given so far. Further confirmation from nmr or ir would be necessary to conclusively say *Unknown 11* was hexanal and not *B*, *C* or *D*. Later on when you have studied Section 9.7 on hydrocarbons you might be able to deduce more about which C_4H_9 structural isomer is present in such a compound.

****

SAQ 9.5d   Fig. 9.5j shows the spectrum of an aldehyde *Unknown 12*. Interpret the spectrum and identify *Unknown 12*.

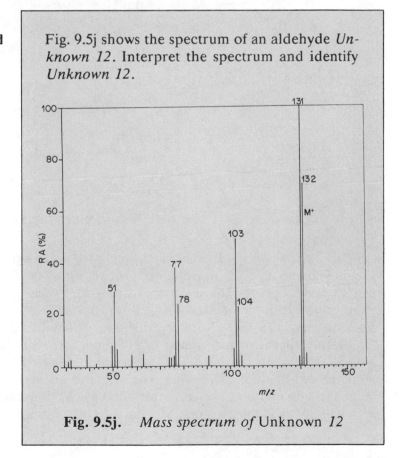

Fig. 9.5j.   *Mass spectrum of* Unknown *12*

Response

Unknown 12 is 3-phenylprop-2-en-1-al (cinnamaldehyde) C_6H_5CH $=CHCHO$. The small number of relatively intense high mass ions show at once that this is an aromatic aldehyde, confirmed by the intense $(M-1)^+$ which is the base peak. This means that this ion is especially stabilised and must be conjugated with the benzene ring which is clearly indicated by the presence of m/z 77. Hence units of C_6H_5 and CHO must exist, leaving 26 amu to be accounted for. This cannot be CN, as the relative molecular mass would then be odd (nitrogen rule), so it must be C_2H_2. This makes sense because it would contain a double bond and conjugate the $-CHO$ to the benzene ring. The main fragmentations are shown below.

Did you notice there were two parallel processes involving loss of CO, then $HC \equiv CH$? One starts from $C_6H_5CH=CH\overset{+}{C}O$ and ends with $C_6H_5^+$ and is fairly obvious, the other involves loss of CO from $M^{\ddagger}$ and leads to m/z 78, an isomer of (benzene)‡. You should have been alerted to this by the intensity of m/z 104 relative to 103. It is too intense to be a ^{13}C isotope peak of the latter. A six-centred transition state can be written for this, and it would be termed a McLafferty process, as mentioned in Section 9.5.1, even though the $C=O$ is not directly involved. It is remarkable how often even simple molecules like this can surprise us with the variety of their EI-induced fragmentations, but you must always bear in mind the large excess energy available to some $M^{\ddagger}$ ions. Expect the unexpected!

****

SAQ 9.5e	What peak(s) would you look for in the mass spectrum of 4-methyl-3-pentanone, $CH_3CH(CH_3)COCH_2CH_3$, to distinguish it from its isomer 4-methyl-2-pentanone (Fig. 9.5m (*ii*))?

Response

The three principle ions of 4-methyl-2-pentanone, Fig. 9.5m (ii) are m/z 85 and 43 caused by the two possible α-cleavages, and m/z 58, the 'McL' ion of a methyl ketone. The two analogous ions expected from 4-methyl-3-pentanone by α-cleavage would be m/z 71 and 57 instead.

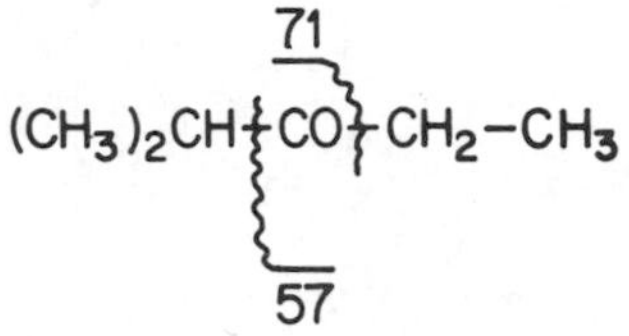

$$(CH_3)_2CH \overset{71}{\underset{57}{|}} CO \overset{}{|} CH_2-CH_3$$

m/z 43 would also be expected for this isomer, formed directly from M^{+} or by loss of :C=O from the isopropyl acylium ion, m/z 71. In this case it is $(CH_3)_2CH^{+}$ rather than CH_3CO^{+} as in Fig. 9.5m (ii), but you could not expect to distinguish these in a low resolution spectrum. Hence it is *not* a point of difference. But 4-Methyl-3-pentanone has *no* γ-H so it would *not* show an even mass 'McL' peak – a prime point of difference.

* *

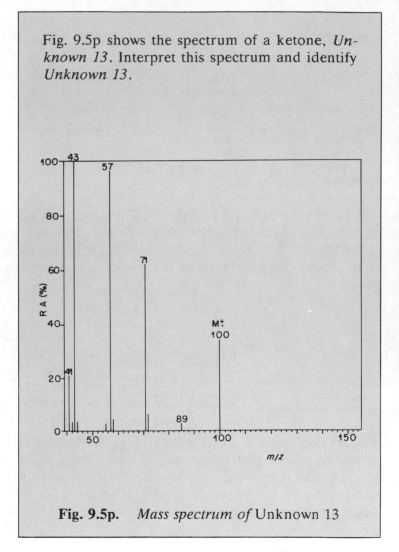

SAQ 9.5f   Fig. 9.5p shows the spectrum of a ketone, *Unknown 13*. Interpret this spectrum and identify *Unknown 13*.

Fig. 9.5p.   *Mass spectrum of* Unknown 13

Response

Unknown 13 is 3-hexanone, $CH_3CH_2COCH_2CH_2CH_3$, isomeric with 2-hexanone, Fig. 9.5l. The M^+ at m/z 100 shows it to be a saturated ketone, even if you did not spot the similar M_r to 2-hexanone. There are *no* even mass ions of any consequence in the spectrum, therefore either there are *no* γ-H, or they are in a CH_3 group which does not release them in the 'McL' process. You have seen this sometimes happens in propyl groups so would suspect this could be the case here. So the α-cleavage ions are $CH_3CH_2CO^+$, m/z 57, and $CH_3CH_2CH_2CO^+$, m/z 71 and the base peak is $CH_3CH_2\overset{+}{C}H_2$ derived from m/z 71 by loss of $:C{=}O$. $(M - CH_3)$, m/z 85 is present but of lower RA, typical of methyl compounds without branches.

****

SAQ 9.5g   Fig. 9.5q shows the spectrum of a ketone, *Unknown 14*. Interpret this spectrum and identify *Unknown 14*.

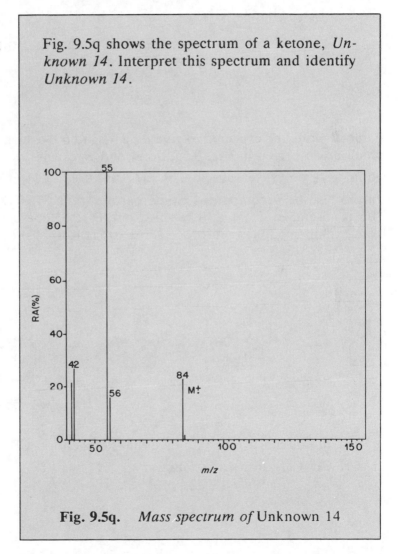

Fig. 9.5q.   *Mass spectrum of* Unknown 14

Response

Unknown 14 is cyclopentanone. $M^{\ddagger}$ is m/z 84 which is two less than the M_r expected for a saturated acyclic ketone. Hence there must be a ring or a double bond present. Possible double bond isomers are:

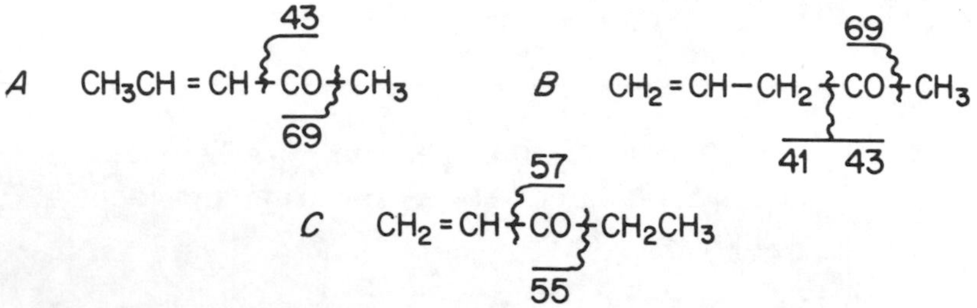

$$A \quad CH_3CH = CH \overset{43}{\underset{69}{\nmid}} CO \nmid CH_3 \qquad\qquad B \quad CH_2 = CH - CH_2 \overset{69}{\underset{41\ \ 43}{\nmid}} CO \nmid CH_3$$

$$C \quad CH_2 = CH \overset{57}{\underset{55}{\nmid}} CO \nmid CH_2CH_3$$

A and *B* would be expected to give m/z 43 and 69, α-cleavage ions, which are not seen in Fig. 9.5g. So *A* and *B* can be ruled out. *C* would give an α-cleavage ion of m/z 55, $CH_2=CH-\overset{+}{C}O$, but it should also be accompanied by its partner m/z 57, $CH_3CH_2\overset{+}{C}O$. This is absent, so *C* can also be ruled out. m/z 55 could be formed from cyclopentanone by a process analogous to that shown in Fig. 9.5n.

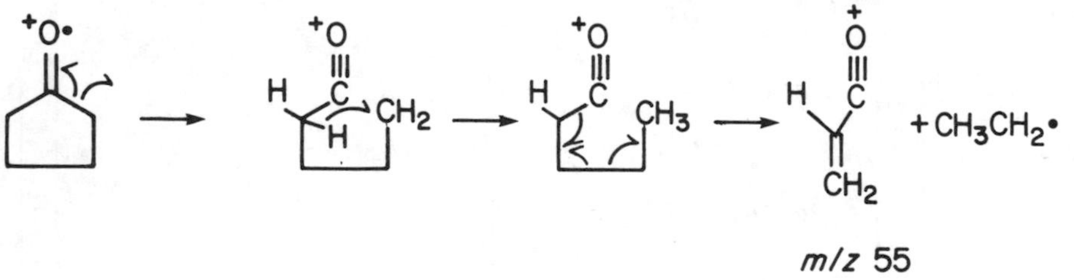

m/z 55

m/z 42 in this spectrum could be either ketene ions, $CH_2=C=O^{\ddagger}$ or $\cdot CH_2CH_2\overset{+}{C}H_2$, formed as follows.

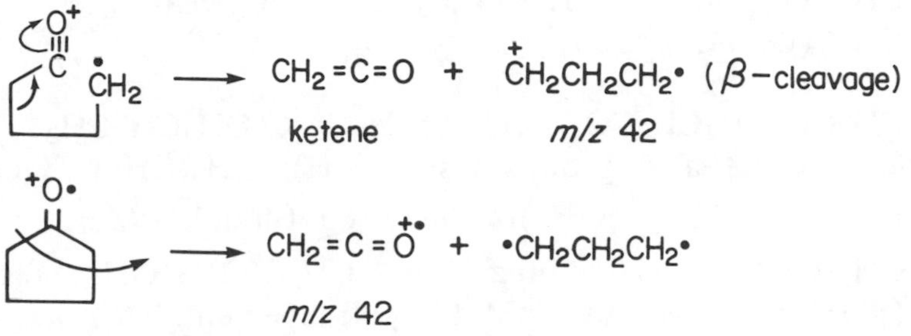

$$CH_2=C=O \quad + \quad \overset{+}{C}H_2CH_2CH_2^\bullet \quad (\beta-\text{cleavage})$$

ketene m/z 42

$$CH_2=C=\overset{+\bullet}{O} \quad + \quad {}^\bullet CH_2CH_2CH_2^\bullet$$

m/z 42

****

SAQ 9.5h

Which acylium ion base peak, 'McL' and 'DHT' ions would you predict for each of the following esters?

(*i*) $CH_3CH_2CO.OCH_2CH_2CH_3$;

(*ii*) $C_6H_5CO.OCH_2CH_2CH_3$;

(*iii*) $CH_3CH_2CO.OCH=CHCH_3$;

(*iv*) $CH_3CO.OCH_2CH_2CH=CH_2$;

(*v*) $CH_3CH_2CH_2CH_2CO.OCH_2CH_3$.

Response

(*i*) CH_3CH_2CO+, m/z 57; $CH_3CH_2COOH^+$, m/z 74 (weak); $CH_3CH_2C(OH)_2^+$, m/z 75.

(*ii*) $C_6H_5CO^+$, m/z 105; $C_6H_5COOH^+$, m/z 122; $C_6H_5C(OH)_2^+$, m/z 123.

(*iii*) $CH_3CH_2CO^+$, m/z 57; no 'McL' or 'DHT' ions as the acid group has no γ-H, and the alkyl group has a double bond.

(*iv*) CH_3CO^+, m/z 43; $CH_3COOH^{\ddagger}$, m/z 60 (weak); $CH_3C(OH)_2^+$, m/z 61.

(*v*) $CH_3CH_2CH_2CH_2CO^+$, m/z 85; $H_2C{=}C(OH)CH_2CH_3^{\ddagger}$, m/z 88 (as the acid group has a γ-H); $CH_3(CH_2)_3COOH^{\ddagger}$, m/z 102; $^{\cdot}CH_2C(\overset{+}{O}H_2)O$, m/z 60 (double 'McL' peak); $CH_3(CH_2)_3C(OH)_2^+$, m/z 103; $^{\cdot}CH_2C(\overset{+}{O}H_2)OH$, m/z 61 ('DHT') followed by a 'McL' rearrangement. This one is as complicated as an ester can get – did you spot most of these possibilities?

****

SAQ 9.5i   Fig. 9.5v shows the mass spectrum of an ester, *Unknown 15*. Interpret this spectrum and suggest a structure for *Unknown 15*.

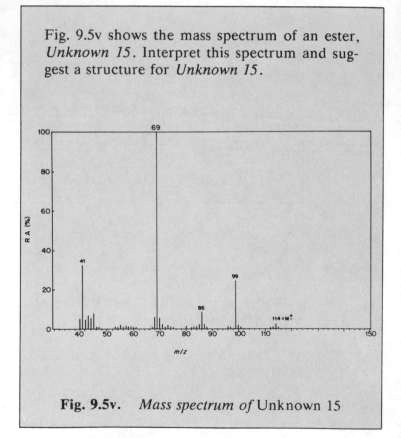

Fig. 9.5v.   *Mass spectrum of* Unknown 15

Response

Unknown 15 is ethyl 2-butenoate, $CH_3CH{=}CH{-}CO.OCH_2CH_3$. You might also have suggested its double bond isomer $H_2C{=}CHCH_2CO.OCH_2CH_3$, but structures reversing the acyl and alkyl groups such as

$$CH_3CH_2CO.OCH_2CH_2CH{=}CH_2$$

or

$$CH_3CH_2CO.OCH{=}CHCH_3$$

would either have other acyl ion base peaks or 'McL'/'DHT' peaks (see SAQ 9.5h).

By comparison with Fig. 9.5s (*ii*) you can see that *Unknown 15* has a similar structure but with a double bond since the M_r is two less. The base peak m/z 69 must be a RCO^+ ion and this too is 2 amu less than the m/z 71 base peak in Fig. 9.5s (*ii*), so the double bond must be in this R group. Possible structures are $CH_3CH{=}CHCO^+$ or $H_2C{=}CHCH_2CO^+$, but the relatively high intensity of the acylium ion would lead you to suspect that it might be conjugated and select the former structure. Loss of CO gives the m/z 41 peak:

$$CH_3CH{=}CH{-}\overset{+}{C}O \rightarrow CH_3CH{=}CH^+ + CO$$

m/z 86 and 87 are the 'McL' and 'DHT' ions derived from the loss of the ethyl group with the 'McL' ion the more intense, as is usual for an ethyl ester.

****

SAQ 9.5j Fig. 9.5w shows the mass spectrum of an ester, *Unknown 16*. Interpret this spectrum and suggest a structure for *Unknown 16*.

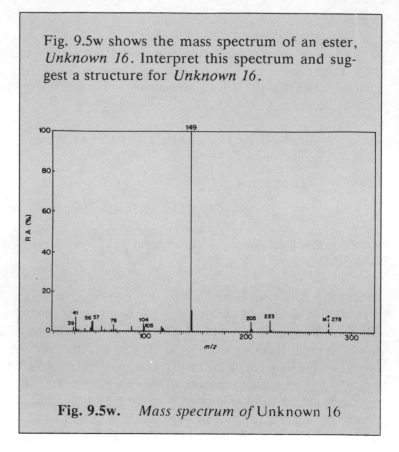

Fig. 9.5w.   *Mass spectrum of* Unknown 16

Response

Unknown 16 is butyl phthalate. The interpretation of the spectrum is shown below. If you said the ester could have been any of the three possible butyl isomers where the R group is primary, secondary, or tertiary, or even structures involving any two of these isomers these were equally satisfactory answers. The mass spectrum alone cannot easily differentiate between them. However, in practice, commercial plasticisers would be symmetrical as these are easiest to manufacture, but both dibutyl and di-isobutyl phthalates are in common use. The spectrum shows the expected m/z 149 dominating the fragments, with minor ions due to its precursor m/z 205 and its fragments m/z 121, 105, 104 and 76. Evidence for the butyl groups appears at m/z 223 ('DHT' ion from one of them – note the 'McL' peak expected at m/z 222 is completely suppressed here) and peaks at m/z 57 and 56, $C_4H_9^+$ and $C_4H_8^+$.

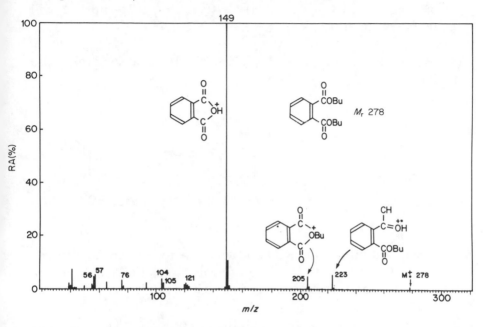

Mass Spectrum of Dibutyl Phthalate

****

SAQ 9.5k

Which ions would you expect to feature prominently in the mass spectra of:

(*i*) $(CH_3)_2CHCO_2H$;

(*ii*) 2-methylbenzoic acid;

(*iii*) 3-methoxybenzoic acid;

(*iv*) $CH_3(CH_2)_4CO_2H$.

Response

(*i*) The expected ions will be: m/z 43, $(CH_3)_2CH^+$ base peak; and m/z 45, CO_2H^+. $(CH_3)_2CHCO^+$, m/z 71 and $M^{\ddot{+}}$ will be fairly small peaks.

(*ii*) $M^{\ddot{+}}$, m/z 136 and $(M - H_2O)^+$, m/z 118 because of the ortho effect:

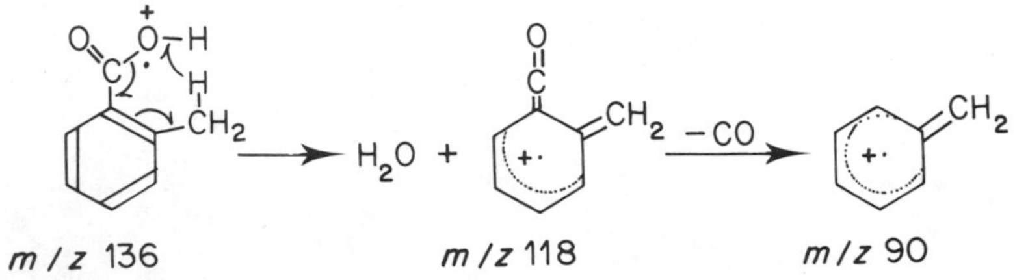

m/z 118 would further lose CO and give rise to m/z 90. The tropylium ion m/z 91 would also be quite intense as the acid is also a tolyl derivative and would tend to lose $\cdot CO_2H$ quite readily.

(*iii*) As this is a 3-substituted acid, there would be no ortho effect, so the fragmentation would be normal, ie

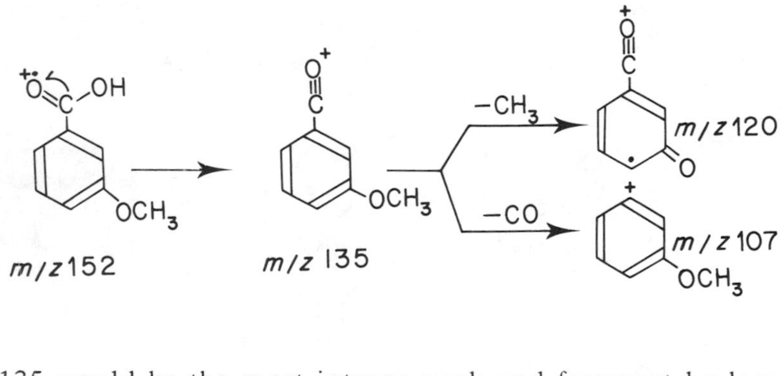

m/z 135 would be the most intense peak and fragment by loss of CH_3 and CO as shown, to give fairly intense m/z 107 and 120 peaks.

(*iv*) The 'McL' peak $CH_2{=}C(OH)_2^{+\cdot}$, m/z 60, would be most intense, m/z 45, CO_2H^+ should be reasonably clear, and β-cleavage as in esters would give $+CH_2CH_2CO_2H$ m/z 73.

There would also be C_nH_{2n+1} ions at m/z 29, 43, 57 and 71 from the pentyl group though these would be fairly weak.

****

SAQ 9.51

> The mass spectrum of an amide C_4H_9NO (isomeric with butanamide, Fig. 9.5y (*ii*)) shows a base peak of m/z 30, with intense ions at m/z 43 and 72. What is the structure of this isomer?

Response

Answer: $CH_3CONHCH_2CH_3$

m/z 30 is typical of an ethanamide CH_3CONHR or propanamide $CH_3CH_2CONHR_1$ structure. The possibilities are therefore $CH_3CONHCH_2CH_3$ or $CH_3CH_2CONHCH_3$ – both give m/z 30.

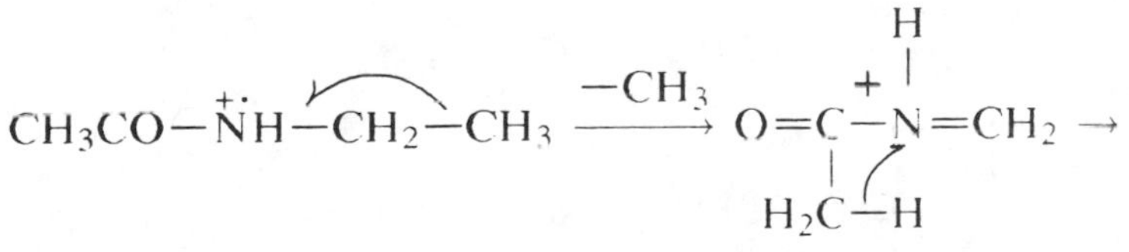

$$CH_3CO-\overset{+\cdot}{N}H-CH_2-CH_3 \xrightarrow{-CH_3} O=\overset{+}{C}-\overset{\overset{\displaystyle H}{|}}{N}=CH_2 \rightarrow$$
$$\underset{H_2C-H}{}$$

$$O=C=CH_2 + H_2N^+=CH_2$$
$$m/z\ 30$$

$$CH_3CH_2CO\overset{+}{N}H-CH_2-H \xrightarrow{-H\cdot} O=C-\overset{\overset{\displaystyle H}{|}}{N}{}^+=CH_2 \rightarrow$$
$$\underset{\underset{CH_3}{|}}{H-C-H}$$

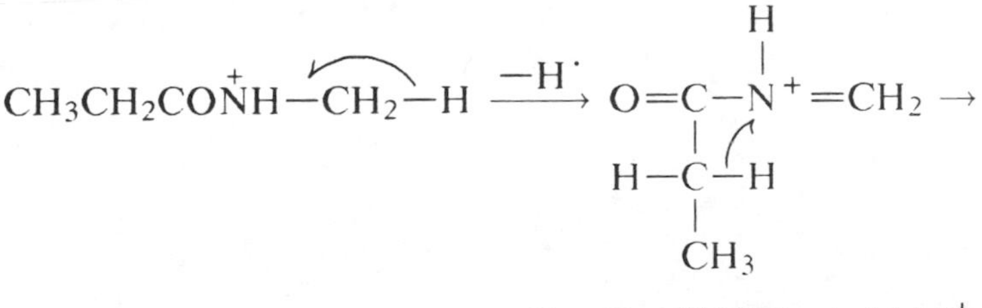

$$O=C=CHCH_3 + H_2N^+=CH_2$$
$$m/z\ 30$$

The difference lies in the α-cleavages

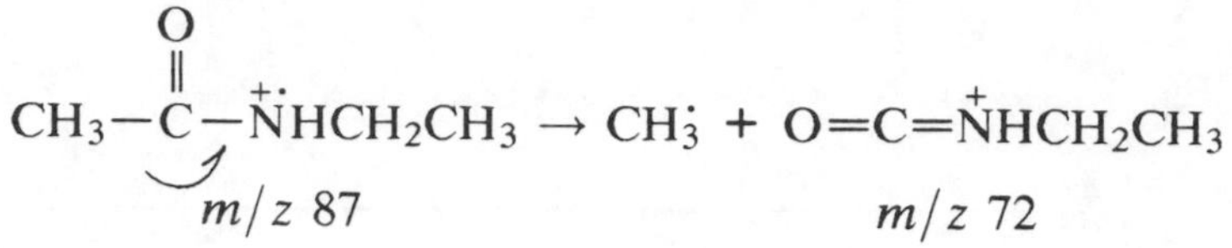

$$CH_3-\overset{\overset{\displaystyle O}{\|}}{C}-\overset{+\cdot}{N}HCH_2CH_3 \rightarrow CH_3^+ + O=C=\overset{+}{N}HCH_2CH_3$$

m/z 87 $\qquad\qquad\qquad\qquad$ m/z 72

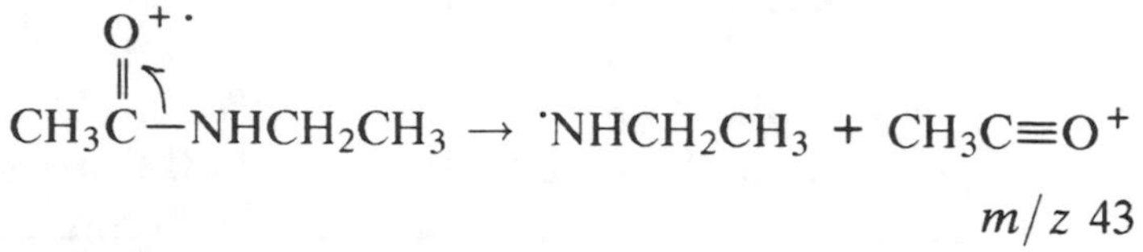

$$CH_3\overset{\overset{\displaystyle O^{+\cdot}}{\|}}{C}-NHCH_2CH_3 \rightarrow {}^{\cdot}NHCH_2CH_3 + CH_3C{\equiv}O^+$$

$\qquad\qquad\qquad\qquad\qquad\qquad\qquad\qquad$ m/z 43

$CH_3CH_2CONHCH_3$ would give m/z 58 and 57 by these processes.

****

SAQ 9.5m

Which typical ions would you expect to find in the mass spectra of:

(*i*) $CH_3(CH_2)_4\,CONHCH_3$;

(*ii*) $CH_3CONH(CH_2)_4CH_3$;

(*iii*) $CONH_2$;

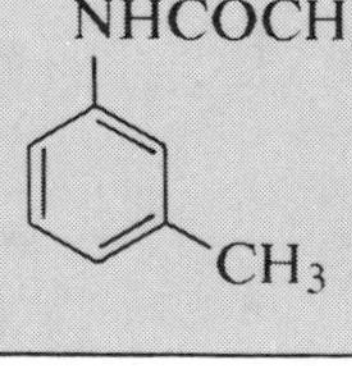

   Cl

(*iv*) $NHCOCH_3$

   CH_3

Response

(*i*) $\overset{+}{\text{OH}}$
         |
    $CH_2\!=\!C\!-\!NHCH_3$, m/z 73 ('McL' peak);

    $O\!=\!C\!=\!\overset{+}{N}HCH_3$, m/z 58;

    $^+CH_2CH_2CONHCH_3$, m/z 86; $H_2\overset{+}{N}\!=\!CH_2$, m/z 30.

(*ii*) CH_3CO^+, m/z 43; $CH_3CO\overset{+}{N}H\!=\!CH_2$, m/z 72; $H_2\overset{+}{N}\!=\!CH_2$, m/z 30 (probably base peak). There is no 'McL' from this compound.

(*iii*) M^{+}, m/z 155/157, 3 : 1; 4-ClC$_6$H$_4$CO$^+$, m/z 139/141, 3 : 1, base peak; 4-ClC$_6$H$_4^+$, m/z 111/113, 3 : 1; C$_6$H$_4^+$, m/z 76.

(*iv*) M^{+}, m/z 149; (M − CH$_2$CO), m/z 107.

****

SAQ 9.5n   Would you expect to observe any ortho effects in either ArNHCOR or ArCONHR compounds?

Response

Answer Yes, in both cases.

The general pattern of ortho effects involves:

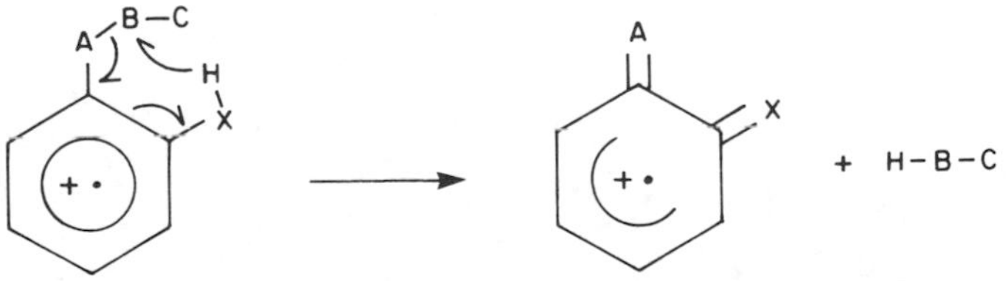

There is one obvious way in which an ortho-substituted ArNHCOR compound could rearrange by a six-centred process;

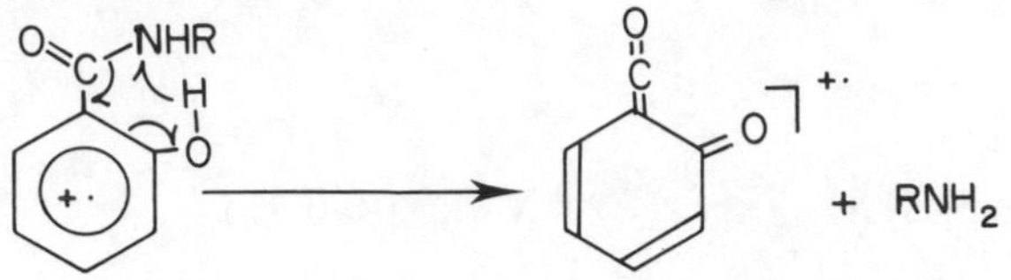

m/z 120

and another for the ArCONHR isomer:

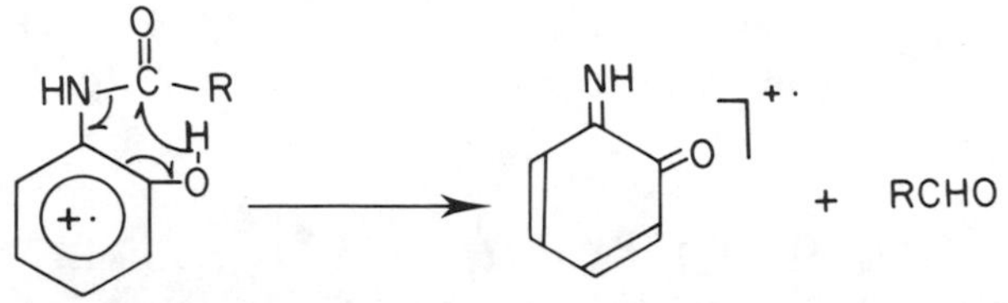

m/z 107

so it would be something to look out for in such compounds.

****

| SAQ 9.5o | Fig. 9.5z shows the mass spectrum of *Unknown 17*, another amide isomeric with butanamide. Suggest its structure. $\longrightarrow$ |

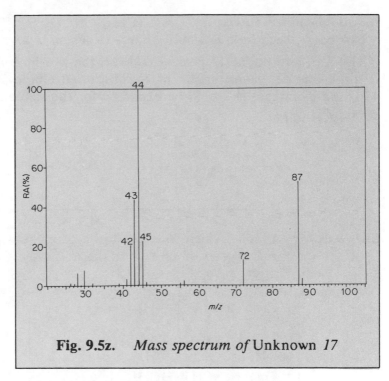

Fig. 9.5z. *Mass spectrum of* Unknown *17*

Response

Unknown 17 is N,N—dimethylethanamide, $CH_3CON(CH_3)_2$.

The base peak is $(CH_3)_2N^+$, m/z 44, and the next most intense ion, m/z 43, is CH_3CO+, both formed by simple α-cleavages. Although m/z 44 is also characteristic of $OC\!=\!\overset{+}{N}H_2$ formed from a primary amide, the absence of odd mass 'McL' peaks rules out a long acid chain in *Unknown 17*. m/z 72 shows that $CH_3^{\cdot}$ loss occurs fairly readily from $M^{+\cdot}$, so confirming the CH_3CO group:

$$CH_3-\overset{\overset{\textstyle O}{\|}}{\underset{\smile}{C}}-\overset{+\cdot}{N}(CH_3)_2 \rightarrow CH_3^{\cdot} + O\!=\!C\!=\!\overset{+}{N}(CH_3)_2$$

$$m/z\ 72$$

If you concluded *Unknown 17* was $(CH_3)_2CHCONH_2$ this was not a bad guess, because this would give a strong m/z 43 due to the $(CH_3)_2\overset{+}{C}H$, and no 'McL' peak. However, the $M^{\ddagger}$ in *Unknown 17* is quite intense, characteristic of N-substituted amides, and m/z 43 is not as intense as would be expected for the stable secondary $(CH_3)_2\overset{+}{C}H$ ion.

****

SAQ 9.6a   Predict which ions would be prominent in the mass spectrum of each of the following isomeric amines:

(*i*) $CH_3(CH_2)_5NH_2$;

(*ii*) $CH_3NH(CH_2)_4CH_3$;

(*iii*) $CH_3CH_2NHCH(CH_3)CH_2CH_3$;

(*iv*) $(CH_3)_2CHN(CH_3)CH_2CH_3$;

(*v*) $(CH_3CH_2)_3N$.

Could they be distinguished from their mass spectra alone?

Response

(*i*) The expected fragmentation routes are:

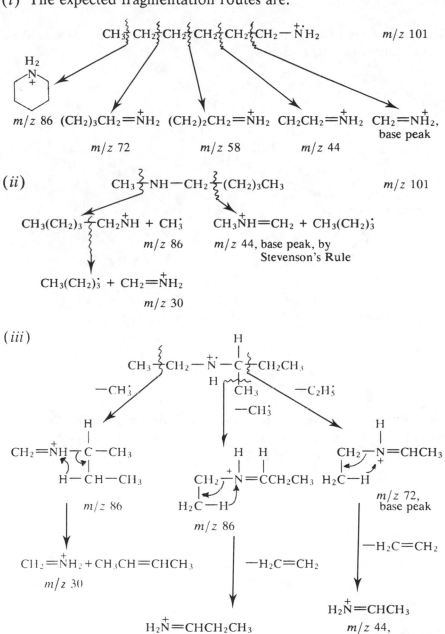

In this case there are *two* possible m/z 86 ions depending on which $CH_3^{\cdot}$ is lost from $M^{+\cdot}$. Both can eliminate alkenes giving m/z 30 or m/z 58.

Since $CH_3CH_2^{\cdot}$ loss is most favoured from $M^{+\cdot}$, these ions would be less intense than m/z 72 and 44.

(*iv*)

$$CH_3 \overset{\text{\scriptsize{}}}{\underset{}{{-}}} \overset{CH_3}{\underset{\overset{|}{H}}{C}} - \overset{+\cdot}{\underset{\underset{CH_3}{|}}{N}} - CH_2 \overset{\text{\scriptsize{}}}{\underset{}{{-}}} CH_3$$

$-CH_3^{\cdot}$ $-CH_3^{\cdot}$

A
$$CH_3CH{=}\overset{+}{N}{-}CH_2$$
$$\quad\quad\quad CH_2$$
$$\quad H$$

m/z 86

B $CH_3{-}C$ CH_2
$$\quad H_2C{-}H$$

m/z 86

$\downarrow$ $-CH_2{=}CH_2$

$$CH_3CH{=}\overset{+}{N}CH_3$$
$$\quad\quad H$$

m/z 58. base peak

$-CH_3CH{=}CH_2$

$$CH_3\overset{+}{N}H{=}CH_2$$

m/z 44

Here again there are two possible ions with m/z 86, *A* and *B*. *A* would be favoured because it has the most substituted double bond, leading to m/z 58 as base peak.

(*v*)

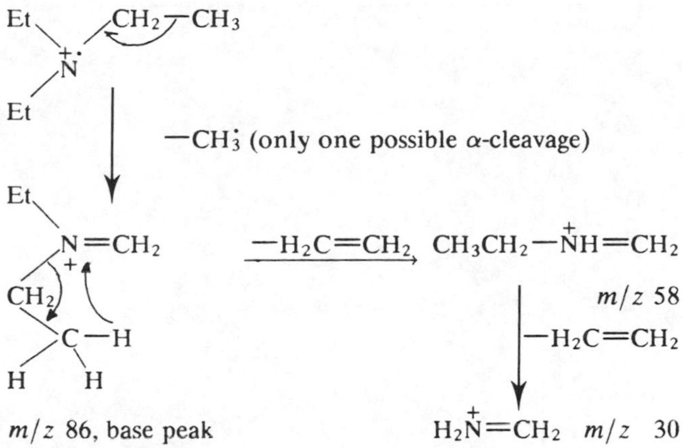

$$Et \diagdown \overset{+\cdot}{N} \diagup CH_2{-}CH_3$$
$$Et \diagup$$

$$\downarrow \quad {-}CH_3^{\cdot} \text{ (only one possible } \alpha\text{-cleavage)}$$

$$Et \diagdown$$
$$\overset{+}{\underset{CH_2}{N}}{=}CH_2 \qquad \xrightarrow{-H_2C=CH_2} \quad CH_3CH_2{-}\overset{+}{N}H{=}CH_2$$
$$\overset{\textstyle C-H}{\underset{\textstyle H \quad H}{\diagup}}$$

$$m/z \ 86, \text{ base peak} \qquad\qquad m/z \ 58$$

$${-}H_2C{=}CH_2$$

$$\downarrow$$

$$H_2\overset{+}{N}{=}CH_2 \quad m/z \ 30$$

Note that these five isomers may be distinguished either by the fact that they give different base peaks, or combine the same base peak with different prominent ions.

****

SAQ 9.6b

Fig. 9.6b shows the mass spectrum of a tertiary amine, *Unknown 18*. Suggest a structure for this amine. $\longrightarrow$

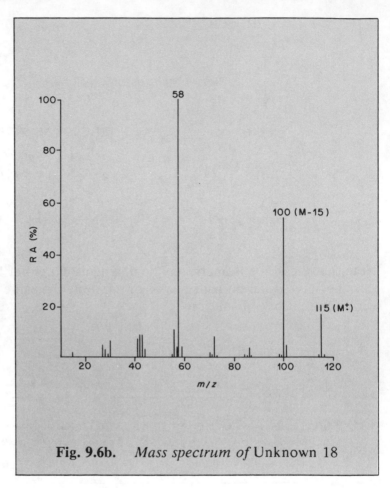

Fig. 9.6b.   *Mass spectrum of* Unknown 18

Response

Unknown 18 is N-methyldiisopropylamine, $[(CH_3)_2CH]_2NCH_3$.

Compared with the examples in SAQ 9.6a, the relative molecular mass is 15 amu higher, so the compound must be a saturated heptylamine. The large $(M-CH_3)$ peak indicates the ready loss of a β-methyl group which is typical of a $N-CH(CH_3)R$ structure. The

m/z 58 ion must be formed from m/z 100 by alkene elimination, with H transfer, so the alkene lost must be propene ($C_3H_6 = 42$ amu). The partial structure indicated for m/z 100 is therefore:

$$
\begin{array}{c}
CH_3 \\
R_1 \quad | \\
C{=}N{-}CH{-}CH_3 \\
R_2 \quad \overset{+}{} \quad | \\
CH_2 \\
H
\end{array}
\quad \rightarrow \quad
\begin{array}{c}
R_1 \\
C{=}\overset{+}{N}HCH_3 \\
R_2
\end{array}
$$

$$m/z\ 100 \qquad\qquad\qquad m/z\ 58$$

It follows that $R_1 = H, R_2 = CH_3$ in the m/z 58 ion. If either or both of the isopropyl groups were propyl, then the original α-cleavage would take place by $CH_3CH_2^+$ loss to give m/z 86, by Stevenson's Rule:

$$
(CH_3)_2CH{-}\overset{\overset{\displaystyle CH_3}{|}}{\underset{+\cdot}{N}}{-}CH_2{-}CH_2{-}CH_3 \longrightarrow (CH_3)_2CH{-}\overset{\overset{\displaystyle CH_3}{|}}{\underset{+}{N}}{=}CH_2 + CH_3CH_2^+
$$

$$m/z\ 86$$

$$-CH_3CH{=}CH_2$$

$$CH_3\overset{+}{N}H{=}CH_2$$

$$m/z\ 44,\ \text{base peak}$$

The base peak would then be m/z 44, *not* 56.

****

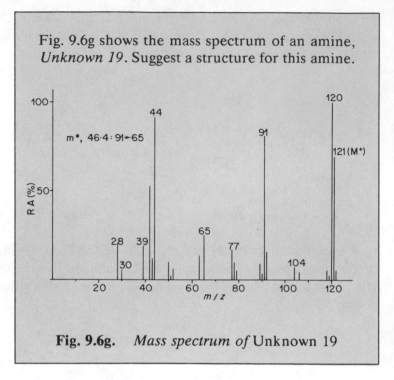

Fig. 9.6g shows the mass spectrum of an amine, *Unknown 19*. Suggest a structure for this amine.

Fig. 9.6g.   *Mass spectrum of* Unknown 19

Response

Unknown 19 is N-methylbenzylamine, $C_6H_5CH_2NHCH_3$.

The spectrum shows intense ions at (M-1), m/z 120; 91 and 44. There are also significant ions at m/z 77 and 65, and a m^* for 91 → 65. This strongly suggests a benzyl group, rather than an aniline derivative such as $C_6H_5NHCH_2CH_3$ or $C_6H_5N(CH_3)_2$. The m/z 44 is $CH_2=\overset{+}{N}HCH_3$ arising from the loss of $C_6H_5^{\cdot}$. This is as expected from Stevenson's Rule. Loss of a benzylic hydrogen from M$^{+\cdot}$ would give $C_6H_5CH=\overset{+}{N}HCH_3$ which is of course stabilised by resonance with the benzene ring and may indeed be a methylamino-substituted tropylium ion:

m/z 120

****

SAQ 9.6d Fig. 9.6h shows the mass spectrum of an aromatic amine of formula $C_6H_6N_2O$, *Unknown 20*. This compound shows a strong absorption in its IR spectrum at 1700 cm^{-1}. Suggest a structure for this compound.

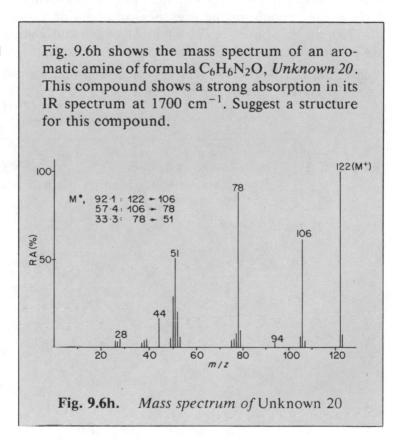

Fig. 9.6h. *Mass spectrum of* Unknown 20

Response

Unknown 20 is nicotinamide,

If you realised it was a pyridyl amide, but not where the amide group was attached, this was fair enough. As we have said before, it is difficult to tell 3- and 4-substituted aromatics apart by mass spectra alone. M^+ loses 16 amu to give m/z 106 ($m^* = 92.1$), a loss shown by few compounds except primary aromatic amides which fragment off $\cdot NH_2$ leaving a stable aroyl ion $ArCO^+$. This further loses 28 amu ($m^* = 57.4$) which would be CO rather than $H_2C=CH_2$ as the ir confirms the presence of an amide $C=O$ frequency at 1700 cm^{-1}. The m/z 78 is typical of pyridyl compounds and fragments as expected to m/z 52 and 51. Finally the weaker but significant m/z 44 is $O=C=\overset{+}{N}H_2$ typical of primary amides, as well as $-CH_2NHCH_3$ amines (*Unknown 19*).

****

SAQ 9.7a   Fig. 9.7b shows the mass spectrum of a hydrocarbon *Unknown 21*. Identify the hydrocarbon.

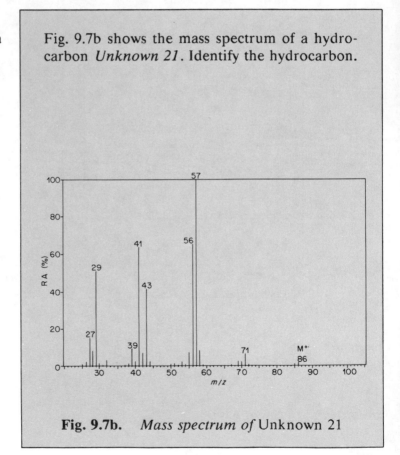

Fig. 9.7b.   *Mass spectrum of* Unknown 21

Response

Unknown 21 is $CH_3CH_2CH(CH_3)CH_2CH_3$, 3-methylpentane.

$M^{\ddagger}$ at m/z 86 shows it to be a hexane isomer and the m/z 57 base peak is very intense compared to either 71 or 43, showing that the chain is branched so as to give preferential cleavage at C_3—C_4. The m/z 57 ion must be either

$$CH_3\overset{+}{C}HCH_2CH_3 \text{ or } (CH_3)_3C^+, \text{ indicating}$$

$$\begin{array}{c} CH_3 \\ | \\ CH_3CH_2CHCH_2CH_3 \text{ or } (CH_3)_3CCH_2CH_3, \end{array}$$

both of which would lose $CH_3CH_2^{\cdot}$ preferentially (Stevenson's Rule).

If you got either of these, well done. We can tell the difference between them by considering the $(M - CH_3)^+$ peak, m/z 71. This is quite small in *Unknown 21* (compare *Unknown 22*) showing that the loss of $CH_3^{\cdot}$ is not favoured. This would be the case for 3-methylpentane as the resulting m/z 71 is $(CH_3CH_2)_2CH^+$, a *secondary* ion. In the alternative structure, 2,2-dimethylbutane, the loss of $CH_3^{\cdot}$ would give $(CH_3)_2\overset{+}{C}CH_2CH_3$, a *tertiary* ion. This would give a more intense m/z 71.

****

SAQ 9.7b   Fig. 9.7c shows the mass spectrum of a hydrocarbon *Unknown 22* isomeric with *Unknown 21*. Identify this isomer.

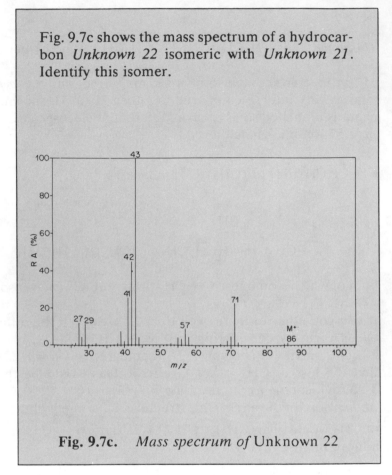

Fig. 9.7c.   *Mass spectrum of* Unknown 22

Response

Unknown 22 is $(CH_3)_2CHCH_2CH_2CH_3$.

The intense base peak at m/z 43 coupled with the very low C_4 ion at m/z 57 strongly suggests a $(CH_3)_2CH$-end group. Another secondary carbocation is formed at m/z 71 by loss of a CH_3, giving $CH_3CH(CH_2)_2CH_3$, so m/z 71 is more intense than expected for a saturated hydrocarbon. Another possibility you might have suggested is $(CH_3)_2CHCH(CH_3)_2$. The mass spectrum cannot entirely rule this one out, so it is an acceptable answer, but note that m/z 57 *is* present, though weak, and it is difficult to see how this structure could give such an ion.

****

SAQ 9.7c

Fig. 9.7f shows the mass spectrum of a hydrocarbon, *Unknown 23*. Suggest a structure for this compound and explain the main features of the spectrum.

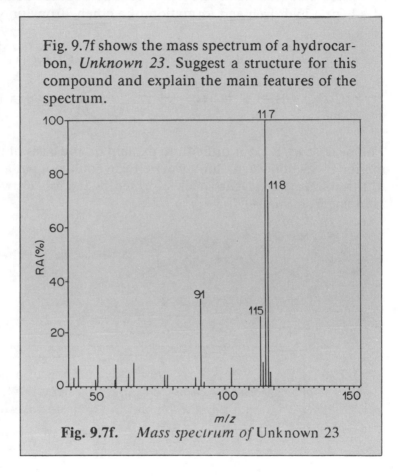

Fig. 9.7f. *Mass spectrum of* Unknown 23

Response

Unknown 23 is 3-phenylpropene (allylbenzene), $C_6H_5CH_2CH{=}CH_2$.

The presence of m/z 91 shows it to be a benzyl compound, leaving 27 amu to be assigned. This must be C_2H_3 and the compound has a double bond in the side chain. Therefore, it must be $C_6H_5CH_2CH=CH_2$. The isomer $C_6H_5CH=CHCH_3$ clearly could not fragment to give m/z 91 without H transfer, which would not be expected from an unsaturated sidechain. The (M-1) base peak would be due to loss of a benzyl hydrogen and formation of a ethenyltropylium ion:

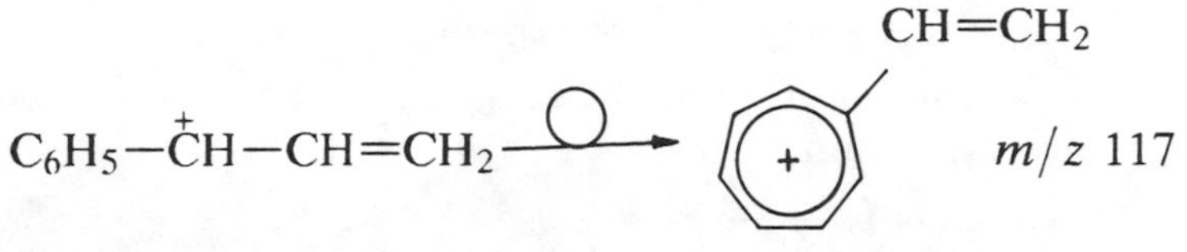

The peak at m/z 115 is difficult to explain on the basis of the information in Section 9.7, so don't worry if you could not put a structure to this ion. It is a bicyclic species formed by the loss of two further H atoms from m/z 117:

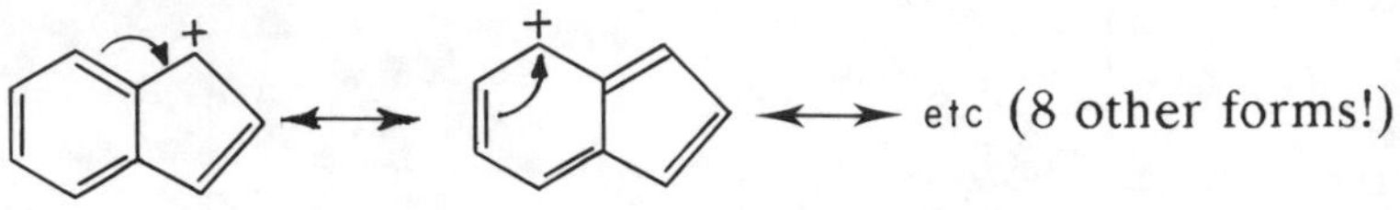

etc (8 other forms!)

****

SAQ 9.7d	Fig. 9.7g. shows the mass spectrum of a hydro-carbon, *Unknown 24*. Suggest a structure for this compound and explain the main features of the spectrum. $\longrightarrow$

**SAQ 9.7d
(cont.)**

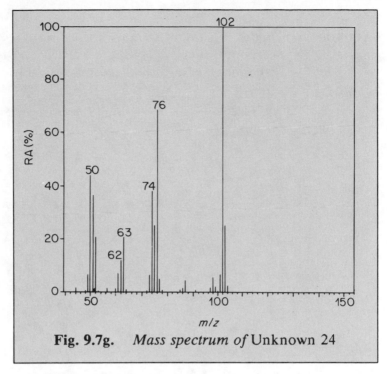

Fig. 9.7g. *Mass spectrum of* Unknown 24

Response

Unknown 24 is phenylethyne (phenylacetylene), $C_6H_5C{\equiv}CH$.

Since the benzene ring is clear from ions at m/z 76, 51, 50, although m/z 77 is weak, subtracting C_6H_5 from the relative molecular mass of 102 leaves 25 amu, which can only be $-CH{\equiv}CH$. (M-1) is weak here, but the $M^{\ddagger}$ is highly stabilised by conjugation with the $-C{\equiv}CH$ group, and the loss of the alkyne H is not very favoured. The main fragmentation is by loss of C_2H_2 to m/z 76, then by C_2H_2 again, to m/z 50. Likewise, m/z 63 is formed by loss of C_3H_3 from $M^{\ddagger}$. The other peaks are formed by various H losses from these ions.

****

SAQ 9.7e   Fig. 9.7h shows the mass spectrum of a hydrocarbon, *Unknown 25*. Suggest a structure for this compound and explain the main features of the spectrum.

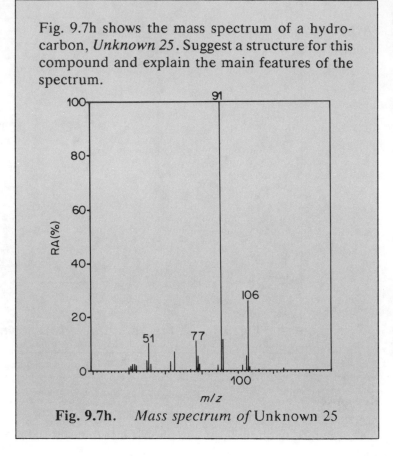

Fig. 9.7h.   *Mass spectrum of* Unknown 25

Response

Unknown 25 is ethylbenzene, $C_6H_5CH_2CH_3$.

Compared with the previous two unknowns, $M^{\ddot{+}}$ is 2 and 4 amu higher, so this is a saturated alkylbenzene. It could possibly be one of the dimethylbenzenes, but since the base peak is m/z 91, it must contain $C_6H_5CH_2-$. Hence it can only be ethylbenzene (all the methylbenzenes give m/z 105).

****

SAQ 9.7f	Consider the isomers of the hydrocarbon, $C_{10}H_{14}$, a benzene derivative. Their mass spectral behaviour divides them into three main groups. What are these groups? Would you be able to positively identify any of the isomers from its mass spectrum alone?

Response

$C_{10}H_{14}$ hydrocarbons could be tetramethylbenzenes, dimethylethylbenzenes, propyl and isopropylmethylbenzenes, and the four butylbenzene isomers. There is no need to put down all the structures because they divide into three main groups as follows:

Group	Mass Spectral Behaviour	Compound
1	Show $(M - CH_3)$ tropylium ions $(m/z\ 119)$ *but* NO 'McL' peaks	tetramethylbenzenes dimethylethylbenzenes t-butylbenzene methylisopropylbenzenes
2	Show $(M - CH_3CH_2)$ tropylium ions, $(m/z\ 105)$ and a 'McL' peak $(m/z\ 106)$.	methylpropylbenzenes 2-methylpropylbenzene
3	Show $(M - C_3H_7)$ tropylium ions $(m/z\ 91)$ and a 'McL' peak $(m/z\ 92)$.	butylbenzene 2-methylpropylbenzene

As you can see, most of the isomers fall into Group 1 where identification would clearly need careful confirmation. Since there are two compounds in each of Groups 2 and 3, you would not be able to uniquely identify *any* one of these isomers by its mass spectrum alone!

****

SAQ 9.8a   Which fragment would you expect to be most prominent in the mass spectra of:

(*i*) $4\text{-FC}_6\text{H}_4\text{CHO}$ 4-fluorobenzaldehyde;

(*ii*) $4\text{-FC}_6\text{H}_4\text{CH}_3$ 4-fluorotoluene;

(*iii*) $4\text{-IC}_6\text{H}_4\text{Br}$ 4-bromoiodobenzene;

(*iv*) $3,4\text{-Br}_2\text{C}_6\text{H}_3\text{Cl}$
                                                          3,4-dibromochlorobenzene;

(*v*) $4\text{-CH}_3\text{C}_6\text{H}_4\text{CH}_2\text{Br}$
                                                          4-methylbenzylbromide?

Response

(*i*) $4\text{-FC}_6\text{H}_4\text{CHO}$ would behave as an aromatic aldehyde, ie like benzaldehyde but 18 amu higher. Hence, its principle fragments would be due to loss of $H^\cdot$ followed by loss of CO to give

$$4\text{-FC}_6\text{H}_4\text{CO}^+, \; m/z \; 123, \text{ and } 4\text{-FC}_6\text{H}_4^+, \; m/z \; 94$$

There would be very little loss of $F^\cdot$ so m/z 105 and 76 would be negligible.

(*ii*)   Here again, $4\text{-FC}_6\text{H}_4\text{CH}_3$ would behave like toluene, $C_6\text{H}_5\text{CH}_3$, which loses an $H^\cdot$ from its M^+ ion to give the tropylium ion, $C_7\text{H}_7^+$. Hence, we would expect the fluorotropylium ion, $C_7\text{H}_6\text{F}^+$, m/z 109 to be intense, and for it to lose $C_2\text{H}_2$ to give m/z 83.

(*iii*) $4\text{-IC}_6\text{H}_4\text{Br}$ would lose $I^\cdot$ first to give BrC_6H_4^+, m/z 155/157, which would lose $\overset{\cdot}{\text{Br}}$ to give $C_6\text{H}_4^+$, m/z 76 and HBr to give m/z 75. IC_6H_4^+, m/z 203 would be very small.

(*iv*) 3,4-$Br_2C_6H_3Cl$ would lose both Br consecutively before Cl is lost, hence the most prominent fragment ions would be the 189/191/193 cluster of $BrC_6H_3Cl^+$ (3:4:1 ratio, compare Fig. 7.3f) and the 110/112 cluster of $C_6H_3Cl^+$ (3:1 ratio). $Br_2C_6H_3^+$, the 233/235/237 cluster (1:2:1 ratio, compare Fig. 7.3e) would be present, but much less intense.

(*v*) 4-$CH_3C_6H_4CH_2Br$ is both a benzyl halide and a tolyl derivative. Hence, it would be expected to give two tropylium ions, $CH_3C_7H_6^+$, m/z 105, by loss of Br˙, and the tropylium ion itself, m/z 91, by loss of ˙CH_2Br. The m/z 105 would be the more intense owing to the weakness of the C—Br bond compared with the C—CH_3 bond.

Note that in all of these examples there are no special features expected arising from the substitution patterns of the benzene rings – all the possible isomers would give the same ions, but in differing RA.

****

SAQ 9.8b   Which types of aliphatic halocompounds are likely to give significant $(M - HX)^{+\cdot}$ ions in their mass spectra?

Response

Only fluoro- and chloroalkanes show significant loss of HX, because the C—X bond is fairly strong inhibiting the loss of X˙, and the α-cleavage ions $RCH\overset{+}{=}X$ are not very stable when X is the highly electronegative F or Cl.

****

SAQ 9.8c | In the mass spectrum of fluoroethane, M_r 48, the base peak m/z 47 is CH_3CHF^+ while $(M - CH_3)^+$, m/z 33 is only 30% RA. Why are these RA values anomalous?

Can you explain them?

Response

Because of the strength of the C—F bond, neither F nor HF is lost readily, so this leaves α-cleavage as the only possible fragmentation route of CH_3CH_2F. You would expect CH_3^+ to be lost far more readily than $H^{\cdot}$ by Stevenson's Rule, but clearly, $H^{\cdot}$ is lost three times more readily. This anomaly can be explained by the effect of the strongly electronegative F on the stability of the α-cleavage ions $CH_3\overset{+}{C}HF$ and $\overset{+}{C}H_2F$. Because the former has a $+I$ $+M$, CH_3 group to help to stabilise the positive charge on the carbon, it is formed more readily. This anomaly is only found for fluorocompounds because of the reluctance to lose $F^{\cdot}$ or HF and the great electronegativity of F.

****

SAQ 9.8d | Which would you expect to give the most intense R^+ ion peak, an aliphatic primary or branched chain compound?

$$(R—X)^{+\cdot} \rightarrow R^+ + X^{\cdot} \ (X = Br, I)$$

Response

This hinges on the relative stability of the carbocations formed. The order of stability is tertiary > secondary > primary, so if it were secondary or tertiary, then the *branched compound* would give a more intense R^+ ion. If the branch were elsewhere in the molecule you would not expect much difference as the R^+ ion would still be primary. If you had forgotten the stability order of aliphatic carbocations, revise Section 8.2.

Note that fluoro- and chloroalkanes do not give very intense R^+ ions (but see SAQ 9.8e below).

****

SAQ 9.8e

> If a chloro- (or fluoro-) alkane were branched at the carbon bearing the chlorine (or fluorine) would you expect the intensity of the $(M - Cl)$ (or $M - F$) ion to increase, decrease, or stay the same compared to the primary isomer? For which type of compound and isomer would you expect to see the most intense $(M - X)$ ion?

Response

Because carbocations increase in stability as you go from primary to secondary, and secondary to tertiary, branching at the carbon bearing Cl or F will enhance the cleavage of the C—X bond leading to secondary and especially tertiary R^+ ions, so these $(M - X)$ species will *increase* in intensity with greater branching. Since the C—Cl bond is weaker than the C—F, the compounds which will show the most intense $(M - X)$, R^+, ions are *tertiary chloroalkanes*. In these compounds loss of Cl˙, HCl and α-cleavage of the alkyl groups all compete and give ions of significant intensities.

****

SAQ 9.8f   Apart from the M^+ ions, the mass spectra of chloro- and iodoethane show one very significant difference. Can you suggest what that would be?

Response

Iodoalkanes in general fragment by loss of $I^{\cdot}$ to give R^+ ions. Iodoethane gives m/z 29 by this process.

$$CH_3CH_2 \overset{+\cdot}{\frown} I \rightarrow I^{\cdot} + CH_3CH_2^+ \qquad\qquad m/z\ 29$$

On the other hand, they do not readily lose HI, so m/z 28, $\dot{C}H_2 - C\overset{+}{H}_2$ would not be very intense. Chloroalkanes do lose $Cl^{\cdot}$, so CH_3CH_2Cl would also give m/z 29, but they lose HCl preferentially, giving $\dot{C}H_2 - C\overset{+}{H}_2$ in this case. The essential difference is the presence of an intense m/z 28 in the spectrum of CH_3CH_2Cl, which is more intense than m/z 29.

* *

SAQ 9.8g

Fig. 9.8c shows the mass spectrum of *Unknown 26*, a haloalkane. Suggest a structure for this compound. Assign structures as far as you can to the ions m/z 77/79, 63/65, 57, 56 and 41.

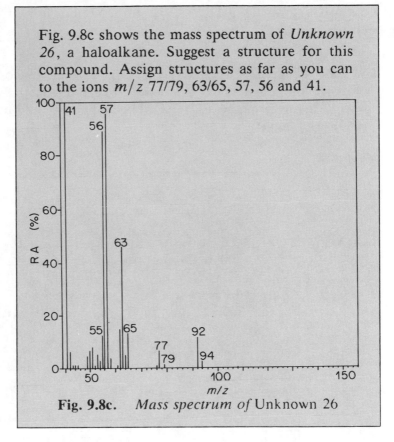

Fig. 9.8c. *Mass spectrum of* Unknown 26

Response

Unknown 26 is 2-chlorobutane, $CH_3CHClCH_2CH_3$, M_r 92/94.

The $3:1$ ratio shown by m/z 92/94 clearly shows this is a monochloroalkane. Two other Cl-containing ions appear at m/z 77/79 and 63/65, showing the losses of CH_3^+ and $CH_3CH_2^+$, respectively. This, coupled with the relatively high RA of M^+ shows that *Unknown 26* is a branched compound with CH_3 and CH_3CH_2 groups attached to the C—Cl carbon. $CH_3CH_2^+$ is lost preferentially to give m/z 63/65 (50%) rather than m/z 77/79 (10%).

Loss of Cl˙ gives $CH_3\overset{+}{C}HCH_2CH_3$, m/z 57, and loss of HCl gives $CH_3\overset{+}{C}H\overset{.}{C}HCH_3$, m/z 56.

The base peak m/z 41 is $CH_3CH{=}\overset{+}{C}H$ formed by loss of $CH_3^.$ from the latter. Note that *Unknown 26* cannot be the isomer 2-chloro-2-methylpropane, $(CH_3)_3CCl$ because this could not give $(M - CH_3CH_2)^+$ ions.

****

SAQ 9.8h   Fig. 9.8d. shows the mass spectrum of *Unknown 27*, a polyhalo-compound. Which halogen(s) are present in *Unknown 27*? Suggest a structure for *Unknown 27*. Why cannot *Unknown 27* be conclusively identified from its mass spectrum alone?

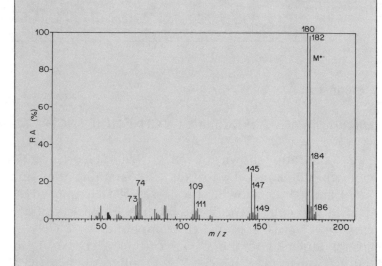

Fig. 9.8d.   *Mass spectrum of* Unknown 27

Response

Unknown 27 is 1,3,5-trichlorobenzene.

The intensity ratios of the

M : M + 2 : M + 4 : M + 6 are approximately 27 : 27 : 9 : 1

as required for Cl_3 (Fig. 7.3g). Successive losses of $Cl^{\cdot}$ occur to give the clusters at m/z 145/147/149 and 110/112, but note that in accordance with the even-electron rule the cluster at m/z 145 more readily loses HCl to give the intense 109/111 pair of ions. Finally C_6H^{+}, $C_6H_2^{+}$ and $C_6H_3^{+}$; m/z 73, 74 and 75 are formed by loss of the final chlorine either as $Cl^{\cdot}$ or HCl. This mass spectrum alone gives no information about the substitution pattern. It could be *any* of the trichlorobenzenes from this evidence.

****

SAQ 9.8i

Fig. 9.8e shows the mass spectrum of *Unknown 28*, a haloalkane. Which halogen is present in *Unknown 28*? Suggest a structure for this compound, and assign structures to the ions m/z 77, 70, 55 and 43 consistent with your structure for *Unknown 28*. $\longrightarrow$

SAQ 9.8i
(cont.)

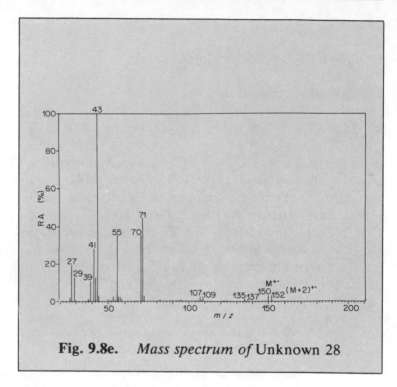

Fig. 9.8e. *Mass spectrum of* Unknown 28

Response

Unknown 28 is 3-methylbromobutane, $(CH_3)_2CHCH_2CH_2Br$.

The fact that the M^{+} is a doublet of m/z 150/152, and $(M\text{-}CH_3)$ and $(M\text{-}(CH_3)_2CH)$, m/z 135/137 and 107/109 are also doublets indicates that the halogen is Br. $(M\text{-}Br)$, m/z 71 and $(M\text{-}HBr)$, m/z 70 are also consistent with this. If R^{+} is m/z 71, $R = C_5H_{11}$. Since neither $(M\text{-}CH_3)^{+}$ nor $(M\text{-}(CH_3)_2CH)^{+}$ are very intense ions, it is unlikely that the structure has any α-branches so it must be either 1-bromopentane or 3-methylbromobutane. The high intensity of the m/z 43 ion points to the structure containing the $(CH_3)_2CH$ group rather than $CH_3CH_2CH_2$. You would also expect the latter to lose $CH_3^{.}$ and give the m/z 135/137 bromonium ion, which $(CH_3)_2CHCH_2CH_2Br$ cannot do:

$$CH_3-CH_2 \overset{\longrightarrow Br^+}{\underset{CH_2-CH_2}{\diagdown}} CH_2 \rightarrow CH_3^{\cdot} + \langle Br^+ \rangle \quad m/z \ 135/137$$

but these ions are very weak indeed.

The suggested structures for the hydrocarbon ions, which are consistent with *Unknown 28* being 3-methylbromobutane are as follows.

$$(CH_3)_2CH \overset{\frown}{-}CH_2 \overset{\curvearrowleft}{-}CH_2^+ \rightarrow CH_2{=}CH_2 + (CH_3)_2\overset{+}{C}H,$$

m/z 71 $\qquad\qquad\qquad\qquad\qquad m/z$ 43

$$CH_3-\overset{\overset{\displaystyle CH_3}{|}}{\underset{\underset{\displaystyle H}{|}}{C}} \overset{\curvearrowright}{-}\overset{\cdot}{C}H{-}\overset{+}{C}H_2 \rightarrow CH_3^{\cdot} + CH_3CH{=}CH{-}\overset{+}{C}H_2$$

m/z 70 $\qquad\qquad\qquad m/z$ 55

****

SAQ 9.9a

A compound X shows base peak m/z 43, other intense ions at m/z 29, 41, 57, 71 and weak ions at m/z 30 and 75. The $M^{\pm}$ is not discernable. X is known to be a nitro compound and could be:

A: $CH_3(CH_2)_4NO_2$

B: $CH_3CH_2CH(CH_3)NO_2$

C: $(CH_3)_2CHCH(CH_3)NO_2$

D: $(CH_3)_2CHCH_2CH_2NO_2$

Which do you think it is?

Response

X is $(CH_3)_2CHCH(CH_3)NO_2$ [C].

Points to note:

(*i*)   A secondary NO_2 compound would give a very unstable

$$M^{+\cdot} \rightarrow (CH_3)_2CH\overset{+}{C}H(CH_3),\ m/z\ 71,$$

which would further fragment to m/z 43, $(CH_3)_2CH^+$.

(*ii*)   It cannot be [B] as this would give m/z 57, not m/z 71.

(*iii*) [A] would give the same R^+ ions as [C], but its 'McL' peak is $CH_2NO_2H^+$, m/z 61. X must have a CH_3 group on the α-carbon to give m/z 75 as its 'McL' peak, $CH_3CHNO_2H^+$.

(*iv*)   [D] would give all the ions listed *except* the 'McL' peak at m/z 75. Like [A], [D] would give m/z 61.

This problem illustrates the importance of small but structurally significant ions such as 'McL' peaks.

****

SAQ 9.9b   A compound Y has $M^{+\cdot}$ 151. It shows base peak m/z 91, and other intense ions at m/z 30, 51, 60, 65. There are weak ions at m/z 77, 105 and 121. Deduce which of the structures E to H is the most likely structure for Y:

E: $C_6H_5CH_2CH_2NO_2$

F: $4\text{-}CH_3CH_2C_6H_4NO_2$

G: $4\text{-}CH_3C_6H_4CH_2NO_2$

H: $C_6H_5CH_2CH_2ONO$

Response

Y is $C_6H_5CH_2CH_2ONO$ [H]

Points to note:

(i) Y is not an aromatic nitrocompound because the sequence (M − NO—CO) is missing (no m/z 93).

(ii) If it were [G], (M − NO$_2$) m/z 105 would be expected as base peak as this would be the highly stabilised methyltropylium ion (Section 9.7). This leaves [E] or [H] as possibilities.

(iii) [E] is an *aliphatic* nitrocompound so would fragment to give $C_6H_5CH_2CH_2^+$ if the NO$_2$ group carried the charge, and $C_6H_5CH_2^+$ if the benzene ring did.

Both should be quite intense and m/z 91 *would* be its base peak.

However, aliphatic nitrocompounds do *not* give m/z 30 or 60. Hence Y must be the nitrite isomer of [E], [H].

****

SAQ 9.9c

A compound Z has M$^{\boldsymbol{\cdot}}$ 151, but shows (M-1) RA 2%. Its base peak is shown, by a metastable, to lose 28 amu to give 93, which further loses 28 amu to give 65. The m/z 104 ion is 30% RA and loses 28 amu to give m/z 76. A close inspection of the spectrum shows m/z 134, 3% and 135, 1%. Suggest a structure for Z.

Response

Z is 2-nitrobenzaldehyde.

The sequence (M − NO−CO), m/z 151 → 121 → 93, and the further loss of CO from m/z 93 to 65 is very typical of a nitroaromatic. The aldehyde group is shown by loss of H to m/z 150, which then loses NO_2 to give m/z 104. Since this still contains the CO of the aldehyde it is not surprising it further fragments to m/z 76. Thus far we have one of the nitrobenzaldehydes, rather than an ethylnitrobenzene which would give an intense m/z 105 tropylium ion. The 2-nitro substitution is shown by the (M − OH) ion at m/z 134. Though weak, this is a distinctive ortho effect.

****

SAQ 9.10a

Compound V has M$^+$ m/z 126 (82%), 111 (100%), 83 (36%), 57 (16%), 45 (18%) and 39 (25%) with m* 62.1 and 39.1. All the ions listed here have (M + 2) peaks of about 5%.

V could be:

A CH_3CO-

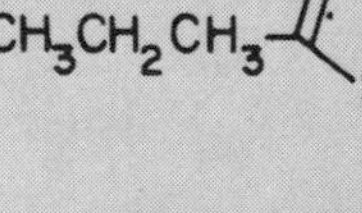

B $-CO_2CH_3$

C $CH_3CH_2CH_3-$

D $(CH_3)_2CH-$

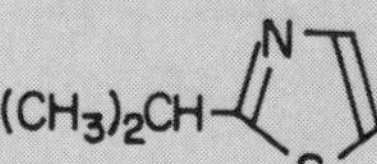

Which do you think it is?

Response

Compound V is 2-acetylthiazole, A.

V cannot be B as this has no sulphur, clearly shown to be present by the $(M + 2)$ peaks. In any case, B would give

$$\text{(furan)} - \overset{+}{C}O \quad m/z \ 95 \text{ as the base peak, by } \alpha\text{-cleavage}$$

of the $O{=}C{-}OCH_3$ bond.

C would show easy β-cleavage to give the thiazole tropylium ion m/z 97 as base peak.

D would lose CH_3 to form the methyltropylium ion of thiazole, m/z 111, $C_4H_4S\overset{+}{C}HCH_3$, which would further fragment by loss of $HC{=}CH$ to m/z 85, not by loss of CO as shown by the m^* at m/z 62.1.

The fragmentation of 2-acetylthiazole is:

m/z 126 $\xrightarrow{-CH_3^\cdot}$ m/z 111 $\xrightarrow[*62.1]{CO}$ m/z 83

$$\overset{+}{HC}{=}S \quad m/z \ 45$$

$$\xrightarrow[*39.1]{-C_2H_2} \quad m/z \ 57$$

$$\xrightarrow{-CS} \quad m/z \ 39$$

****

SAQ 9.10b   *Compound W has M^{+} m/z 95 (100%), 67 (21%), 41 (23%), 40 (26%), 39 (50%) with m^* 47.3 and 22.7. W could be:* $\longrightarrow$

SAQ 9.10b
(cont.)

E

F

G

H Either *F* or *G*

Suggest which of the options *E–H* you think is most likely.

Response

Compound *W* is either 2- or 3-hydroxypyridine (*H*)

The molecular ion loses 28 amu (m* 47.3) which must be C_2H_4 or CO to give m/z 67, then m/z 67 loses a further 28 amu to give m/z 39, ie loss of CH=NH. 2,4-dimethylpyrrole might fragment *once* by loss of CH=NH, but on the second occasion would be expected to lose its homologue, CH_3C=NH. Also dimethylpyrroles would readily lose H from one of the CH_3 groups to form a methylpyrrole tropylium ion m/z 94, or a CH_3 to give the pyrrole tropylium ion m/z 80.

Compound *W* does neither of these so must be a hydroxypyridine. This behaves like a phenol would, losing CO to give m/z 67, which is the M^+ of pyrrole or a similar species. This fragments by losses of HC≡CH, HCN and CH=NH to give the ions at m/z 41, 40 and 39, respectively. You cannot be sure which positional isomer is present from the mass spectrum alone.

$$m/z\ 67 \qquad\qquad m/z\ 39$$

****

SAQ 9.10c

Compound X has $M^{+\cdot}$ m/z 110 (44%), 95 (100%), 67 (5%), 43 (13%), 39 (14%), 38 (4%), m^* 82.1, 47.3, 22.7. X could be:

Which do you think it is?

Response

X is 2-acetylfuran, [furan structure]—COCH$_3$ (J)

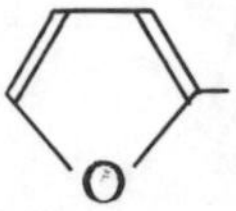

I, thiophenylethene, is easily eliminated because (i) no (M + 2) peak is mentioned, (ii) no thiophenyl ion m/z 83 is present, (iii) HC=S$^+$, m/z 45 is missing.

K is an aromatic aldehyde and would be expected to show (M − H) and (M − CHO) ions, m/z 109 and 81 as well as possibly (M − CH$_3$), m/z 95. Note m/z 43 *is* consistent with 5-methylfurans which eliminate CH$_3$CO instead of HCO from the ring. This leaves J and L, both of which would be expected to fragment very similarly by loss of CH$_3$ to the ArCO$^+$ at m/z 95, then loss of CO to m/z 67, and give CH$_3$CO$^+$ as the alternative α-cleavage fragment of the ketone. The only difference comes in the fragmentation of the Ar$^+$ ions. Fig. 9.10b shows that furans fragment by loss of CHO and imidazoles by loss of HCN. X fragments further to m/z 39, a loss of 28 amu, not 27, and also shows a small m/z 38 due to loss of CHO, from m/z 67. Hence it is a furan.

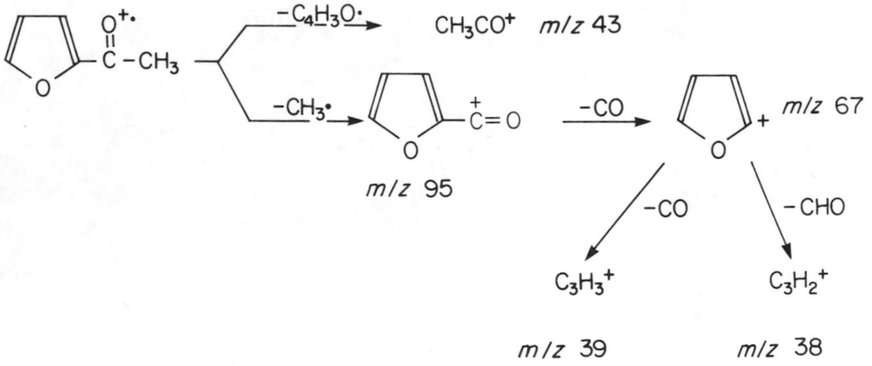

SAQ 9.10d | A compound Y contains one sulphur atom and its mass spectrum shows the following major ions: m/z 90 (100%) [$M^{\ddot{+}}$]; 75 (97%); 49 (42%); 48 (75%); 47 (39%); 43 (68%); 41 (77%) and 27 (27%). m/z 90, 75, 49, 48 and 47 contain a sulphur atom. Compound Y could be:

M $CH_3CH_2CH_2CH_2SH$

N $CH_3SCH_2CH_2CH_3$

O $(CH_3CH_2)_2S$

P $CH_3SCH(CH_3)CH_3$

Which do you think it is?

Response

Compound Y is methylisopropyl sulphide (P)

The primary thiol (M) would be expected to give ions corresponding to α-cleavage ($CH_2{=}\overset{+}{S}H$, m/z 47), loss of H_2S ($CH_3CH_2CH{=}CH_2^{+\cdot}$, m/z 56) and M $-$ (H_2S + $CH_2{=}CH_2$) m/z 28, as well as R^+ ions m/z 29, 43 and 57. Of these Y shows only m/z 47 and 43, so it is not (M).

(N), methylpropyl sulphide would be expected to give $CH_3\overset{+}{S}{=}CH_2$ as base peak at m/z 61 by loss of the $CH_3CH_2^{\cdot}$ group, and also $CH_3\overset{+\cdot}{S}H$ by loss of the $CH_3CH_2CH_2^{\cdot}$ group with β-H transfer to sulphur, ie m/z 48. This latter ion *is* present, indicating the presence of CH_3S- in the structure, but the all-important m/z 61 is absent. Hence, the isopropyl group is the remaining option, (P), or the symmetrical $(CH_3CH_2)_2S$, (O), which would also be expected to lose one of the CH_3 groups to give $CH_3CH_2\overset{+}{S}{=}CH_2$, m/z 75 as base peak, as observed.

You can distinguish between (O) and (P) by two things (i) presence of $RSH^{+ \cdot}$ and (ii) presence of an intense m/z 43. The isopropyl isomer will give $CH_3SH^{+ \cdot}$ and the diethyl compound $CH_3CH_2SH^{+ \cdot}$ m/z 62, and $(CH_3)_2\overset{+}{C}H$, m/z 43 would be expected from an isopropyl compound, as observed. The fragmentations of, Y, methylisopropyl sulphide are as follows:

$$CH_3\overset{+ \cdot}{S}-\overset{\overset{\displaystyle CH_3}{|}}{CH}-CH_3 \quad \rightarrow \quad CH_3^{\cdot} \; + \; CH_3\overset{+}{S}{=}CHCH_3 \qquad m/z \; 75$$

$$CH_3-\overset{+ \cdot}{S}{-}CH(CH_3)_2 \quad \rightarrow \quad CH_3S^{\cdot} \; + \; (CH_3)_2\overset{+}{C}H \qquad\qquad m/z \; 43$$

$$CH_3-\overset{+ \cdot}{S}{-}\overset{\overset{\displaystyle H-CH_2}{}}{\underset{\underset{\displaystyle CH_3}{}}{CH}} \quad \rightarrow \quad CH_3SH^{+ \cdot} + CH_3CH{=}CH_2$$

$$m/z \; 48$$

$$CH_3-\overset{+}{\underset{\cdot}{S}}{-}CH\begin{matrix} H \\ \diagdown \\ \diagup CH_2 \\ | \\ \diagdown CH_2 \\ \diagup \\ H \end{matrix} \quad \rightarrow \quad CH_3S\overset{+}{H}_2 + C\overset{\cdot}{H}_2{-}CH{=}CH_2$$

$$m/z \; 49 \; (\text{also as } m/z \; 41)$$

$$CH_3-\overset{+ \cdot}{S}{-}CH(CH_3)_2 \rightarrow CH_3S^+ \; + \; (CH_3)_2CH^{\cdot}$$

$$m/z \; 47$$

****

SAQ 9.10e

A compound Z contains one sulphur atom and its mass spectrum contains the following major ions: m/z 138 (100%) [M$^+$]; 123 (65%); 110 (66%); 109 (24%); 65 (21%); 52 (23%); 45 (32%). m/z 138, 123, 110, 109, 45 contain the sulphur atom.

Z could be:

Q $C_6H_5CH_2SCH_3$;

R $4\text{-}CH_3C_6H_4SCH_3$;

S $4\text{-}CH_3CH_2C_6H_4SH$;

T $C_6H_5SCH_2CH_3$;

Which do you think it is?

Response

Compound Z is ethyl phenyl sulphide, $C_6H_5SCH_2CH_3$, (T)

Isomer (Q) is a benzylic thioether. This would be expected to give an intense m/z 91, and the usual fragments (m/z 65, 51, 39) from that. It would not give a particularly stable (M $-$ CH$_3$)$^+$ ion as Z shows.

Isomer (R) is more plausible because loss of CH$_3$ here could give a stable thiotropylium ion with a methyl substituent. This should then lose C$=$S to give m/z 79, but this is not found in the spectrum of Z. Loss of CH$_3$S would also be expected but m/z 91 is not found. Hence (R) is unlikely. Isomer (S) is a thiophenol. It would also be expected to lose CH$_3$ to give a sulphhydryl tropylium ion, and lose CS from that, but the lack of m/z 79 throws doubt on this structure. It should also lose HS$^\cdot$ direct from M$^+$, and perhaps HC$=$CH, but m/z 105 and 112 are not found.

The fragmentations of *Z* are:

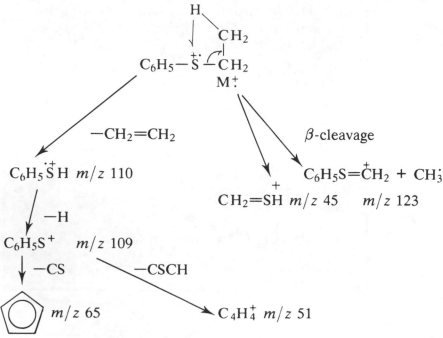

Apart from m/z 123 and 45, the ions in this spectrum result from the formation of the M$^{\ddot{+}}$ of thiophenol (m/z 110) and its daughter ions. The (110-C$_2$H$_2$) and (110-SH) ions are present in the spectrum but are less than 10% RA.

****

SAQ 10.1a	Suggest derivatives which will be suitable for the gc-ms analysis of each of the following classes of compound: (*i*) fatty acids; (*ii*) amino acids and peptides; $\longrightarrow$

SAQ 10.1a (cont.)

> (*iii*) carbohydrates;
>
> (*iv*) phenols;
>
> (*v*) 2-hydroxyacids;
>
> (*vi*) steroid hormones (contain C=O and OH groups);
>
> (*vii*) fruit essences (contain alcohols, carbonyl compounds and esters).
>
> In some cases there is a choice – select which derivative you think would be best from the ms point of view.

Response

(*i*) Methyl esters are most suitable for acids. TMS derivatives could be used but they increase M_r by 72 amu per acid group which may be disadvantageous.

(*ii*) Amino acids and peptides are best methylated to give the methyl esters from the free $-CO_2H$ groups, and either acetylated or trifluoroacetylated on the free $-NH_2$ groups. Neither alone is sufficient for good gc-ms. TMS treatment would derivatise both $-CO_2H$ and $-NH_2$, but gives a large mass increment (144).

(*iii*) Permethylation (OCH_3 derivatives) is most suitable because of the number of $-OH$ groups in the carbohydrate. The use of acetates, trifluoroacetates and TMS derivatives would push M_r up to levels beyond the usual range of the gc-ms. These latter groups are sometimes used with monosaccharides which have a smaller number of $-OH$ groups.

(*iv*)   There is a real choice here unless the phenol has a high M_r – any of the —OH derivatives in Fig. 10.1b could be used as phenols are easily derivatised. So —OCH_3, —$OSi(CH_3)_3$, —$COCH_3$ and —$OCOCF_3$ could be used. At a pinch the lower M_r —OCH_3 would be best.

(*v*)   Any of the three boronate derivatives should be your choice here, as a 2-hydroxyacid would form a 5-membered cyclic boronate derivative.

(*vi*)   Steroid hormones have C=O groups so your choice should fall on the O-methyloxime derivatives, perhaps using TMS for good measure for any —OH groups present.

(*vii*) Fruit essences usually contain mainly esters which do not require derivatising, plus alcohols and limited amounts of aldehydes and ketones. The best strategy would be a TMS derivatisation for the alcohols, to help reduce tailing on the column and improve gc separation, while providing observable M^{+} ions.

****

SAQ 10.1b	In what ways are gas chromatography and mass spectrometry compatible analytical techniques?

Response

(*i*)   *Volatility* – both can cater for gaseous, liquid and solid samples which have high vapour pressures.

(*ii*)   *Sensitivity* – mass spectrometers have high sensitivities, they can detect down to pg of sample, adequate for capillary column gc, which is where it counts.

(*iii*) *Temperature* – the operating range of gc columns is very similar to that of ms inlet systems and sources.

(*iv*) *Speed* – Mass spectrometers exist which can scan fast enough to produce several spectra even across a narrow gc peak emerging from a capillary column.

(*v*) *Chromatogram Output* – A simple TIC device enables mass spectrometers to produce a mass chromatogram which is very similar to an FID or other gc detector chromatogram. Alternatively, the data system can reconstruct a TIC trace by summation of all the ion peak intensities recorded in each scan. Splitting of the sample stream is unnecessary.

****

SAQ 10.1c	In what ways are gas chromatography and mass spectrometry incompatible techniques?

Response

Pressure difference is the main problem. Lesser problems are column bleed; the need to derivatise some samples, eg alcohols; and cost.

****

SAQ 10.1d	How have the difficulties been overcome to marry the two instruments successfully?

Response

Packed columns can be coupled *via* a jet separator. You should have briefly described this, and explained that it works by differential expansion of a jet of carrier gas and sample into a low vacuum chamber at high speeds. The lighter carrier gas molecules diffuse outwards and miss the opening into the mass spectrometer inlet positioned about 1 mm away. Capillary columns, though more expensive, can be directly coupled to the ion source owing to their lower flow-rates and the high capacities of modern pumping systems. Bonded phases reduce the bleed problem to acceptable levels. There are now available a wide choice of suitable derivatives to enable gc-ms of more polar and thermally sensitive compounds to be done successfully. There is nothing to help reduce the high cost. Not all applications will justify it, but it is a fact that installations of gc-ms have increased ten-fold in the last ten years!

****

SAQ 10.1e	A number of factors are considered important for choosing a column for gc. Which do you think is the most important for gc-ms?

Response

Degree of column 'bleed'. In most gas chromatographs 'bleed' leads only to an elevated baseline which may limit sensitivity but has no other drawback. In gc-ms the 'bleed' enters the ion source and may give a background which completely obscures important sample ions. Deposit of 'bleed' on source components will contaminate it, leading to defocusing, arcing across insulation and over-frequent shutdowns for cleaning.

****

SAQ 10.1f	Think back to how a magnetic mass spectrometer is scanned. What do you think is the best way of carrying out SIM? Are quadrupole analysers suitable for SIM work?

Response

Accelerating voltage switching is used. The magnetic mass spectrometer equation is

$$m/z = B^2 r^2 / 2V$$

Normally B is scanned while r and V are constant. However, over a small mass difference, it is easier to switch from one accelerating voltage to another to focus selected ions (some machines have provision for eight ions to be monitored sequentially) while the magnetic field is held constant. This avoids the delay for recycling the magnet and enables more rapid switching between the selected ions, and is the method usually adopted provided the m/z lie within a range of 30% or so. Quadrupoles focus ions by a suitable combination of rf and dc fields applied to their four rods in pairs. These can be switched from one selected pair of values to another very rapidly so a quadrupole is an excellent SIM analyser too.

****

SAQ 10.1g	It is noticeable in the components of ylang-ylang oil that some are esters, apparently arising from the combination of alcohols and acids also present in the oil. How would you investigate in more detail which alcohols and acids were present and explore the possibility that they are actually formed by the hydrolysis of the natural esters during the extraction of the oil from the flowers by steam distillation?

Response

Compounds containing —OH groups in general are difficult to gc, and their mass spectra show weak or non-existent M^+ peaks (see Section 9.1, 9.5.5). It would be best to take samples of the oil and derivatise the —OH groups in order to enhance their M^+ and, therefore, identify them by the mass shift and fragmentation patterns obtained. Methylation would convert —CO_2H to —CO_2CH_3 (but not —OH to —OCH_3) and would distinguish any acids, while trimethylsilyation would convert all —OH to —$OSi(CH_3)_3$ with a mass shift of 72 amu per —OH reacted. Only OH-containing components would show retention shifts and M^+ shifts, which should be characteristic.

It is quite likely that some of the —OH compounds are artifacts produced in the steam distillation of the plant material. There are two possible strategies.

(i) Obtain some of the flowers and perform a cold extraction with an organic solvent such as propanone then analyse, as above, to see if the same —OH derivatives are still present. Not easy unless you are in the 'trade', or going on a winter sun holiday to the Indian Ocean! But a feasible strategy even on a small amount of plant material because gc-ms only needs about 10^{-4} g for a complete analysis.

(ii) Take a sample of the extracted oil, already analysed for —OH compounds, and steam distil again. A repeat of the derivatisation and gc-ms would reveal whether the concentration of the already identified alcohols and acids changed after steam distillation.

****

SAQ 10.2a | What are the fundamental requirements for a lc-ms system?

Response

(*i*) An interface capable of handling at least 1 cm^3 min^{-1} of an aqueous buffer solution, ie *ca* 200 cm^3 min^{-1} of vapour.

(*ii*) A mass spectrometer ionisation system capable of providing molecular ions, or MH$^+$ and structural information on large, non-volatile and thermally labile molecules.

****

SAQ 10.2b | What coincidental developments have aided lc-ms?

Response

(*i*) The development of the 'soft' methods of ionisation such as CI, FD and FAB which enable large polar molecules to give intense recognisable pseudomolecular ions and, hopefully, some structurally useful fragments. Without these the direct coupling of lc and ms would achieve little in the way of useful analytical information.

(*ii*) The discovery by Vestal that ionic compounds gave positive ions when their solutions were sprayed into the ion source.

(*iii*) The development of microbore lc columns which reduce the problem of vapour volume to more reasonable levels for conventional pumping systems.

****

SAQ 10.2c | What is the chief problem in coupling liquid chromatographs directly to mass spectrometers?

Response

The solvent volume used in typical lc separations is 0.5–5 $cm^3 min^{-1}$. This translates into gas volumes of 100–3000 $cm^3 min^{-1}$, one or two orders of magnitude greater than gc flows. The efficient removal and concentration of the sample is beyond the mass spectrometer pumping capacity if such flows are passed directly into the ion source. The presence of buffers only adds to this problem.

****

SAQ 10.2d | What methods are used to overcome the problem of direct coupling of lc with ms and to what extent are they successful?

Response

(*i*) Direct liquid introduction splits the lc effluent into two streams, about 1% being admitted to the mass spectrometer as a jet into the ion source. The vaporised solvent then acts as a reagent gas for CI. This has the disadvantage that 99% of the sample is not examined, only a limited range of solvents can be used, and only relative molecular mass information can be obtained when operating in the CI mode. With the advent of microbore columns this method might be used more widely but it lacks sensitivity and many compounds do not give MH^+.

(*ii*) Thermospray (TS) can take flows up to a few cm^3 min^{-1} directly into an ion source rather like a CI source (Fig 10.2b). Provided at least 10% water is present in the solvent and ammonium salts as buffers the electric field generated as the charged droplets shrink in the vacuum causes MH$^+$ ions to be formed. These are sampled by means of a cone inserted into the beam and admitted to the mass spectrometer analyser, which can be a magnetic or quadrupole type. This is a 'soft' ionising technique rather like FD or CI and will readily produce ions from very polar biological molecules and charged species of high relative molecular mass. Thus it extends the mass range and type of compounds which can be examined by mass spectrometry very significantly. Some very polar molecules give multiply protonated molecular ions which analyse at m/z M_r/n, where n is the number of protons acquired. In general the mass spectra are very comparable to CI spectra, but more sample is required on column to give similar MH$^+$ intensities. Structurally related ions are sometimes, but not always, obtained. On balance, TS is a very promising interface design but it is not, at the moment, a universal solution to the problems posed by the lc solvent.

(*iii*) Moving belt interfaces can also take flow rates up to a few cm^3 min^{-1} but here the idea is to evaporate the solvent before the belt carries the sample into the ion source chamber. Provided this can be achieved efficiently for a given solvent mixture, conventional ionisation methods, such as EI, CI and, most importantly, FAB can be used to give M$^+$, MH$^+$ species directly off the belt. The matrix required for FAB can be introduced into a washbath situated where the belt leaves the vacuum system, so is constantly renewed while remaining buffer salts and sample residues are removed, thus reducing the 'memory' of previous samples. By spraying the lc effluent onto the belt at an angle much more efficient evaporation can be achieved, and transfer efficiences are quite good (40–80%). The method gives good results for drugs and other biomolecules of high relative molecular mass, down to the ng level, which is better than current thermosprays.

Units of Measurement

For historic reasons a number of different units of measurement have evolved to express quantity of the same thing. In the 1960s, many international scientific bodies recommended the standardisation of names and symbols and the adoption universally of a coherent set of units—the SI units (Système Internationale d'Unités)—based on the definition of five basic units: metre (m); kilogram (kg); second (s); ampere (A); mole (mol); and candela (cd).

The earlier literature references and some of the older text books, naturally use the older units. Even now many practicing scientists have not adopted the SI unit as their working unit. It is therefore necessary to know of the older units and be able to interconvert with SI units.

In this series of texts SI units are used as standard practice. However in areas of activity where their use has not become general practice, eg biologically based laboratories, the earlier defined units are used. This is explained in the study guide to each unit.

Table 1 shows some symbols and abbreviations commonly used in analytical chemistry. Table 2 shows some of the alternative methods for expressing the values of physical quantities and the relationship to the value in SI units.

More details and definition of other units may be found in the *Manual of Symbols and Terminology for Physicochemical Quantities and Units*, Whiffen, 1979, Pergamon Press.

Table 1 *Symbols and Abbreviations Commonly used in Analytical Chemistry*

Å	Angstrom
$A_r(X)$	relative atomic mass of X
A	ampere
E or U	energy
G	Gibbs free energy (function)
H	enthalpy
J	joule
K	kelvin ($273.15 + t\,°C$)
K	equilibrium constant (with subscripts p, c, therm etc.)
K_a, K_b	acid and base ionisation constants
$M_r(X)$	relative molecular mass of X
N	newton (SI unit of force)
P	total pressure
s	standard deviation
T	temperature/K
V	volume
V	volt ($J\ A^{-1}\ s^{-1}$)
$a, a(A)$	activity, activity of A
c	concentration/ mol dm^{-3}
e	electron
g	gramme
i	current
s	second
t	temperature / °C
bp	boiling point
fp	freezing point
mp	melting point
$\approx$	approximately equal to
$<$	less than
$>$	greater than
e, $\exp(x)$	exponential of x
$\ln x$	natural logarithm of x; $\ln x = 2.303 \log x$
$\log x$	common logarithm of x to base 10

Table 2 *Alternative Methods of Expressing Various Physical Quantities*

1. **Mass (SI unit : kg)**

$$g = 10^{-3} \text{ kg}$$
$$mg = 10^{-3} \text{ g} = 10^{-6} \text{ kg}$$
$$\mu g = 10^{-6} \text{ g} = 10^{-9} \text{ kg}$$

2. **Length (SI unit : m)**

$$cm = 10^{-2} \text{ m}$$
$$\text{Å} = 10^{-10} \text{ m}$$
$$nm = 10^{-9} \text{ m} = 10\text{Å}$$
$$pm = 10^{-12} \text{ m} = 10^{-2} \text{ Å}$$

3. **Volume (SI unit : m^3)**

$$l = dm^3 = 10^{-3} \text{ m}^3$$
$$ml = cm^3 = 10^{-6} \text{ m}^3$$
$$\mu l = 10^{-3} \text{ cm}^3$$

4. **Concentration (SI units : mol m^{-3})**

$$M = \text{mol } l^{-1} = \text{mol dm}^{-3} = 10^3 \text{ mol m}^{-3}$$
$$mg \, l^{-1} = \mu g \text{ cm}^{-3} = ppm = 10^{-3} \text{ g dm}^{-3}$$
$$\mu g \, g^{-1} = ppm = 10^{-6} \text{ g g}^{-1}$$
$$ng \text{ cm}^{-3} = 10^{-6} \text{ g dm}^{-3}$$
$$ng \text{ dm}^{-3} = pg \text{ cm}^{-3}$$
$$pg \, g^{-1} = ppb = 10^{-12} \text{ g g}^{-1}$$
$$mg\% = 10^{-2} \text{ g dm}^{-3}$$
$$\mu g\% = 10^{-5} \text{ g dm}^{-3}$$

5. **Pressure (SI unit : N m^{-2} = kg m^{-1} s^{-2})**

$$Pa = Nm^{-2}$$
$$atmos = 101\ 325 \text{ N m}^{-2}$$
$$bar = 10^5 \text{ N m}^{-2}$$
$$torr = mmHg = 133.322 \text{ N m}^{-2}$$

6. **Energy (SI unit : J = kg m^2 s^{-2})**

$$cal = 4.184 \text{ J}$$
$$erg = 10^{-7} \text{ J}$$
$$eV = 1.602 \times 10^{-19} \text{ J}$$

Table 3   *Prefixes for SI Units*

Fraction	Prefix	Symbol
10^{-1}	deci	d
10^{-2}	centi	c
10^{-3}	milli	m
10^{-6}	micro	μ
10^{-9}	nano	n
10^{-12}	pico	p
10^{-15}	femto	f
10^{-18}	atto	a

Multiple	Prefix	Symbol
10	deka	da
10^{2}	hecto	h
10^{3}	kilo	k
10^{6}	mega	M
10^{9}	giga	G
10^{12}	tera	T
10^{15}	peta	P
10^{18}	exa	E

Table 4   *Recommended Values of Physical Constants*

Physical constant	Symbol	Value
acceleration due to gravity	g	9.81 m s^{-2}
Avogadro constant	N_A	$6.022\ 05 \times 10^{23} \text{ mol}^{-1}$
Boltzmann constant	k	$1.380\ 66 \times 10^{-23} \text{ J K}^{-1}$
charge to mass ratio	e/m	$1.758\ 796 \times 10^{11} \text{ C kg}^{-1}$
electronic charge	e	$1.602\ 19 \times 10^{-19} \text{ C}$
Faraday constant	F	$9.648\ 46 \times 10^{4} \text{ C mol}^{-1}$
gas constant	R	$8.314 \text{ J K}^{-1} \text{ mol}^{-1}$
'ice-point' temperature	T_{ice}	$273.150 \text{ K exactly}$
molar volume of ideal gas (stp)	V_m	$2.241\ 38 \times 10^{-2} \text{ m}^3 \text{ mol}^{-1}$
permittivity of a vacuum	ϵ_0	$8.854\ 188 \times 10^{-12} \text{ kg}^{-1} \text{ m}^{-3} \text{ s}^4 \text{ A}^2 \text{ (F m}^{-1})$
Planck constant	h	$6.626\ 2 \times 10^{-34} \text{ J s}$
standard atmosphere pressure	p	$101\ 325 \text{ N m}^{-2} \text{ exactly}$
atomic mass unit	m_u	$1.660\ 566 \times 10^{-27} \text{ kg}$
speed of light in a vacuum	c	$2.997\ 925 \times 10^{8} \text{ m s}^{-1}$